AF553630

FISHING TECHNIQUES

ENCYCLOPAEDIA OF FISH AND FISHERIES

Vol. III

FISHING TECHNIQUES

By

Dr. Arvind N. Shukla

School of Studies of Zoology & Biotechnology

Vikram University

Ujjain

DISCOVERY PUBLISHING HOUSE PVT. LTD.

NEW DELHI-110 002

First Published-2008

ISBN 978-81-8356-303-1 (Set)

Published by:

DISCOVERY PUBLISHING HOUSE PVT. LTD.
4831/24, Ansari Road, Prahlad Street,
Darya Ganj, New Delhi-110002 (India)
Phone: 23279245 • Fax: 91-11-23253475
E-mail: dphbooks@rediffmail.com
dphtemp@indiatimes.com

Printed at:

Sachin Printers, Delhi

Preface

The present title "*Encyclopaedia of Fish and Fisheries*" has been designed for undergraduate and postgraduate students of all Indian Universities. The text of the present title is organized on some major areas like fishing techniques, fish behaviour, and fish physiology, fisheries of important Indian forms. Fishes inhabit every kind of aquatic environment, and their wide distribution resulted in many different designs for their special mode of life. The present title, with a skilful mix of breadth combined with detail, reviews what is known about fishes. Biological inter-relationship are fully discussed, and fishes with features of special interest are highlighted, whether they are economically significant or not. The present text brings together many scattered observations as well as the results of recent investigations. It is hoped that the general biologist, the zoologist and the Icthyologist will find wealth of exciting information in it.

The present title gives a comprehensive overview of the fishery science and also offers diverse information on the subject of fisheries to make the readers up-to-date with the latest in this field where proliferation of scientific information has not only been fast but enormous too in the last decade. Written in scholarly yet easy to understand language. It combines the usefulness of a reference work with the readability of a browsing book, perfect for any one attuned to the splendor of our natural world. It is hoped that the book will prove indispensable to fisheries organizations, lecturers and the young zoologists engaged in teaching and research.

In the preparation of this book large number of books and research papers have been consulted. So no authenticity is claimed.

The author expresses his gratitude to Mr. Wasan and staff of M/s Discovery Publishing House for their whole hearted co-operation in the publication of this book.

The author tried hard to be accurate and upto date in statement and realises the impossibility of completely avoiding errors therefore, the author will greatly appreciate having his attention called to any questionable statement.

Author

Contents

1

INTRODUCTION

The recognition of the need for new and improved methods in fishing gear research has resulted in some promising new tools and techniques. In connection with some midwater trawling gear research studies, the United States Fish and Wildlife Service has demonstrated the systematic application of underwater television and divers using self-contained breathing *apparatus* and cameras as practical fishing gear research *instruments*.

UNDERWATER TELEVISION

Since 1950, the U.S. Fish and Wildlife Service has conducted experimental trials with one- and two-boat midwater trawls. Results from earlier Service work had shown economical and operational advantages -in the single boat type, and its continued development was favoured for possible use by American fishing vessels. In May 1954 a Service research vessel carried out some *preliminary* trials with the *Larsson Phantom Trawl* in North Pacific offshore waters. Results of fishing performance were inconclusive. The gear was then shipped to the Services Gear Research and Development Unit at Miami, Florida, for further test and evaluation, using underwater television and other experimental methods.

Here Service engineers had successfully adapted industrial type. closed-circuit television for direct observations of fishing gear in action [4]. By means of progressive experiment and *refinement* of equipment a *remotely* controlled *submersible* vehicle was developed for the television cameras.

In November 1954, a joint cruise of the Service research vessels *Oregon* and *Pompano* made the first practical use of underwater television in fishing gear research, with the Phantom Trawl as the subject to be viewed. While under tow in clear Gulf Stream waters, the trawl warps, doors and net opening were observed by a television camera streamed between the *Oregon's* towing warps. A second television camera was streamed from the *Pompano* to make simultaneous lateral observations of the gear. Still picture and kinescope recordings were made of the entire operation.

The first tests were made with the gear rigged as originally received from Sweden: cotton trawl, hydrofoil floats and depressors, aluminium floats, hydrofoil doors and 30 fathom manila towing legs with danlenos. Previous trials had established that this gear did open to an estimated 30 feet or more, but that it was unsatisfactory to tow the trawl at 3.0 knots and even below this speed. The heavy drag caused continued tearing at sudden speed variations or full vessel power[3]. With 50 fathoms of cable from vessel to net, the trawl was nbw seen well below the propellor wash at speeds below 3 knots. At a speed of 2.0 knots, the monitor screen revealed the trawl doors, warps, and net mouth to be quite stable in the water with the trawl in an approximately 20 feet square opening, and thc headline rising in a gentle arc. A close-up camera lens revealed the terrific stress on the twine in a rigidly appearing trawl mouth. Most of the patent trawl floats had been cleared only with difficulty and those still fouled were seen to produce uneven stress on the twine. While the patent hydrofoil floats on short pennants performed well, a slight tendency to oscillate suggested turbulence particularly at wing tips. This was remedied by tying off the floats close-up. Round aluminium float performance seemed most satisfactory in comparison. Increases in speed up to the 3.0 knot limitation resulted in the estimated 30 feet horizontal spread with, however, *a closing* of vertical opening. This had been suspected in previous sounding tests, but not confirmed until registered on the television monitor.

In another trial the 30 fathom manila towing legs and danlenos were replaced by $_{s}$ in. wire cable and dandy-line gear to facilitate handling. Observers noted no appreciable change iii net performance except that the trawl fished slightly deeper in the water and tended to close at the lower wings. This was remedied by adding one fathom of cable at each lower towing leg. No marked losses of horizontal or vertical spread could be attributed to the removal of the danlenos. There followed various substitutions and arrangements of floats, leads and depressors, in an effort to attain the desired vertical trawl opening at 2.0 to 3.0 knot speeds. With the use of adequate weights and depressors, the desired vertical opening was obtained, but at the sacrifice of horizontal spread in each case. Although the hydrofoil trawl doors had handled and performed well, they were replaced with standard type trawl doors of equal and of larger sizes in an attempt to improve performance. Apparently there was no appreciable change in the trawl opening.

Most noticeable to all observers during these trials was the evidence of continued heavy drag and extreme "sweep back" of head, breast and leadlines at all test speeds. Completely absent during all television observations was any appearance of the net acting as an inflated body containing a volume of water exerting a pressure on the twine.

It was indicated to observers that, due to the excessive drag, the trawl could not be opened to its maximum designed displacement without some structural modification.

DIVING SLED AND CAMERA OBSERVATIONS

Through 1955, experimental trials with the modified Phantom Trawl continued in clear Florida Gulf Stream waters, subject to observations by underwater television, surface water glasses and divers using a towed diving sled and camera gear. A diving sled was made by converting a tubular steel ambulance litter for use when underwater television observations were not feasible[6]. This device was equipped with elevator controls to permit manoeuvrability. It provided a large degree of comfort and protection for the pilot-observer and cameraman diving team while under tow from the research vessel. The diving sled permitted first-hand observations .and photography of all aspects of the trawl performance from various angles, to detailed inspections in actual contact with the fishing gear.

During some 20 towing observations at speed and depth ranges proximating the previous television trials, various trawl modifications and accessory rigs were considered separately and in combination. Since in the previous trials the trawl performance had not been seriously affected by open or closed codend, emphasis was placed on reducing resistance at the trawl mouth. Progressively the net wings were shortened, body length was reduced and trawl hanging and headline increased to enlarge the proportion of the net square dimension to the length. These were effected with no real improvement apparent to observers. The first indication that the gear was opening in a better manner was not obtained until head, breast- and leadlines were reduced by about one-third in diameter and the number of floats and weights reduced by one-half.

When quarter doors or kites of equal area were affixed at the four wings in the same angle of attack as the trawl doors, greater trawl opening was achieved but drag and turbulence on wings greatly increased. In a similar manner, triangular pieces of heavy canvas attached to the wing extremities failed to assist trawl opening due to low attack angle made necessary to reduce drag.

Trials were continued with the net modified to about two-thirds the original size and with trawl doors approximately twice the area of the patented hydrofoil doors. For the first time this net was seen to open to an estimated 90 per cent. of the possible mouth aperture at a 2.0 knot towing speed. At this speed, a bulge or lump in the top square was still seen to cause some uneven distribution of tension on the body twine. This was remedied by removal of two centre aluminium floats and replacement at wing tips. The midwater trawl now was completely inflated as if exerting a pressure from the centre outwards to all points on the twine. Meshes were revealed as taking the desired diamond shape under tension from every angle. That there was an apparent pressure effect within the trawl was determined by physical contact by the diving team who found the net rigid yet malleable-and capable of partially supporting a 50 pound lead weight placed upon the twine. With reduced towing speed below 2.0 knots, there was evidence that the net opened, more as a result of resistance to the water than as the result of lift by the floats. This was accompanied by some apparent transfer of towing strain to rib lines. Strong evidence had now been presented showing the need to restrict future midwater trawl construction to materials of higher strength and lower drag characteristics. Also, this demonstrated the need for float or lead and depressor rigs permitting better trawl opening. Recognised as of vital importance was the extreme need for symmetry in design, exactness in measurement, and careful sewing detail to ensure perfect twine balance.

DISCUSSION OF RESULTS

Within the frame of reference provided by the underwater determinations, an all-nylon 40 foot square opening trawl was constructed. Particular attention was given to choice of materials, twine sizes and dimensions, geometric configuration and proportions to permit performance of the net as a turgid elastic body. Final adjustments to trawl and accessory gear were again subject to underwater inspection. The resulting trawl design has performed well in operation from Service research vessels in North Pacific and North Atlantic offshore waters, in open swells and moderate seas with 25-30 m.p.h. winds. In fishing trials from vessels of 200 to 750 h.p. at speeds of between 2 and

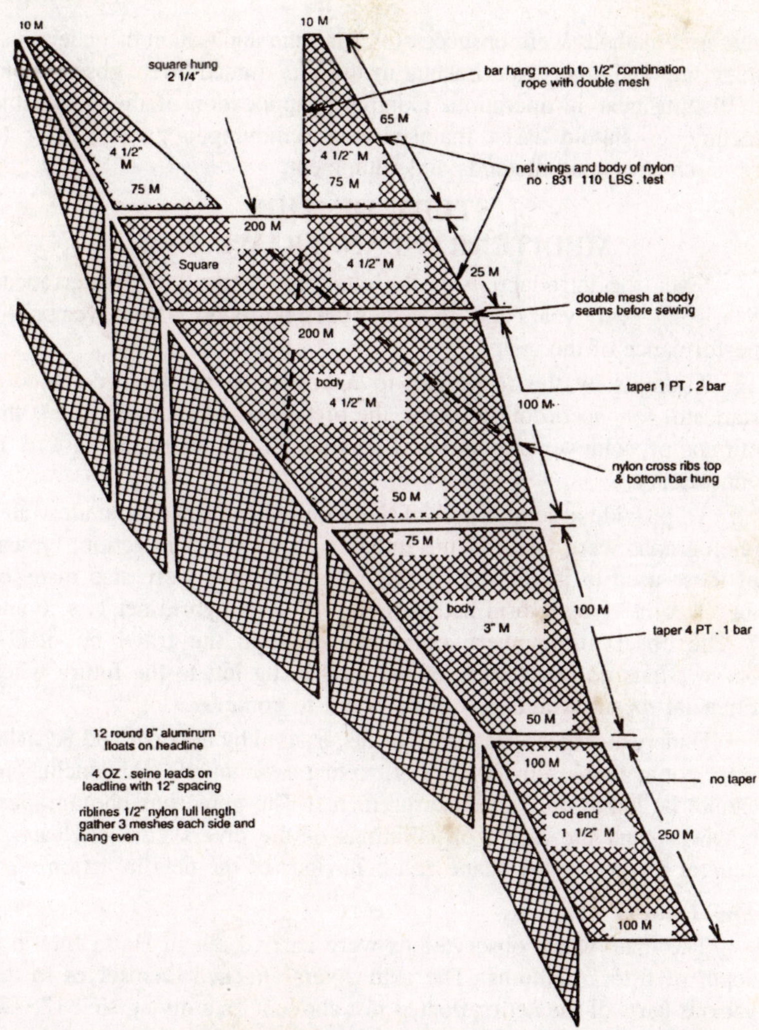

approximately 5 knots, individual trawls have withstood over 100 tows without serious damage attributed to excessive drag from towing strain in taking catches varying from 5,000 to an estimated 20,000 pounds.

There remain a number of variables to be examined and the trawls currently in use are, of course, subject to continued refinement. It is noted also that the general design features compare closely with those of recently introduced Canadian and European midwater gears.

For more than four years, the experiments-using conventional gear research methods-were unsuccessful, until the application of underwater television and other experimental methods permitted direct observations of fishing gear in operation. Continued application of these valuable techniques should assist materially in removing a great barrier to research in fishing methods and equipment.

STUDY OF THE MEDITERRANEAN TRAWL NET

Since the introduction of the otter trawl into the Mediterranean Sea, some thirty years ago, there has been no report whatsoever on the performance of the gear.

Attempts by the fishermen to improve the net have come to a standstill for, according to them, the present trawl gear represents the climax of achievement and any change in its technical pattern is unnecessary.

To provide at least a part of the required information, underwater photographs were taken of the Italian type trawl net in action, typical of those used in Israel and southern Italy. Studies were also made of the" Tzofia" type hybrid net, and the "B" type hybrid net 1, s, 6 and 7 The observations were concentrated upon the trawl net itself, observations on other parts of the gear being left to the future when financial means would permit the work to continue.

Underwater measuring instruments, as used by De Boer and Scharfe, were not available for this survey, so that assumptions and conclusions cannot be based on exact measurements. The numerous photographs, however, and the direct observations of the divers give, at least, a general picture of the shape and behaviour of the nets in action.

The Trawler

The underwater observations were carried out in Haifa Bay at a depth of 6 to 7 fathoms. The skin divers attached themselves to the various parts of the net or floated just above it on a diving sled 12, 13.

The observed gear was towed by the F.R.V. *Hatzvi, a* Danish built wooden boat of Scandinavian type with 120 h.p. engine and converted to stern trawling.

The Gear (Other Than Net)

The differences between the examined and the commercial trawl gear were as follows :

The warp was of 10 mm. θ steel wire, 100 m. long, and appeared to be proportional to the depth (under normal conditions 250 m. at 20 fathoms and 200 at 15 fathoms are used in this area). The otter boards

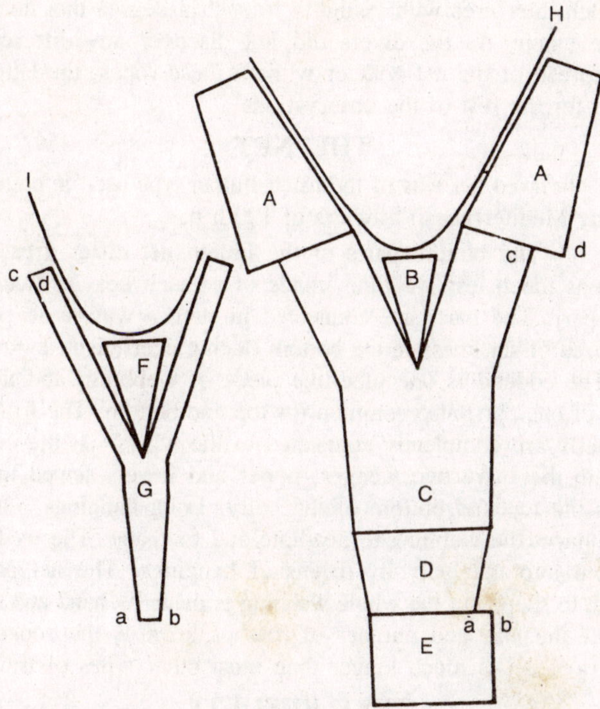

Fig. 1.1 : Italian trawl net for 120 h.p. vessels.

A ≐ wing, 10-11 m. long, 140 meshes reduced to 90, 100 mm. stretched.
B = upper wedge, 5 in. long, 80 meshes reduced to 20, 80 mm. stretched.
C = belly, 15 m. long, 220 meshes in each part, at the connection to the wing, 400 meshes at the throat connection, 54 to 50 mm. stretched.
D = throat, 400 meshes reduced to 360 meshes, 40 mm. stretched, 4 m. long.
E = codend, 5 m. long, 400 meshes, 40 to 50 mm. stretched.
F = lower wedge, 5 m. long, 80 meshes reduced to 20, 100 mm. stretched.
G = lower body, 21 to 22 m. long, 60 meshes in each part, at the connection to the wing, 40 to 60 meshes at the codend, 50 to 60 mm. stretched.
H = headline, 14 to 16 mm. a hemp rope, 28 m.
I = footrope, 40 mm. o hemp rope, 34 m. The small letters: a, b, c, d, show how the lower body is connected to the rest of the net.

were 160 × 90 cm. and typical for vessels of the *Hatzvi* size in the local trawl fishery. The sweeplines of 22 mm. a combination rope were shortened from 210 to 100 m. The mudropes, which are several pieces of thick rope whipped together, 5 to 7 m. long and placed in the gear between the sweeplines and the wings of the net, form an integral part of the Mediterranean trawl gear. They were only used during one of the tows. The divers found that they raise a cloud of mud which interferes with visibility to such a degree that no records could be taken. As the divers did not discover any differences in fishing spread of the net with or without these ropes, the latter were removed for the rest of the observations".

THE NET

The observed net was of the usual Italian type used in commercial fishing by Mediterranean trawlers of 120 h.p.

The top and bottom parts of the Italian net differ greatly. The bottom has much less webbing, made of a much heavier twine, than the top part. The parts are connected in such a way as to permit a high degree of slackness in the bottom (lacing coefficient: c = 0.85 to 0.90). The codend is one tube-like piece of webbing, and the wing consists of one part, also common for top and bottom. The front edges of the belly are completely connected to the wings. At the centre of the mouth there are two wedges, upper and lower, sewed into long clefts in the top and bottom of the belly. Long hangings, (30 to 45 cm.), connect the webbing to headline and footrope. The wedges are also sewn into the belly by means of hangings. The net parts are never cut to shape but the whole webbing is made by hand and tapered, to reduce the size and number of meshes towards the codend. The Italian trawl net is much longer than most other types of trawl nets.

TEST RESULTS

The Opening Width

The method used in finding the approximate distance between the otter doors was based on measuring the distance between the warps at two particular points on board the vessel. Knowing the exact length of the warp, the spread was then calculated. This method, although not exact, suffices to obtain a rough appraisal of the fishing spread.

Calculations showed that the distance between the boards of the standard gear of *Hatzvi* varies from 60 to 70 m., and in the examined gear from 40 to 50 m.[16] To assess the angular difference between the net's wings in both cases, a comparison can be made between the isosceles triangles formed by the wings and the sweeplines and the line joining the two otter boards.

(1) Common gear: Distance between the boards—60 m. Length of the arm—220 m. (this includes the sweepline, the mudrope and the wing of the net). Therefore the sine of $\frac{\alpha}{2}$ will be 0.14 and the angle itself will be about 16 degrees (α = angle between the two equal arms).

(2) Experimental gear: Distance between the boards—40 m.; the arm—112 m. (this includes the shortened sweepline and the wing of the net). The sine of $\frac{\alpha}{2}$ equals 0.18 and the angle itself about 21 degrees.

The difference between the common and the experimental gear, regarding the angles between the wings, calculated in this simplified way, is about 5 degrees. There is no reason to believe that this value makes a great difference in practice, causing basic changes in the shape and behaviour of the net in action.

All measurements and observations were carried out while the codend was empty. The current opinion of local fishermen is that a big catch in the codend increases the total resistance of the net. It is understandable that if the codend fills with mud, stones or any other heavy load, as often happens, the total resistance of the gear increases, the fishing spread decreases and in some cases the towing speed decreases as well. The reaction may be quite different if the codend fills with fish only. De Boer, who measured the trawl gear under water, by means of special instruments, expresses the opinion that a big catch may cause a decrease instead of an increase of the pull in the legs of a trawl but a normal catch would not influence the tension.

Influence of Bottom Conditions

All observations were made on a sandy bottom. Various beliefs and ideas about the influence of the type of the sea bottom on the behaviour of the trawling gear exist among the fishermen. The most common opinion is that on a soft muddy bottom the sweeplines dig into the mud, and move mole-like under its surface.

In spite of the differences between the actual fishing conditions and the conditions under which the study was made, the author believes that the over-all picture obtained of the Italian trawl net in action is sufficient to allow a preliminary analysis of the problem.

The Shape of the Trawl Net in Action

Some facts concerning the shape of the net in action may be determined from the photographs and observations made by the divers.

(1) The curve of the footrope is much wider than the curve of the headline. This suggests that the overhang of the net is much smaller in action than it appears to be when the net is spread on shore.

(2) The wing forms a triangular wall, with its concave surface inwards. The height and the cross section of this wall increase

from the danleno towards the junction with the body.

(3) The cross section of the net's body is ellipse-like, becoming more and more circular in form towards the codend.

(4) The after part of the codend forms a swollen ball, the diameter of which is larger than the diameter of the cylindrical throat.

(5) The fishing height, measured by the divers at the centre of the headline, is 90 to 120 cm.

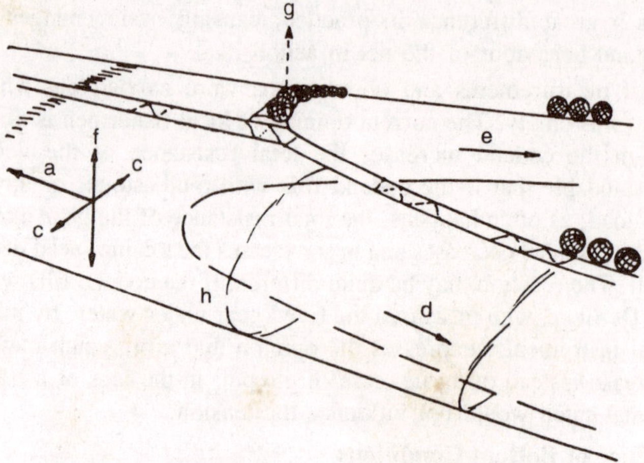

Figure 1.2 : Combined photograph-drawing of the Italian trawl net in action.

a = *total net resistance.*
c and f = *spreading forces.*
d and e = *towing forces.*
g = *lifting force of the floats.*
h = *connection between the wing and the body.*

The Forces Acting on the Net

Forces acting on the Atlantic type trawl gear have already been calculated theoretically and studies of various types of trawls are being made in some countries. But up to the present no report has been received on attempts to measure these forces in the Mediterranean trawl gear. However, the experience of the fishermen and netmakers, the direct underwater observations, and the knowledge of the construction of the Italian trawl net, can provide an explanation in general terms.

In the horizontal plane, three main forces are acting:

(1) *Total resistance of the net "a"* opposes the movement of the gear.

(2) *Towing force "e", "d"*, acts at an angle *a* to the direction of movement.

(3) *Spreading force "c"* acts perpendicular to the direction of the movement and is a result of the action of the water stream on the webbing.

Although we cannot deal with the actual values of these forces because of the absence of any measured data, it is clear that the resulting force acts along the net, parallel to the movement. This force will be called in the following *"the parallel force"*.

The spreading force acts on the webbing of the net from inside out in all directions, causing the net's body to expand. The action of all these forces together with the lifting force, if any, of the spherical floats, creates the final shape of the Italian trawl net while fishing. As whirlpools are formed by the net moving through the water, it is possible that additional mutual influences exist between them and the shape of the net.

The Behaviour of the Gear when Towing is Started

The divers found that when the net is released and the whole gear is loose, the height of the vertical opening reaches 4 or more metres, caused by the lifting force of the floats. But as soon as towing is started, the vertical opening of the net decreases rapidly and the tightened headline comes down to a height of about 1 m. As the trawl begins to move, a stream of water through the net body presses in all directions, which causes the body to swell. The after-end of the codend opens up into a ball and, being constructed from heavy webbing, causes a serious

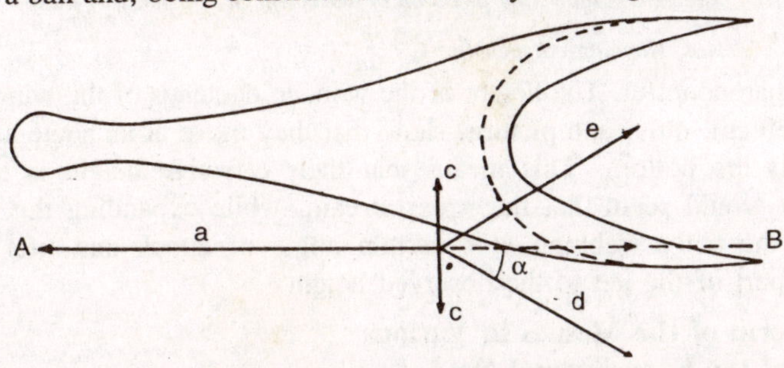

Figure 1.3 : Scheme of forces acting on an Italian trawl net. a, c, e and d as above. AB =parallel force, α = the angle at which the towing forces act.

resistance. While the towing speed increases, the parallel force increases and the lifting effect of the spherical floats decreases. As the Italian trawl net has no sidelines the elongation of the body, due to the resistance of the codend, is limited only by the webbing of the upper part. As the parallel force increases, the meshes close and the body of the net consequently becomes longer. The result is that the net body narrows and forms a flat, long funnel shape.

The Lifting Force of Spherical Floats

The headline of the experimental net was equipped with common spherical glass floats. The lifting force of these floats at usual towing speeds of about 3 knots seems to be almost negligible. This feature of the spherical floats has been described by various authors.

The Opening Height

The headline of the Italian trawl net lifts at regular towing-speeds to a height of 90 to 120 cm. The efficiency of the spherical floats is

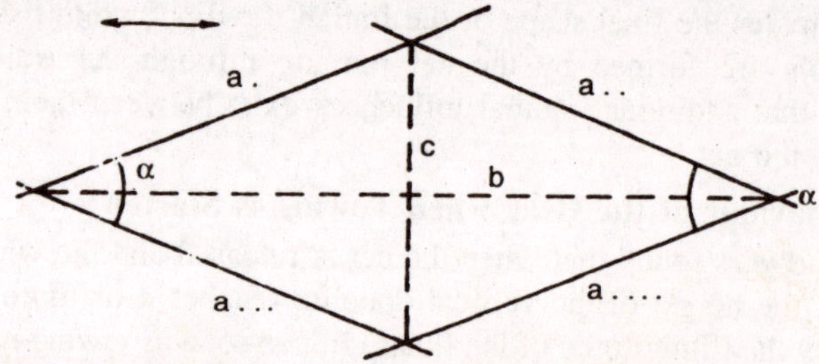

Figure 1.4 : Shape of a net mesh in working position.
a.,a..,a.... a bars of the mesh.
b = mesh length.
c = mesh width.
α = mesh angle.
The arrow shows the direction of movement. a . – a . . = mesh size. Hanging coefficients: $c_1 = \frac{b}{2a}$; $c_2 = \frac{c}{2a}$

more than doubtful. The height of the wooden danlenos of the wings is 40 to 60 cm. only, and pictures show that they move at an acute angle towards the bottom. This means that their effective height is even less. It would seem that the water stream, while expanding the net, meets the belly webbing at a certain angle of attack and lifts the upper part of the net to the observed height.

The Form of the Meshes in Various Parts of the Experimental Net

Due to the angle taken by the camera, the angles of the meshes, as measured on a photograph, are not quite accurate. Those which appeared to be undistorted by the perspective were measured but the results are still valid only for a rough comparison of the shape of the webbing in the various parts of the net.

The length of the mesh bar is the only unchanging factor if the stretching of the twine caused by towing tension is disregarded.

All other qualities of the mesh (the length, the width, the mesh angle) change according to the action of outside forces, so that trigonometric and geometric formulae can be used in designing nets and analysing their action. For this purpose the hanging coefficients are convenient and in common use, for calculating the relations between the webbing and the ropes. Hanging $c_1 = \frac{b}{2a}$, coefficient gives the relation between the mesh length (b) and the mesh size (2a). The coefficient $c_2 = \frac{b}{2a}$, represents the relation of the mesh width to the mesh size. The bigger b, the bigger is c_1, the smaller c, and the mesh angle. The relation between these coefficients and the mesh angle can be calculated trigonometrically, or by means of measuring the angles on drawings. A table was published by Baranov.

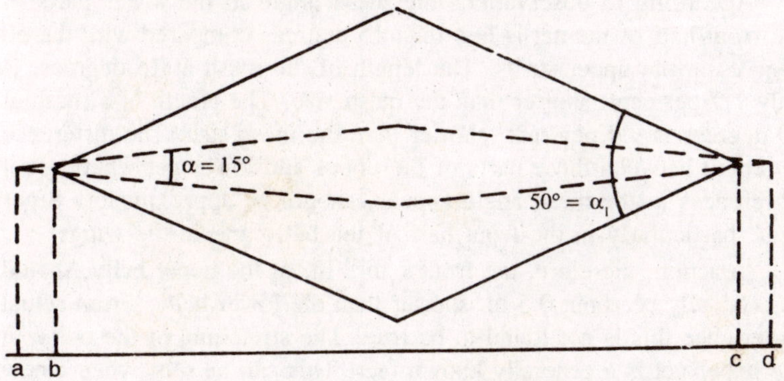

Figure 1.5 : Relation between coefficient of hanging c_1 and mesh angle x ad=mesh length at c_1=0.99; α=15° degrees. be =mesh length at c_1=0.91; α=50° degrees. abed-the difference between the mesh length in both cases.

As seen from the underwater observations and photographs the mesh angles in the Italian trawl net vary from wide open to completely closed. In the after part of the upper belly, and in the throat, the meshes are almost completely closed. The meshes in the central and the front parts are closed in the sides and open in the upper belly. In the central sectors of the belly and the upper parts of the wings, the mesh angle seems to be not less than 40 degrees, while in some places it reaches 80 degrees. The width of the webbing strips, the meshes of which are open, is at least 50 rows counting from the top edge downwards. Theoretically this fact should cause a shortage in the upper part of the webbing in relation to the mid and lower parts.

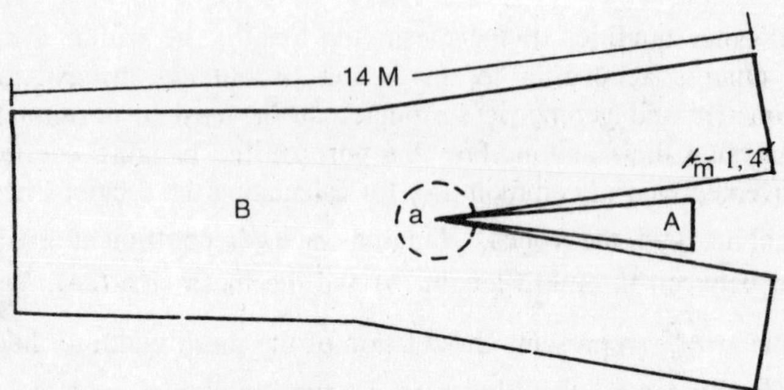

Figure 1.6 : The upper belly with the wedge adjusted into the cleft. A = upper wedge; B = belly; a = distorted sector of webbing.

The Strain in the Upper Part of the Net

According to observations the mesh angle in the lower parts of the front half of the net is less than 15 degrees compared with the 50 degrees in the upper strips. The length of the mesh at 15 degrees, is only 1.7 per cent. shorter than the mesh size. The length of a mesh at 50 degrees is 9.4 per cent. shorter than the mesh size. The difference in length between those parts of the upper and lower net where great differences in the mesh angle appear, should be approximately 8 per cent. particularly in the front half of the belly and in the wings.

In action, therefore, the front 7 to 8 m. of the upper belly, should theoretically be about 0.5 m. shorter than the lower belly. From actual experience this is not found to be true. The stretching of the twine in the upper net is a generally known fact. This can be seen when a new Italian trawl net is taken ashore for the first time for repairs and adjustment. The belly webbing is found to have lost its original form, and differences between the upper and lower parts are apparent, the former being longer than the latter. The fishermen cut triangular strips from the stretched sectors of the webbing, to give the belly its original rectangular form. This proves that there is a greater stretching force acting on the upper strips than on the rest of the webbing, and this causes an elongation of the meshes. The wide opening of the meshes is not due to a weaker parallel force acting on the top of the net, but to additional spreading forces affecting the sides of the belly.

It could be concluded that the increase in the angle of the meshes in the upper net is almost balanced by the stretching. In action, the strips with wide open meshes would be about 5 to 10 per cent. shorter than the parallel strips. with closed meshes below. However, the actual stretching of the webbing in the top is less than 5 per cent. so that

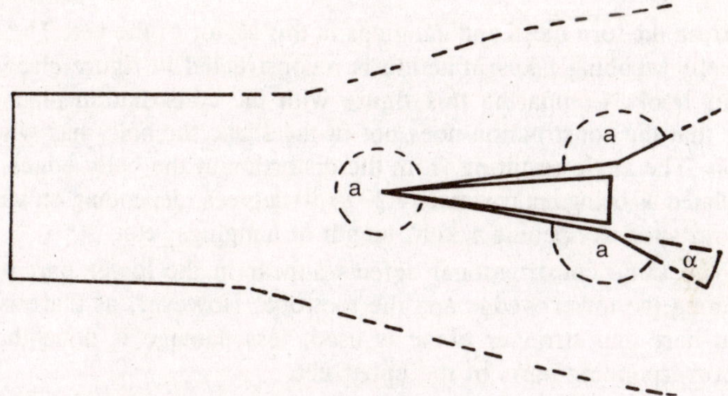

Figure 1.7 : Schematical view of the belly in action.
a = sectors under distorting forces; z = angle at which the belly is broken.

these two phenomena only balance each other in part. It seems, therefore, that in action the upper parti of the belly and of the wings are really a little shorter than the rest of the webbing, although the total strans over the top is stronger.

Consideration of the Suitability of the Italian Net Design

The upper belly of the Italian net is a trapezium which is almost rectangular. In the centre of its front part there is a cleft 6 to 7 m. long. The shape is achieved by gradually reducing the number of meshes per row, and/or by decreasing the mesh size towards the throat. All edges are knitted straight and no shape cutting used. A triangular piece of webbing is laced in the cleft by means of thin lines, in order to enlarge the net mouth. This part is called the upper wedge. While adjusting this wedge into the cleft the natural shape of the belly webbing becomes distorted. Thus, a weak constructional point is produced near the posterior connection of the upper wedge. This place often tears, and the meshes here are always distorted, so that fishermen use various methods to avoid such damage but with relatively poor results. The methods used.are: connecting the wedge to the belly by means of a long line running along the centre of the belly and reaching sometimes to the throat or even the codend, and/or using double twined webbing along the cleft.

In addition to the distortion, two more weak points are created on the upper edges of the belly where the headline, the wedge and the belly meet. The angle is clearly visible in the photograph, although the edge here was constructed straight. The fishermen waste a lot of fishing time repairing the torn mesh and hangings in this sector of the net. The form the belly webbing takes in action is reconstructed in figure elsewhere in this book. Comparing this figure with the construction plan, it is clear that the construction does not fit the shape the belly has when in action. The angle resulting from the distortion at the belly edges, was calculated as being approximately 20 to 30 degrees, depending on various factors, such as opening width, length of hangings, etc.

The same constructional defects appear in the lower part of the net along the lower wedge and the footrope. However, as there is less strain here and stronger twine is used, less damage is done than in the corresponding parts of the upper net.

In most parts of the net, the meshes are closed when the net is in action, so that the webbing partly loses its filtering action. In the after part of the belly, and in the throat, where the body of the net takes the shape of a tube, the meshes are so near to closed that substituting a non-filtering material for net webbing, e.g. canvas, would have only a slight influence on the behaviour of the gear. The poor filtering of the webbing results in increased resistance of the whole net.

The curves of headline and footrope in the sections connected to the wedges, are very broad, especially in the footrope. The fishermen make the hangings of the wedges shortest in the centre (less than 5 cm.), and their length is gradually increased towards the side of the wedge. The divers reported that the difference in length between the central and the side hangings is too great and causes increased tension in the centre of the wedge and some slack, in its sides.

Possibilities for Improvements of the Italian Net

A differentiation ought to be made between the term "improvements in the Italian trawl net" and the term "improvement of the Mediterranean trawl net". In the former, reference is made to small technical improvements in the present net, while conserving the characteristic pattern of its construction and behaviour on the sea bottom. In the latter, this pattern is rejected. The aim of the improvements in both cases is the saving of work and material, decreasing the net resistance to the water, increasing the catch in relation to the towing force invested and saving the young specimens of no commercial importance".

Few changes can be made in the Italian net without deviating from the accepted pattern. The vertical opening is definitely limited because the upper part of the body is 10 to 15 per cent. shorter than the bottom part The meshes in many parts of the net are closed due to the absence of side lines along the body, which would limit the stretching of the webbing.

Reduction of the Amount of Webbing

Decreasing the number of meshes in those parts of the webbing where the mesh is closed, without suitable changes in the parallel force, will not improve the mesh opening. The strain acting on the webbing will continue to keep the meshes closed, with the whole tube of the net body becoming thinner. Since the shape of the net mouth and its dimensions depend mainly on the hydrodynamical features of the body, any change in the size and shape of the body in action must produce parallel changes in the form and behaviour of the net. Therefore, only a limited decrease of the amount of webbing in the throat and the codend can be recommended and care must be taken to avoid exaggeration which can produce negative effects.

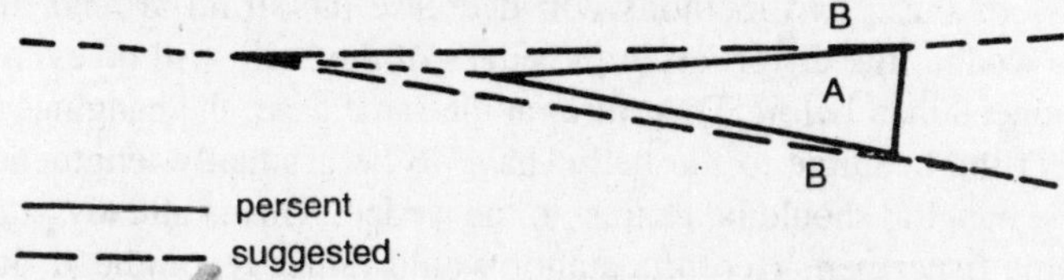

Figure 1.8 : Longer wedge to decrease distortion in belly webbing.
A = wedge; B = belly.

Reduction of the Twine Strength

The fishermen of the smaller trawlers (up to 120 h.p.), have upper parts made from light webbing as it is more efficient despite the fact that the thinner the twine the weaker and shorter living the net. For the bigger trawl nets, the light webbing proved too weak to resist the greater tension of the higher towing speeds of the larger vessels. Substituting the thin webbing by a stronger and heavier one, lowers the lifting abilities of the Italian net and reduces the catch.

Limits of the Italian Pattern

It can be said that in the three most important aspects, namely: (1) saving in the amount of webbing; (2) lengthening the net's life by use of stronger twine without changing the material, and (3) increasing the fishing height, no full satisfactory results can be achieved, as long as the main pattern of the Italian net remains unchanged.

Experiments should be carried out with a shortened body and a flapper in the throat, and with synthetic twine, none of which has yet been tried.

Minor Improvements of Construction

Two different methods are suggested to decrease the damage due to the weak points in the upper edges of the belly and around the posterior connection of the wedge.

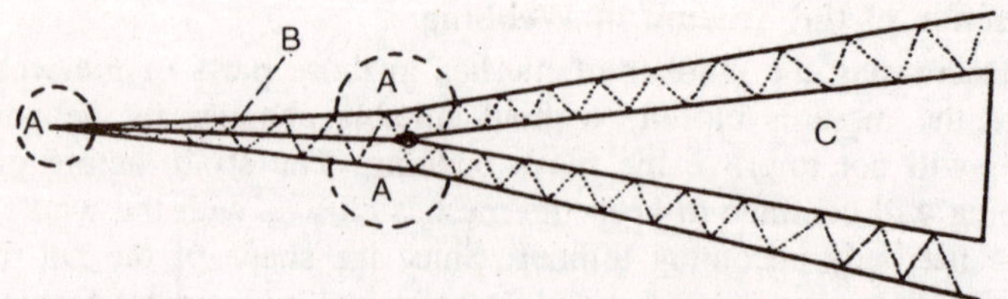

Figure 1.9 : Method of wedge adjustment, to reduce distortions in the belly webbing. A = the weak points performed; B = suggested additional hanging; C = wedge.

The whole wedge is made much longer, and only the cleft is made longer, while the hangings of both sides of the wedge unite behind its end and continue some metres, decreasing their length gradually. Although the use of these two methods will decrease the strain around the end of the wedge, the distortion in the edges of the belly will be even more pronounced than before Therefore, at the same time, the hangings which connect the headline to the belly, have to be gradually lengthened the longest hanging should be nearest to the wedge. This is already practised by some fishermen. Here the author would rather recommend, instead of making the hangings in the critical section longer, that the other hangings along the whole wing be shorter. This method of hanging was already successfully practised. The length of the hangings along the wing should not exceed 20 cm. (instead of 35 to 40 as used generally) but those along the critical section of the belly edge should be gradually lengthened till 40 cm. at the wedge.

The side hangings connecting the wedges to the ropes should not exceed 20 to 25 cm.

A more radical solution would be to cancel the wedges altogether. For this purpose, however, the number of the meshes in the front edge of the belly must be increased by about one hundred. For this purpose a rectangular part of webbing could be removed in order to form the mouth. This would simplify the construction of the Italian net without changing the principle of its action. The weak point around the wedge end disappears together with the wedge and its hangings. After cutting out two small triangular pieces of webbing, a form close to that appearing in action is given to the top of the net mouth.

It must be underlined that such small technical adjustments may improve the economy by reducing time and expense of maintenance but have nothing to do with increasing the efficiency of the Italian trawl net, for they do not change its towing resistance nor the size of its mouth. For more basic improvements the Italian pattern in the construction of a trawl net would have to be rejected.

Considerations About the Suitability of Other Types of Trawl nets

The long experience of making, and fishing with, the Italian trawl net, and the survey described above, has convinced the author that this net cannot form a basis for the development of an improved trawl.

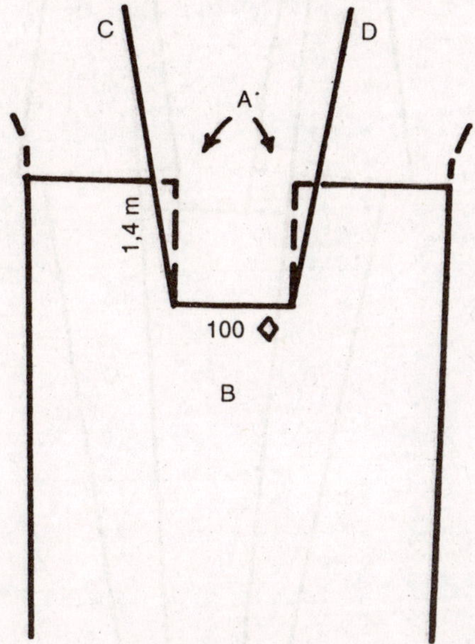

Figure 1.10 : Front edges of upper belly cut to shape. A = triangular pieces to be removed; B = belly; C-D = wings.

Former Experiences

From publications available in Israel we learn that experiments made in the Mediterranean proved that the Atlantic or northern types of nets were in all cases less efficient than the Italian, although in some trials they caught more pelagic fish.

Experimental trawling with the Atlantic gear in 1957 by the South African steam trawler, *Drom Africa,* yielded very poor results, taking into consideration the size of the vessel and her gear and the richness of the north-east Mediterranean where the trawling was carried out"'. Some experiments were made by the late Dr. Lissner, Director of the Sea Fisheries Research Station, but the author did not succeed in getting enough information about them. A Yugoslavian fishing vessel, *Napredak,* arrived in Israel in 1953 to carry out experimental tuna fishing. This vessel, when the expected tuna did not appear, fished with an Atlantic type of trawl net but, although her power was 400

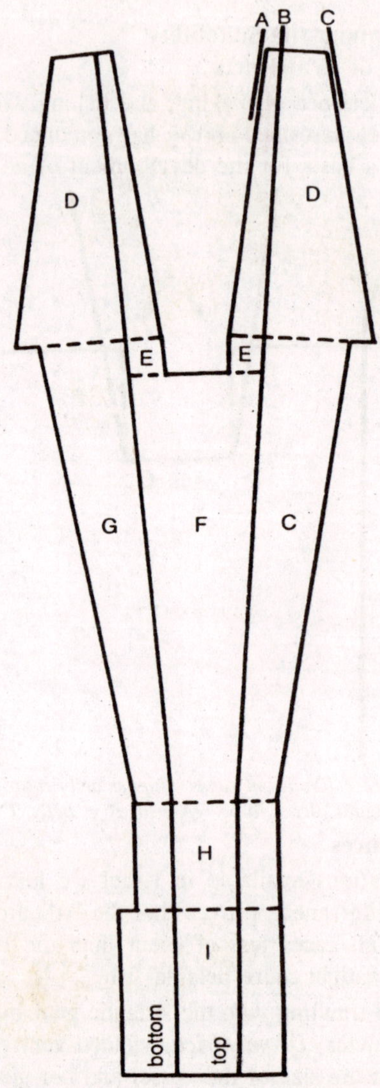

Figure 1.11 : Hybrid net, advanced type, as used on board F/V Shomria (110 h.p.) Upper part. A = headline, 12 mm. o hemp, 28 m. long; B = sideline, 12 mm. a hemp, 2 × 32 m. long; C = footrope, 40 mm. o hemp, 34 m. long; D = wing, 140 reduced to 90 meshes, 110 rows, 100 nun. stretched; E = square, 40 meshes, 24 rows, 50 mm. stretched; F = top, 170 reduced to 120 meshes, 300 rows, 50 min. stretched; G = side, 115 reduced to 80 meshes, 300 rows, 50 mm. stretched; H = throat, 280 meshes, 60 rows, 40 mm. stretched; I = codend, 300 meshes, 100 rows, 46 to 48 mm. stretched. The rows are of full meshes (2 bars).

h.p., the catches were always less than those of the small Israeli trawlers until a net of Italian pattern was used. Experimental fishing was done by the Sea Fisheries Research Station in 1957 in the Bay of Tarsus with Portuguese gear on board the *F/V Lamerchav* (240 h.p.). The results were almost nil, and the trawling was stopped after two tows. At the same time and place, big catches were taken by *Lamerchav* and other Israeli trawlers fishing with their standard gear.

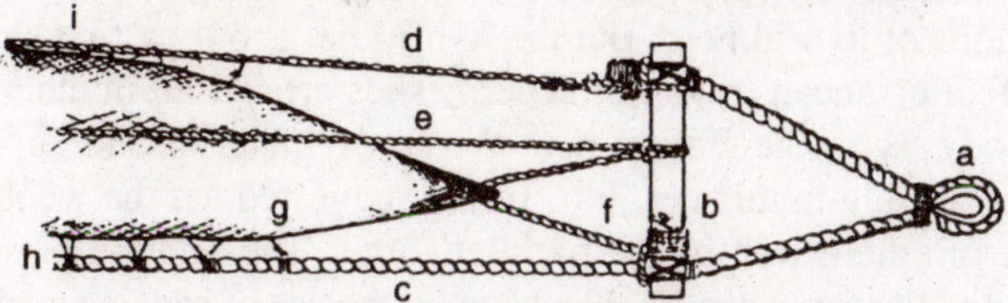

Figure 1.12 : Adjustment of wing tips to the wooden danleno in the hybrid-net. The height of the danleno is 70 to 80 crn. a = thimble; b = wooden danleno; c = footrope; d = headline; e = sideline; f = wing connecting rope; G = wing webbing; h = net to foot rope connection (Italian) method; I = direct net-to-rope connection.

The Mediterranean needs an improved trawl net, which has all the features of the Italian net, is economic to construct and has a fishing height that will ensure catching fish swimming some metres off the bottom. The mesh should be large enough to avoid catching young fish of non-commercial size, as recommended in the survey carried out by the Sea Fisheries Research Station.

The Hybrid Net

During 1956 and 1957 four different vessels fished commercially with hybrid nets as well as with standard nets, and the hybrid nets had *at least the same catches* as the standard Italian nets. The important point is that a net constructed on a different pattern from the Italian can achieve satisfactory results. The main principle of the hybrid net is that the towing forces are acting on the sidelines (seized to the side-seams).

The top (F) is connected to the sidelines with a hanging coefficient c = 0.88 to 0.90, and the sides (G) with a coefficient c= 0.95 to 0.97. The Italian pattern of the bottom parts below the sides is preserved. The loosely connected top part can be lifted by means of, for instance, Philip's trawl-planes, and reaches an opening height of 2.5 m. (twice that of the Italian net), as determined by the divers. And this is achieved with less webbing in the upper part of the net.

The figure elsewhere in the chapter shows the plan of a hybrid net for 100 to 120 h.p. trawler as now used by the F/V *Shontria.*

FURTHER DEVELOPMENTS

The preliminary observations of the Italian trawl net, together with the theory of net construction', show the way for further development. The way suggested by the author is to remove the towing strain from the net webbing by the use of a rope skeleton sewn into the net.

Such constructional improvement will give back to the net webbing its original ability to filter the water and will save a lot of net material without decreasing the net size. Moreover, it will enlarge the mouth opening. These advantages have been proved in fishing with and underwater observations of hybrid nets. But the hybrid net is only a first step to the future Mediterranean trawl net. The meshes are closed in the after part of the body as in the Italian net. Waste of material and superfluous resistance are big in this net, too. In the suggested net the webbing will be free from the action of the parallel force. The relation between the webbing and the ropes can be calculated by the use of hanging coefficients. This will enable the constructor to design a net pattern suitable to the net shape when in action. Furthermore, in such a net all the meshes will be open and less webbing will cover a particular surface.

According to calculations made by the author, at least 64 per cent. of the webbing can be saved in those parts where the mesh angle does not exceed 18 degrees, if adjusted to the side ropes with a hanging coefficient c= 0.87. This coefficient gives a mesh angle of 60 degrees.

The length of the whole net will be assured by means of sidelines preferably made of thin (12 to 14 mm.) combination rope, while the webbing will control only the swelling of the net. Such an improvement leads to: (a) saving of net material; (b) decreasing of resistance, and (c) liberating the top of the net from horizontal strain, thus enabling it to attain a larger opening height. It does not oppose, otherwise, the Italian principle of loose bottom parts. This net can be constructed of stronger (heavier) webbing, because its shape in action will be assured by the rope skeleton and the hydro-dynamic floats, and not only by the action of the water stream on the webbing, as seems to be the case in the Italian trawl net. The new net can be constructed of machine-made webbing.

CONCLUSION

A cheap, strong and efficient trawl net for the Mediterranean could be constructed as a result of technological surveys and experiments. The best way to develop such a net seems to be in rejecting the old pattern and shifting the towing strain from the webbing to ropes.

The proposed pattern could be equally applied to areas outside the Mediterranean where an increase in the fishing height may be of far greater importance.

FACTORS AFFECTING THE EFFICIENCY OF DREDGES

This paper is intended for marine biologists as well as for commercial fishermen. Much of the discussion is of an elementary nature and over-simplified; qualifications should follow most of the statements. However, as dredges have evolved very slowly, and nearly always in an empirical manner, at least in Europe, it will perhaps make a starting point to discuss in a general way some of the problems involved.

Underwater observation of dredges in action shows that scallop dredges with rope warps have a tendency to skip over the bottom' and oyster dredges to slither. Increases in speed of tow above a low level tend to reduce catches. This can be overcome to some extent by fitting depressors or diving plates to the dredges. One such dredge for scallops has already been described[2]. The fitting of teeth to dredge bars improves performance in some cases, but makes the angle of attack of the blades more critical than without teeth. Tooth spacing is also important.

The shape of the catenary of a warp in the water will depend on the drag of the warp which will vary with the material. In general, a wire warp will have a forward and downward catenary, the weight in water being greater than the drag; a rope warp will have a backward and upward catenary, the drag being greater than the weight in water.

In the following discussion, the weight of the warp will be ignored, although it can be seen that a wire warp can increase the effective weight of the dredge and a rope warp can exercise considerable lift.

The normally used warp-depth ratio for dredging is 3:1. The actual length of warp required-disregarding the effect of the warp itself-will be determined by the drag of the dredge which is proportional to the velocity squared ($D=KDV^2$), and by its weight in water. An increase in speed of tow necessitates a much increased ratio of warp to depth.

EFFECT OF DIVING PLATES

With a diving plate, another factor-lift-is introduced, which can act upwards (positive) or downwards (negative). In addition to the lift from a diving plate, induced drag occurs. Thus, there are two sources of drag, that from the dredge itself, called parasitic drag (Dp), and induced drag from the diving plate (Dj). As with drag, lift is proportional to velocity squared ($L=K_L V^2$). It can be seen that although the negative lift from a diving plate increases the warp angle, increase in speed still decreases the warp angle and thus some more warp is required for a given depth. Parasitic drag should be kept as low as possible by the use of as large a mesh as possible and by avoidance of unnecessary large surfaces in the construction.

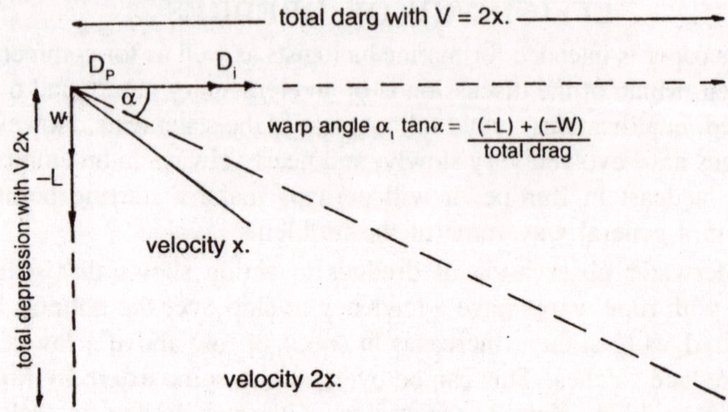

Figure 1.13 : The effect on warp angle of increasing speed of tow with diving plate. The sign of lift and weight show direction.

The lift/drag ratio is at its maximum at rather small angles of attack, the drag increasing, with increased angle of attack, quicker than the lift which, furthermore, increases only up to the point of stall. However, when the parasitic drag is high, the amount of negative lift needed for a given size of diving plate requires a bigger and, therefore, less efficient angle of attack.

A secondary effect of a diving plate is to help material into the bag of the dredge by deflecting upwards the waterflow at the mouth of the dredge. It has been observed that if the angle of attack of the diving plate is too great, more trash is retained. This may be explained by the eddies formed by the partial or complete stall of the plate resulting in a slowing down of the water flow at the mouth of the dredge, with a tendency for trash to accumulate in the front of the dredge bag.

The use of diving plates affects the stability of the dredge while it is in mid-water during shooting. A dihedral angle on the diving plate will give lateral stability about the longitudinal axis. Longitudinal stability is maintained by the point of tow. If negative lift is maintained on the plate with negative dihedral (anhedral) while shooting, righting moments occur when the dredge is disturbed laterally. If the plate gives a positive lift with negative dihedral angle the dredge will be turned over.

When the dredge is disturbed laterally, weight continues to act vertically downward but the negative lift is inclined to the vertical. If the negative lift is split into its vertical and horizontal components, it can be seen that the horizontal component is acting to the right, causing a relative waterflow to the left, which gives positive lift to the left plane and negative lift to the right plane, resulting in righting moments. However, if, as sometimes happens, when the tow is taken from the leading edge of the dredge and the warp is run out with restraint, the diving plate will give positive lift, turning the dredge on to its back.

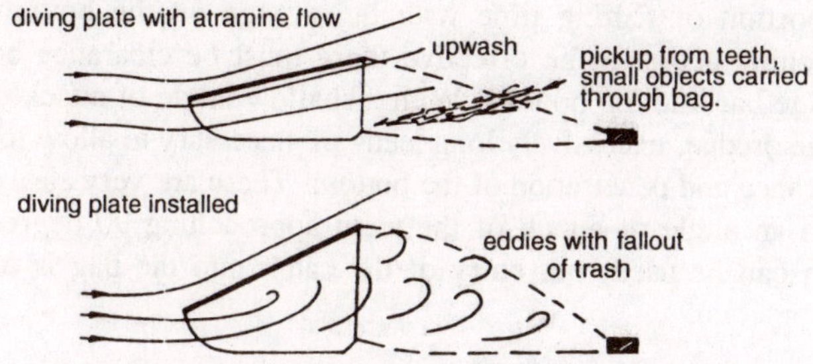

Figure 1.14 : Secondary effect of stalled diving plate. Eddies behind stalled plate can cause accumulation of trash in the dredge bag

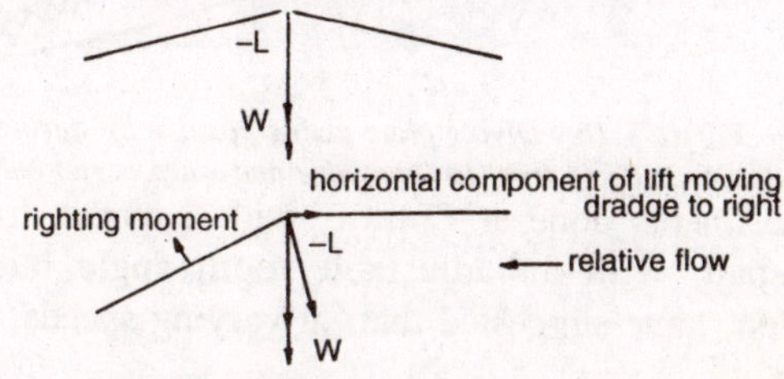

Figure 1.15 : Effect of dihedral. Temporary displacement of plate results in forces righting moments.

A dredge designed for mussel and oyster fishing is shown with two alternative points of tow. When the forward points of tow are used, the dredge has positive 'lift and is unstable in mid-water if checked during shooting. This had the disadvantage that when the depth suddenly increased the dredge left the bottom and turned over. With the after points, a negative lift was always maintained and the dredge was completely stable and could be towed in mid-water without turning over.

THE EFFECT OF TEETH

Scallops normally lie recessed with the flat valve approximately in the plane of the bottom. On escallop dredges the teeth penetrate the bottom and get below the edge of the shell, thus lifting the scallops into the bag. There are, however, considerable secondary effects. It has been shown' that the teeth on scallop dredges give a highly selective effect, operating probably more efficiently than the selective effect of the meshes of the bag. Quantitative information on this effect is limited to date, but from my own data and from those of Mason the suggestion is that the sizes at which 50 per cent. are selected are between 20 per cent. and 50 per cent. larger than the tooth spacing. The teeth, furthermore, are also sifting out trash, i.e. smaller organisms, shells and small stones. This allows a much longer tow with an increased proportion of fishing time with the dredge on the bottom. For this screening action to be effective there must be clearance between the dredge bar and the bottom. With a shallow angle of attack of the teeth of the dredge, excessively long teeth are necessary to allow for sufficient clearance and penetration of the bottom. These are very easily damaged. With an angle of attack of the teeth approaching 90 degrees, shorter teeth can be used, but entry of the catch into the bag is impeded.

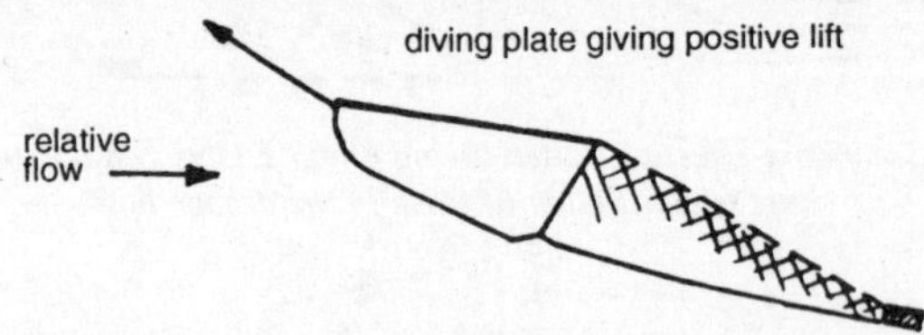

Figure 1.16 : Diving plate giving positive lift during shooting when tow is taken from leading edge and warp is run out with restraint.

Experiments done at Conway with a model dredge of about 0.5 metres span, with an adjustable tooth angle but uniform bottom penetration, have suggested that, at varying speeds, entry into the bag

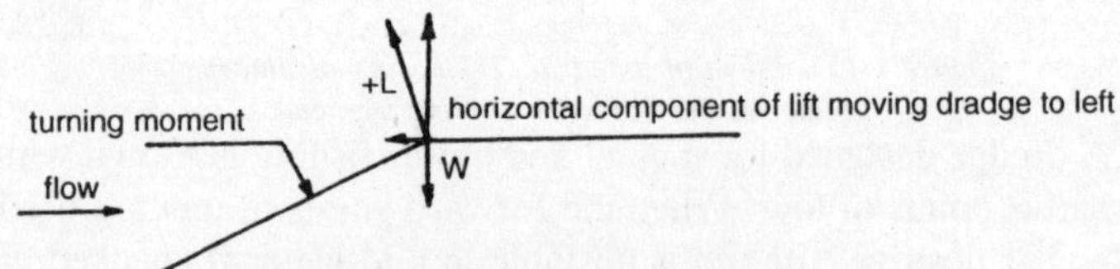

Figure 1.17 : Forces capsizing dredge during shooting when diving plate gives positive lift.

is not much impeded up to angles of 45 degrees. Within limits, higher speeds of tow allow steeper tooth angles to be used before passage of scallops over the teeth is impeded.

ABSOLUTELY EFFICIENCY

The catching efficiency of dredges is low and will vary with the type of bottom. Coral gravel bottoms in Cornwall and near Port Erin, Isle of Man, at depths of about 20 metres, have been observed to be furrowed to a depth up to 10 cm. with 20 to 30 cm. between the crests of the ridges. The passage of a dredge effectively flattens this type of bottom. It can be seen that a dredge will have a low catching efficiency on a furrowed bottom because it will be either skimming the crests of the ridges and missing many of the scallops, or digging into the ridges and catching a lot of trash which will soon fill the dredge. An increase in efficiency will occur as the bottom is flattened by working. (It is often stated by scallop fishermen that catches increase on some grounds after some days of intensive working.)

Dickie[3] measured the efficiency of Canadian dredges by releasing marked scallops and afterwards dredging in the area, assessing efficiency by the estimated number encountered per tow and the actual number caught. He found an efficiency of from 5 per cent. to 12 per cent., depending on the ground worked. Walne assessed the efficiency of a hand oyster dredge on slipper limpets *(Crepidula fornicata)* by comparing the catch of the dredge to the density of limpets as assessed by grab sampling. He found an efficiency of 16 per cent. Shelbourne[4] measured the efficiency of a winch-operated oyster dredge on slipper limpets by distributing a known number of limpets between poles on a clean intertidal area at low water and fishing between the poles at high water. He found this dredge to be 30 per cent. efficient. A special sampling dredge he devised had an efficiency of 60 per cent. However, although it might be quite possible to devise a dredge that was nearly 100 per cent. efficient for a short tow on a soft bottom, it does not follow that this will give the most economical fishing. This is clearly demonstrated by Shelbourne where the ratios of catch per distance run of 3.3, 1.0, 0.6 were obtained for the survey dredge, the winch-operated dredge and hand dredge respectively on oysters. The ratio of output of oysters per man day was 1.0, 1.1, 1.6 for the most, medium and least efficient dredges respectively. This was almost entirely due to the degree of selectivity, the least efficient dredge being most selective and resulting in the highest daily catch of oysters on a "mixed" ground. Where the organisms to be caught are fairly sparse on the bottom, a high selectivity is as important as a high efficiency, otherwise the dredge fills too quickly with unwanted material and so has to be handled more frequently.

A preliminary attempt to measure the efficiency of a traditional scallop dredge was made at Port Erin, Isle of Man, with the aim of using the dredge as a quantitative sampling device and as a standard to which other dredges could be related. Using a self-contained breathing apparatus, the actual density of scallops was measured by collecting all scallops passing between the runners of a sledge towed slowly over a known distance. Dredge hauls of measured distances were made in the same area. If only commercial scallops, 11.5 cm., and upwards, are considered, diving showed these to be present at a mean density of 1 per 40 m. and the dredge covered 250 m. for every one caught, giving an efficiency of 16 per cent. for commercial scallops. The mean density of all sizes of scallops, found by diving on the bed being investigated, was I per 11 m. The dredge covered 130 m. of bottom for every scallop caught, giving an efficiency of 8.5 per cent. The selective effect of the dredge will cause this figure to vary with the sizedistribution of scallops on the bed.

It must be emphasised that the range and scatter of these results was such that the only inference that can safely be made is that dredge efficiency is of a low order, probably between 5 and 20 per cent. with a higher efficiency when working on the larger scallops.

TRAWL GEAR MEASUREMENTS OBTAINED BY UNDERWATER INSTRUMENTS

Film taken by frogmen, showing trawls and seines in action under water, have enabled fishermen for the first time to see the behaviour of their nets on the bottom and the reaction of the fish to the net.

Although helpful, underwater filming does not solve all the problems. It is also necessary to take measurements of the gear in action. For this reason the Netherlands Inspection of Fisheries has, during the past four years, developed several underwater instruments to record the behaviour of the gear in tow.

Extensive experiments have been made on board the Netherlands F.R.V. *Antoni van Leeuwenhoek*. These have led to many improvements in the instruments which, as they were designed for work in the Southern North Sea, only withstand pressures to a depth of 100 metres.

DESCRIPTION OF THE INSTRUMENTS

The instruments can be read to an accuracy of 1 centimetre, half a degree or 1 kilogram.

1. **A spread meter.** This records the distance between the after ends of the otter boards. If the boards are attached to the wings directly (without legs) this corresponds with the horizontal opening of the net.
2. **A clinometer.** This records the tilt of the otter board sideways as well as fore and aft.
3. **A net height meter.** This records the vertical opening of the net (Headline or kite).
4. **An angle of attack meter.** This records the angle of attack of the otter board.

5. **An hydraulic dynamometer.** This records the pull in the legs between the otter board and the net or the pull on the warps on board ship.

SPREAD METER

The spread between the otter boards was first measured by reading the angle between two rods clipped on the warps. At the same time, a length of twine was attached to one of the otter boards, the other end being wound round a small barrel mounted on the other otter board. The barrel was pulled down by springs against two wooden brake cleats. Pull on the twine lifted the barrel from its brakes so that twine unreeled according to the spread of the otter boards and then braked again. In this way, the maximum spread could be measured.

However, the average spread measured by these two methods differed by about 6 metres, so a much more accurate instrument had to be developed.

First it was necessary to determine the curve in a steel wire of 3 mm. diameter and 12, 16, 20 and 24 metres length pulled through the water at speeds of 2, 3, 4 and 5 knots, and carrying longitudinal loads of from 0 to 500 kg. Then graphs were drawn for each speed, giving the difference in lengths at certain loads between the curved wire and a straight line, viz. the actual distance between the otter boards.

Accepting an inaccuracy of 1 per cent. and with an expected spread between the otter boards from 12 to 16 metres, it was found necessary to have a pull of 11 to 15 kg. on the wire when travelling at a speed of 3 knots.

This pull is provided by trawlplanes attached to the free end of the wire. Tests in the towing tank in Wageningen showed that 3 trawlplanes on a 3 mm. wire produced a load of 16 kgs. at 3 knots. In practice, however, this pull was too much for the wire, which had to be renewed frequently because of kinks and flattening where it went over the guiding reels. So, only 2 trawlplanes, which exerted a load of about 10 kg., were used although this increased the error to 2 per cent. in the spread reading.

At a speed of 5 knots, the pull of 2 trawlplanes is 42.2 kgs. and for 1 trawlplane about 20 kg., which is double the amount that can be used for 3 mm. diameter wire. The new spread meter therefore cannot be used for speeds over 3 knots without renewing the wire after every haul or using a stronger wire and recalculating all the data.

The spread meter works as follows: a 3 mm. diameter steel wire is fastened to the aft part of one otter board and passes over guiding reels through a hole in the aft part of the other otter board. The wire then passes (actually in two turns) over a wheel and over another set of guiding reels to the two trawlplanes.

In towing, when the distance between the otter boards increases or decreases, the wheel turns one way or the other and so registers changes in the spread. A recording device is attached to the wheel.

The distance between the otter boards is measured before shooting and added to the distance read on the meter, the total being the spread + 2 per cent.

CLINOMETER

The clinometer mainly consists of a pendulum arm, free to swing in a vertical plane, and a clockwork paper recorder. Changes in the tilt of the otter board are marked by a stylus on the pendulum arm, which carries a weight that can be moved up or down to regulate the sensitivity. Oscillations are damped by attaching the pendulum to a horizontal spindle, which operates a piston rod in two air cylinders. Two vent screws regulate the escape of air in the cylinders.

The apparatus is shown, together with the spread meter, attached to the otter board in operation position to determine the sideways tilt of the otter board. By unscrewing the sole plates, it can be fixed in a position at right angles to measure the fore and aft tilt of the otter board. The range is ± 45 degrees.

NET HEIGHT METER

The height of the headline is established by using a differential manometer to measure the difference in hydrostatic pressure between the bottom of the net and the centre of the headline.

A tank is attached to the footrope. It has a hole at one end while the other is connected to a plastic tube, with an inside diameter of 3 mm. The tube runs along the headline upwards to a paravane which has a buoyancy of 4 kg. and is towed by a steel wire attached to the centre of the headline. A second tank (an ordinary 8 in. float with a hole at the bottom) is fastened to the middle of the headline, and is connected by a short plastic tube to the paravane.

When the net is shot, the water enters and compresses the air in the tanks and in the plastic tubes. The submerged capacity of the tanks is such that they only half fill with sea water.

The two plastic tubes inside the paravane are connected with one pair of bellows each, welded to a steel plate, which moves out of its zero position if the pressure in one pair of bellows is more than in the other. A stylus records the movements. The paravane is shown in figure elsewhere in the chapter, shows it with the head off, disclosing the differential manometer and paper recorder.

As this apparatus measures the difference between the water levels in the two tanks, only that difference can give rise to inaccurate readings. If excess seawater enters one tank by accident, the inaccuracy will not be more than 5 cm. If the tanks are lowered carefully, the maximum error, therefore, will not be more than 10 cm.

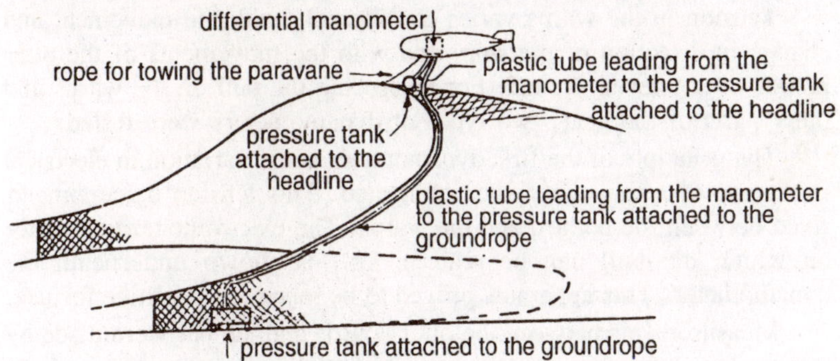

Figure 1.18 : Net height meter. Arrangement of the different parts at the net opening.

The paravane is lowered at the same time as the headline, with the towing wire and plastic tubes connected to both sides. When the paravane is afloat the veering wire is disconnected at a length of 4 metres, a small plastic float being attached to the end. This can be picked up easily when the net is hauled again.

As the first paravane could be used only between depths of 15 and 30 metres, measuring a maximum height of 5 metres, another was developed on the same principles, designed to withstand an outside pressure of 10 atmospheres.

A new type of differential manometer divided in two partitions by a large brass diaphragm on a rubber disc, has been used with this paravane. Movements are marked on a recording drum by a stylus mechanism. The paravane can be used at all depths up to 100 metres, measuring a maximum height of 10 metres.

ANGLE OF ATTACK METER

If a rod is attached to an otter board and suspended to move in horizontal and vertical directions, the free end sliding over the ground will adopt the towing direction of the otter board.

This is the principle underlying the angle of attack meter.

A steel tube, about 2 metres long, and weighted at the end with lead, is connected by a lever with a turning disc operating in a sole plate attached to the otter board.

The registration apparatus mounted on the sole plate. The complete instrument fixed to the otter board ready for operation.

DYNAMOMETER

Tension in the warps varies considerably with the movement and change in direction of the ship, and with the movements of the otter boards over uneven ground. For measuring the pull on the warps and other parts of the gear, two types of dynamometers were tested.

The principle of the first dynamometer is the variation in electrical resistance of a thin wire when elongated. This tension dynamometer fixed between the bollard and the warps. The electronic tension meter on which the pull can be read in kg., is shown underneath the dynamometer. This apparatus proved to be much too sensitive for use.

Measurements between the otter boards and the net were made by using a hydraulic dynamometer attached to the upper leg between the otter board and the wing of the net. This dynamometer, attached to a bracket mounted to the aft side of the otter board, consists of an oil cylinder, with piston rod and piston which compresses the oil in the cylinder. The cylinder is connected with the registration apparatus by a high pressure rubber tube. A plug regulates the sensitivity. The pressure, which is proportional to the pull, is measured by a spiral hollow tube moving a stylus, which is supported to prevent bad recording when the otter board bumps. The recorder is mounted in a vertical position so that ink can be used instead of a pencil, which proved to be too unyielding.

The best results from the instruments mentioned are obtained when trawling over even ground and when the wind force is not more than 3 to 4. These conditions apply to even bigger ships than the *Antoni i·an Leeuwenhoek.*

RESULTS

Instruments 1 to 4 have been used together, or in different combinations, in a number of experiments. The hydraulic dynamometer has been used separately in more recent trials.

The depth was noted during each haul and the ratio of the length of the warps to the depth was determined, while the distance travelled per haul over the ground was calculated to find out the strength of the tide. Wind and sea conditions were noted as well.

A manila trawl, with a headline length of 20.8 m., a footrope of 25.2 m., and with 8 cm. meshes in the codend, was used. The dimensions of the otter boards were 1.10 × 2.10 m. (weight 209 kgs. each). The wings of the net were attached to the otter boards without legs, while 10 trawlplanes of 8in. diameter, 1 m. apart, were attached to the headline, 5 to each side of the middle. The upper tank, plastic tubes and towing wire of the paravane were fixed to the centre of the headline.

Hauls were made always with the running tide and with a constant number of propeller revolutions, giving the ship a fishing speed of 1.7 to 3 knots over the ground.

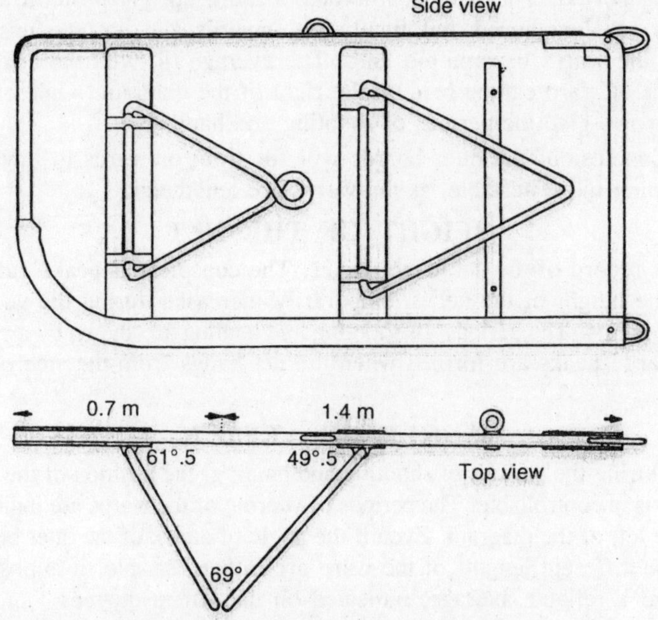

Figure 1.19 : Schema of the otter board used.

Experiments took place from 5th May to 4th July 1953, over smooth bottom between Scheveningen and Katwijk at a mean depth of 17 metres. Hauls lasted one and a half hours, during which the warps were lengthened by veering about 7.5 fathoms every 15 minutes.

SPREAD

For convenience the scale of the total spread has been given at the bottom of the diagram. The distance between the otter boards measured on board was 9.45 m. The zero-position of the stylus before lowering the otter boards is indicated by a thick short arrow. The recording is unsteady at first but becomes nearly constant as the gear settles on the bottom. As the warps are lengthened by about 7.5. fathoms, the spread increases. The periods of shooting and hauling are indicated with arrows on the left of the diagram.

Conclusion: the spread will increase by lengthening the warp.

TILT

A clinometer record for one haul. The first line shows an outward tilt about 16.5 degrees when the otter board is hanging in the gallows. When lowering, the tilt is unsteady, but becomes steady after a short settling period. The otter boards take a more upright position as the warps are lengthened and finally tilt inwards and become unstable when the warps become too long. The average tilt with the different lengths of warp can be read on the right of the diagram, while on top the arrows give the periods of shooting and hauling.

Conclusion: the otter boards will tilt from outwards to inwards, becoming more unstable, as the warps are lengthened.

HEIGHT OF THE NET

A record of the net height meter. The conspicuous peaks indicate that the height of the net is temporarily increased during the veering of the warps. The periods' of shooting and hauling are shown by arrows. The large peaks are formed when the net hangs from the side of the ship.

ANGLE OF ATTACK

During the periods of shooting and hauling, the position of the otter board is uncontrollable. The periods of veering of the warps are indicated on the left of the diagram. Even if the angle of attack of the otter boards for the different lengths of the warp are rather variable, it is possible to read a reliable average, indicated on the left in degrees.

Conclusion: by lengthening the warps, the angle of attack decreases.

The recordings of the tilt, the height of the net and the angle of attack show great peaks when the warps are veered.

It appeared that, during the veering of the warps, the otter boards tilted, on average, 23 degrees inwards, while the angle of attack diminished by 8 degrees and the height of the net increased by 35 cm. The position of the otter boards, drawn in dotted lines, indicate the situation during the veering of the warps, from which the conclusion can be made that there is more slack in the headline, which will be lifted by the floats.

There are peaks also on the other side because of the sudden braking of the warps after veering, in which case the opposite action occurs, but, with gradual braking no peaks appear.

INFLUENCE OF THE INSTRUMENTS ON THE GEAR

The paravane has a buoyancy of 4 kg., equal to that of a spherical float of 8 in. diameter, while the *clinometer* has a *buoyancy* of 0.8 kg. The new *paravane* was given a buoyancy of 8 kg. It may be accepted that these instruments will not affect the gear in any way.

Since the spread meter, the angle of attack meter and the clinometer are mounted on the aft side of the otter boards where, as the film "Trawls in Action" shows, great whirls are created; the water resistance of these instruments is not likely to affect the action of the otter boards. But the angle of attack meter weighs 9 kg. and the spread meter 5 kg. At a speed of 3 knots, 2 trawlplanes pull with a force of 10 kg. on the wire of the spread meter. It had, therefore, to be determined what influence these instruments had on the gear, separately and in combination.

The instruments were distributed over both otter boards in 9 series of hauls.

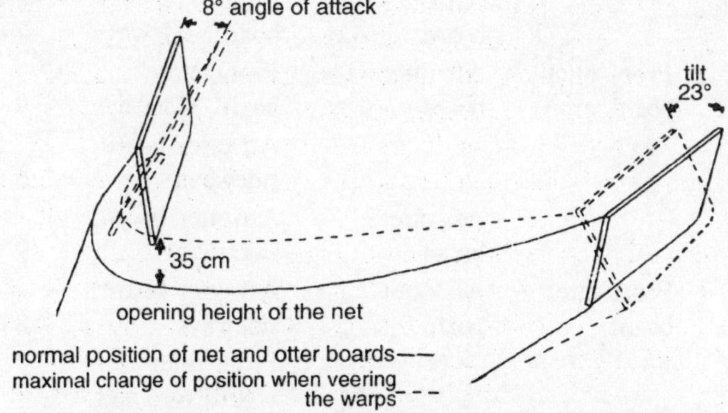

Figure 1.20 : Schematic drawing, explaining the behaviour of the otter boards and the net opening when veering the warps.

Before starting *h* and *i*, the warps had to be renewed and shortened and since differences in lengths of warps produce different readings, series *h* was made a repetition of series *d,* so that *h* and *i* could be compared.

The net height meter and the clinometer were used during all these hauls. A comparison of the data obtained shows that the instruments exerted no influence whatsoever on the operations of the otter boards. Even the strain on the wire of the spread meter had no influence on the angle of attack and the tilt of the otter boards.

However, the height of the net became 2.20 m. when the spread meter was used, compared with 2.12 without it. This corresponds to a decrease in spread of about 52 cm. On an average spread of 12.89, an error of nearly 4 per cent., which must be added to the measured spread. As already mentioned, by using 2 trawlplanes, 2 per cent. has to be subtracted for the bend in the wire, so the actual distance between the aft parts of the otter boards will be the measured distance plus 2 per cent., in this case 12.89 m. + 2 per cent. = 13.15 m.

INFLUENCE OF THE LENGTHS OF THE WARPS

After determining the influence of the instruments on the gear, the average was taken of the results obtained from series a to g, while gives the averages of the hauls in h and i.

Table 1.1

Series	Spread Meter on:	Attack-Angle meter on:	Clinometer on:	Number of hauls:
a.	—	Aft otter board	Fiont otter board, sideways	7
b.	Front otter board	Aft otter board	Front otter board, sideways	9
c.	—	—	Aft otter board, sideways	6
d.	—	Aft otter board	Aft otter board, sideways	9
e.	Frort otter board	Aft otter board	Aft otter board, sideways	8
f.	—	—	Front otter board, fore and aft	2
g.	—	—	Aft otter board, fore and aft	3
h.	—	Aft otter board	Aft otter board, sideways	8
i.	—	Front otter board	Front otter sideways	9

By lengthening of the warps:

1. the angle of attack decreases;
2. the otter board tilts from outward to inward;
3. the otter board tilts from forward to aftward;
4. the spread increases;
5. the height of the net decreases.

We can also say that, if the depth decreases with a fixed length of warp; the same will happen as mentioned above. If the depth increases, the opposite will occur.

RATIO BETWEEN WARP LENGTH AND DEPTH

The effective spreading surface of the otter board is greatest when it takes up a position perpendicular to the bottom, i.e. when the sideways tilt of the otter board is zero.

The small differences between the sets of values are caused by a smaller ratio between the lengths of the warps and the depth, by small errors in the readings due to differences in current, wind force, swell and slight changes in the depth, and by small instrument errors.

However, the results are sufficiently accurate to show that, while fishing at a depth of 17 metres with an upright otter board, the angle of attack is about 28.5 degrees the height of the net just over 2 metres and the spread 12.89 m. + 2 per cent. = 13.15 metres. For an upright position of the otter board the length of the warps must be about 5.1/2 times the depth.

Influence of Legs

Two additional series of hauls have been carried out with the same gear but with a fixed length of the warps.

In the first 7 hauls, legs of 3.60 m. length were used between the otter boards and the net; in the second series of 4 hauls the wings of the net were coupled to the otter boards directly. The tilt of the otter boards was not determined.

These results are represented diagrammatically in figure elsewhere in the chapter (gear with the legs shown by dotted lines).

In fishing with legs, the angle of attack is decreased and the spread of the otter boards is increased, but not the spread of the net, so there is more slack in the headline, and the height of the net is increased by 41 cm.

INFLUENCE OF DIFFERENT TYPES OF FLOATS ON THE OPENING HEIGHT

With the same gear as mentioned above, using 3.70 m. legs between the otter boards and the net, the height of the net without any floats was 1.20 m., or 10 cm. more than the height of the otter board. With 4 to 6 spherical 8 in. floats, the increase in height was not more than 30 cm., while with 8 floats the lift was considerably more, being about 55 cm. (total height 1.75 m.). A kite (with two small glass floats mounted on it) was used with the 8 floats, but it provided no increase in height.

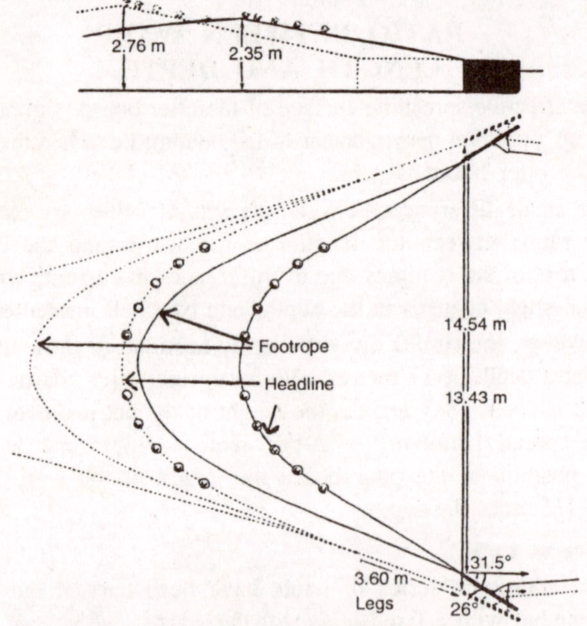

Figure 1.21 : Schematic drawing showing the influence of short legs on the net opening.

Comparative trials were carried out with 10 spherical 8 in. floats, 5 Phillip's Siamese-twin floats and 10 Phillip's trawiplanes, all having a diameter of 8 in. Strips of about 7 cm. width were out off the sides of the half circular plates on the Siamese-twins, giving them a winglike shape, which reduced resistance and increased buoyancy.

The results were:

10 spherical floats ·	height of headline	1 90 m.
5 Siamese-twins ·	• height of headline	2.30 m.
10 trawlplanes ·	height of headline	2.75 m.

PULL ON THE LEGS AND AMOUNT OF CATCH

The hydraulic dynamometer has only been used submerged between the otter boards and the net, and only a limited number of hauls have been made. A diagram in the chapter is given showing a tension of about 180 kg. in the upper leg. During this haul the sea was rather rough, with a wind force 4.

During another haul with very calm weather the pull decreased from 250 to 130 kg. The same happened in 6 consecutive hauls although in earlier and later trials a fairly constant tension was recorded. The reason was found in an enormous number of colony-building Hydrozoa which gradually choked the net during fishing. The net was cleaned before each haul.

On the assumption that the specific gravity of fish is practically the same as the specific gravity of the water, the strain in the legs should be unaffected by the catch, especially if most of the fish are swimming with the net. But eventually the codend becomes choked by the fish and starts to overflow. This may result in a decrease, not an increase, of the pull. Therefore, a dynamometer between the otter board and the net may give an indication of the amount of catch.

This is in contradiction to a claim by an American manufacturer of dynamometers and it is suggested that the rate of catch may be determined *by a decrease in the tension* on the warps or the legs.

TWO-BOAT TRAWLS AND OTTER TRAWLS

The Automatic Net Height Meter

The recorder (A) and the guide part (B) are both equipped with bellows (C_2 and C_1) which are connected by a vinyl pipe (F) filled with oil. The antipressure vinyl pipe (G) equalises the air pressure inside the recorder and guide part in order to avoid temperature effects. The recorder is fixed to the footrope and the guide part to the point to be measured. The two bellows measure the difference in hydrostatic pressure between the measuring point and the footrope which is recorded in the usual way by a pointer (E) writing on a chronograph (D).

Three types of such net height meters have been developed which are all of sturdy construction to stand rough handling, and which are small and light enough to exclude any influence on the net to be measured. They have a working range of 0 to 150 m. water depth, a measuring range of 0 to 7 m. and an accuracy of 5 cm. with a sensitivity of 2 cm.

THE AUTOMATIC FOOTROPE INDICATOR

This instrument consists of a handle which can freely turn on a vertical axle and which, due to the bottom friction of its resistance plate, is kept in towing direction during action of the trawl. The position of this handle is continuously recorded on a chronograph which is attached to the footrope in a water and pressure tight casing. By attaching several of such instruments distributed over the footrope the curvature it takes during trawling and its variations can be determined.

TWO-BOAT TRAWLS

Opening Height

A diagram of the trawl which was tested is given in figure elsewhere in the chapter.

Table given in the chapter shows that the net sinks to the bottom with a speed of about 0.3 m./sec. and the warps are veered at about 3.1 to 4.2 m./sec. It is also shown that before towing really starts the net opening is much higher than during trawling.

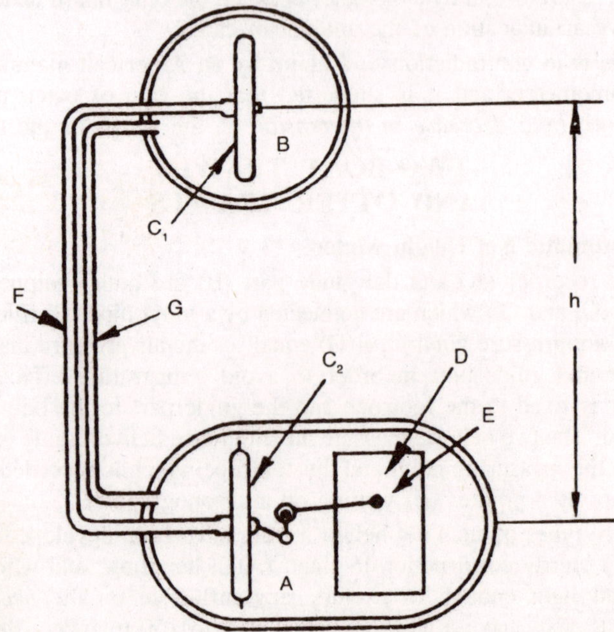

Figure 1.22 : Principle of the recording net height meter. A _Recorder; B=Guide part; C_1=Bellows (guide); C_2=Bellows (recorder); D = Chronograph; E=Pointer; F=Liquid pipe; G=Air pipe.

Table 1.2

Experiment No.	Time from throwing the net overboard until it reahes the bottom	Time for paying out 750 in. warp	Height of net-entrance immediately after reaching the sea bottom	Sea depth
1	3 min.	3 min.	6.28 m.	51 m.
2	3	4	5.55	55
3	3	3	5.70	52
4	3	3	6.28	57
5	3	3.5	6.30	51

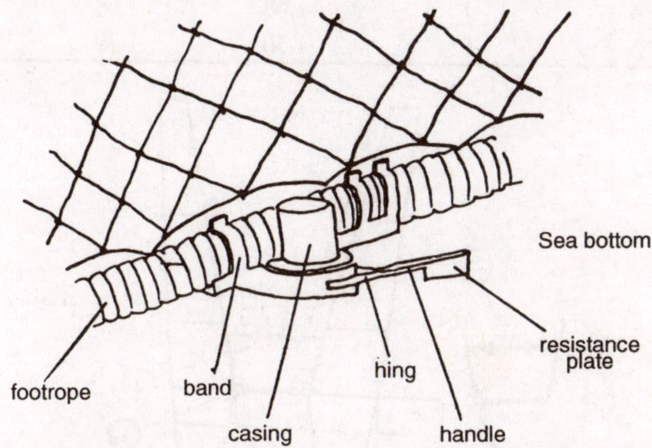

Figure 1.23 : Principle of the recording footrope indicator.

There are two ways of shooting this gear which have been found to influence the time needed from starting towing until the net opening acquires its stable height. With the V-type manoeuvre it takes less time than with the U-type where, furthermore, the opening decreases to a minimum before it becomes stable.

With about 750 m. warp length, 300 to 400 m. distance between the two towing boats is considered to be convenient. Even up to about 500 m. distance no significant effect to the opening height occurred. The variation is due to the changing tidal current influencing the actual towing speed.

To improve the opening height, the construction of the original net was changed as shown in figure elsewhere in this chapter. These changes concern increase of webbing mainly around the net opening, the relation between headline and footrope lengths which shifts the main pull to the footrope, and the flapper attached in such a way that it can open completely during towing.

Table 1.3

Experiment No.	*Time from start of towing to stable net-opening*	*Minimum height of net opening*	*Height of stable net opening*	*Type of setting*
1	12.5 min.	1.43 m.	1.82 m.	U
2	14	1.00	1.55	
3	8	1.48	2.14	
4	1	1.68	1.68	V
5	5	1.30	1.30	

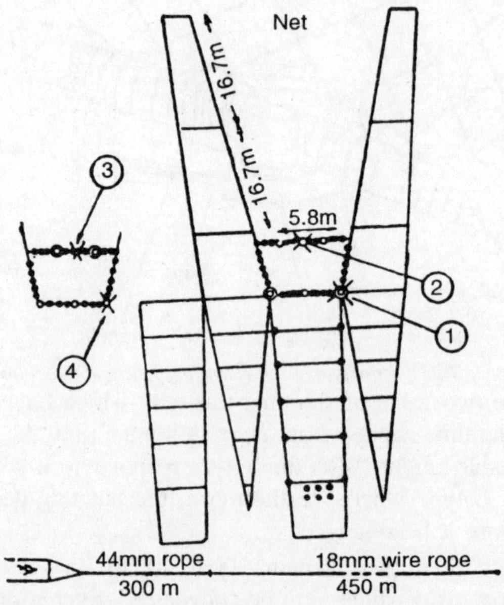

Figure 1.24 : Diagram of two boat trawl tested with indication of the measuring points for the net height meter (1 to 4). O = Glass float, 30 cm. diameter; O= Glass float, 18 cm. diameter; ● =Glass float, 15 cm. diameter; —=Chain.

A comparison of the values of opening height of the original design with the new design proves that a considerable increase of about 1.5 m. or 75 per cent. could be obtained.

With increasing towing speed the opening height decreases considerably at all points measured, e.g. middle of headline bosom, quarter point and middle of wing. The main reason for this is considered to be the decrease in buoyancy of the floats caused by their towing resistance. Without any floats, the opening height was only 0.6 to 0.8 m.

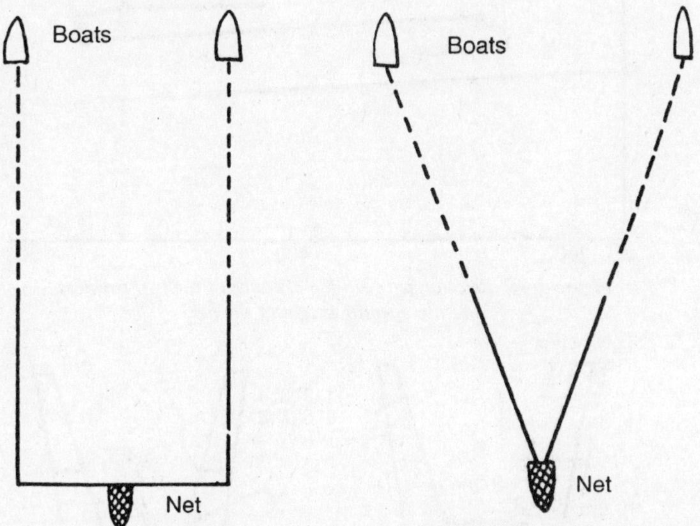

Figure 1.25 : Two different types of shooting a two boat trawl.

The graphs in figure show that the influence of the distance between the trawlers on the opening height increases with the towing speed. With a towing speed of 2.3 to 2.5 knots the most satisfactory distance between the trawlers in regard to the opening height is 300 to 350 m. under the given conditions.

CURVATURE OF THE FOOTROPE

The curvature of the footrope was measured at five points by means of the footrope indicators described below.

It was found that the footrope settles to a stable curvature in 10 to 12 minutes with the U-type of shooting and in 2.5 to 5 minutes with the V-type of shooting, after towing is started.

The angles formed at the five measuring points and the distance calculated accordingly between the respective points for different distances between the trawlers are given in Tables IV and V. The magnitude of the change in distance between the vessels. explains the influence on the opening height discussed above.

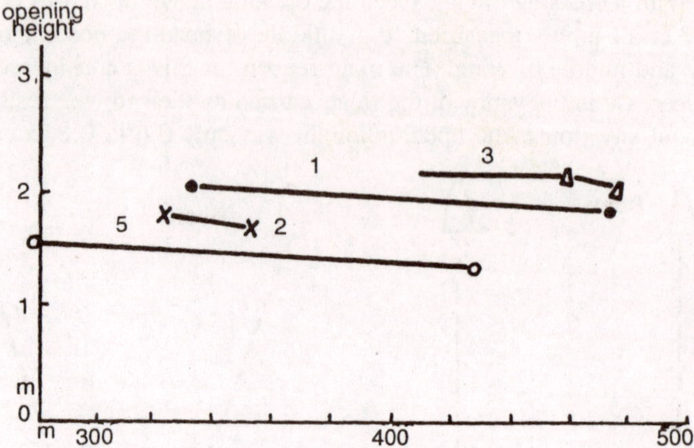

Figure 1.26: Relation between the distance of the two trawlers and the opening height of the net.

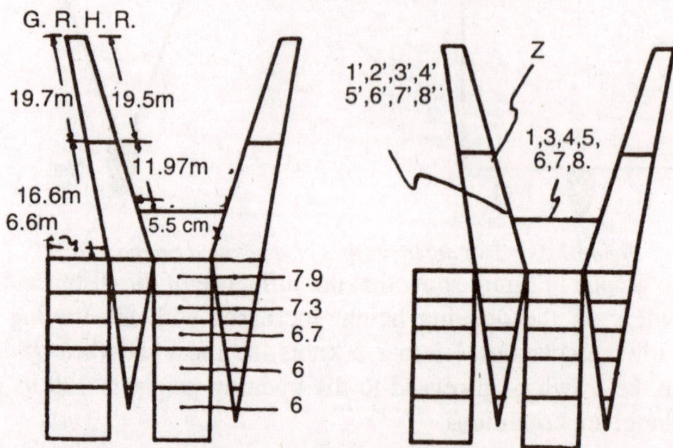

Figure 1.27: Diagram of the improved net for a higher opening with indication of measuring points for different experimental hauls.

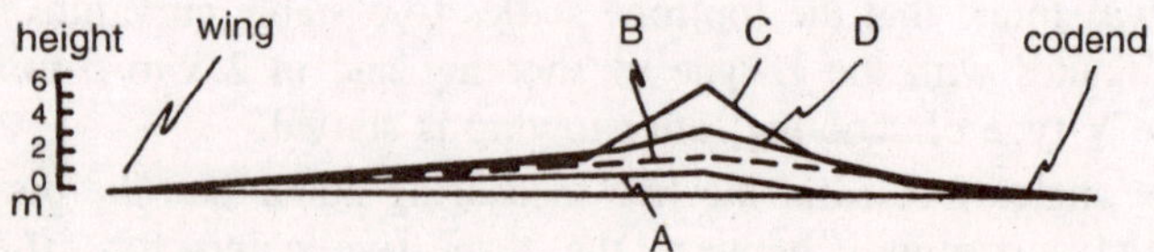

Figure 1.28 : Comparison between the original net and the improved one. A=Net without floats; B=Original net, 2.3 knots towing speed; C=Improved net, 1.7 knots; D=Improved net, 2.3 knots towing speed.

Table 1.3

Ex. No.	*Distance between the boats*	*Towing Speed*		*Height*		
		Boat	*Net*	*The Middle of the headline bosom*	*The quarter point*	*Difference*
1-1'	420 m.	2.2 kt.		3.3 m.	2.15 m.	1.15 m.
	370 „			3.9 „	2.4 „	1.5 „
2-2'	300 „	2.5		The middle point of the wing. 1.7 m.	1.9 „	0.2 „
	500 „			1.7 „	2.0 „	0.3 „
3-3'	400 „	2.2 „		3.4 „	3.0 „	0.4 „
	600 „	2.1 „		3.2 „	2.9 „	0.3 „
	300 „	2.3 „		3.7 „	3.15 „	0.55 „
	500 „			3.5 „	2.95 „	0.55 „
4-4'	280 „	3.5 „		0.65 „	0.35 „	0.3 „
	400 „	3.3 „		0.6 „	0.3 „	0.3 „
	300 „	2.66 „		0.8 „	0.4 „	0.4 „
5-5'	300 „	1.66 „		5.3 „	3.0 „	2.3 „
	350 „	1.6 „		5.7 „	3.05 „	2.65 „
6-6'	320 „	1.6 „		5.5 „	3.1 „	2.4 „
7-7'	530 „	1.66 „	0.3 kt.	5.2 „	2.95 „	2.25 „
	530 „	1.8 „	0.6 „	5.3 „	3.05 „	2.25 „
	530 „	1.86 „	1.0 „	4.85 „	2.80 „	2.05 „
8-8'	380 „			4.3 „	2.6 „	1.7 „
	270 „			5.0 „	2.85 „	2.15 „
	450 „	2.4 „		3.6 „	2.45 „	1.15 „
	350 „			4.45 „	2.8 „	1.65 „
	400 „			3.5 „	2.35 „	1.15 „

Table 1.4

Distance between the boats	*Angle of the Footrope and Towing Direction*				
	a	b	c	d	e
400 m.	26.5°	32.5°	34°	38°	46°
450	35	35	37.5	46	52
500	32	36	38	53	55

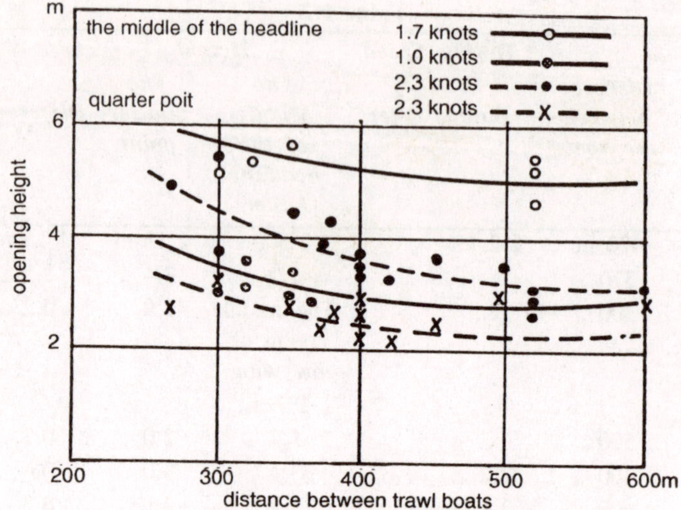

Figure 1.29 : The effect of towing speed on the influence of the distance between the trawlers on the opening height.

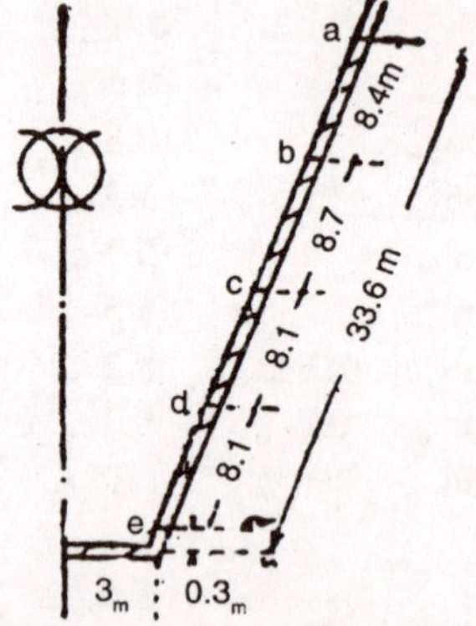

Figure 1.30 : Measuring points for determining the curvature of the footrope.

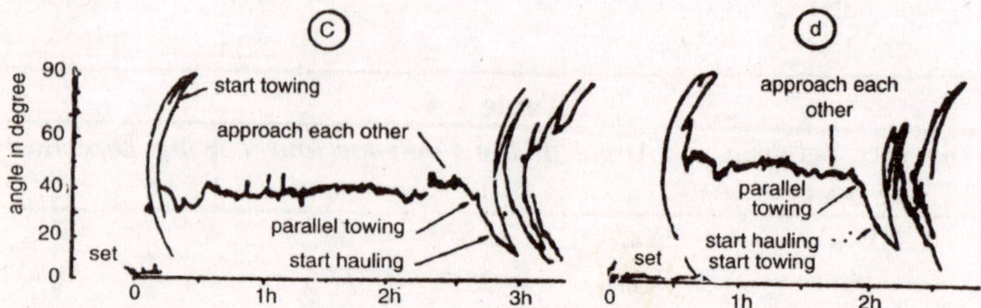

Figure 1.31 : Example for the recordings of the footrope indicators.

Table 1.5

Distance between the boats	*Distance between wing tips*	*Span of foot-rope bosom (6 m. long)*	*Depth of footrope curvature*
400 m.	41.1 m.	5.1 m.	29.3 m.
450	45.6	5.4	27.8
500	48.6	5.5	26.3

Just before hauling the two boats approach each other and then tow for a short while parallel and at about 10 m. distance to force the whole catch into the codend. The changes in the curvature of the footrope during this manoeuvre is shown in Table and figure. For the stages 1, 2, and 3 the distance was 500, 450 and 400 m. respectively. The stages 4, 5, 6 and 7 refer to the approaching, and 8, 9, 10 and 11 to parallel towing with reduced distance (about 10 m.). In addition to the quantitative data given it was found that after about 10 minutes the footrope has settled in a stable curvature according to the reduced distance between the trawlers and that extending the time of parallel trawling to get the warps parallel would be useless.

OTTER TRAWL

Measuring experiments with this trawl by means of net height meter and footrope indicator are being carried out in the Yellow Sea since 1953. The construction of the net in question and the measuring points for different experiments are given in figure elsewhere in the chapter.

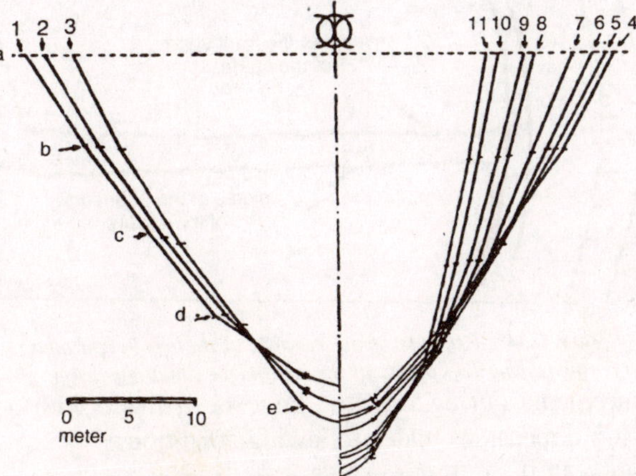

Figure 1.32 : Shape of the footrope, as calculated from the angle measurements, the trawlers keeping different distances.

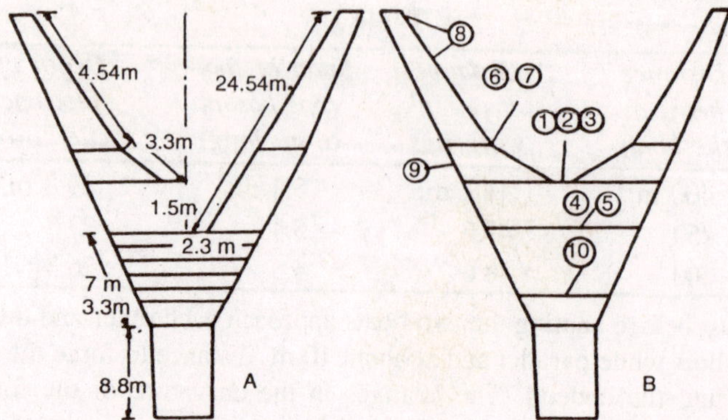

Figure 1.33 : One boat otter trawl net used during the measuring experiments. A=Construction; B=Measuring points for the net height meter referring to the different experiments.

For measuring the opening height an improved meter was used with which two points of the net height could be measured simultaneously. An example of the readings obtained is given in figure.

The behaviour of the trawl and the variation of the distances between different parts, as well as the angles of net, sweep lines and warps during the shooting operation have been studied thoroughly. One example of the configurations drawn according to the measurements obtained. It was, for instance, found that the otter boards reach the bottom first and the net follows some time later.

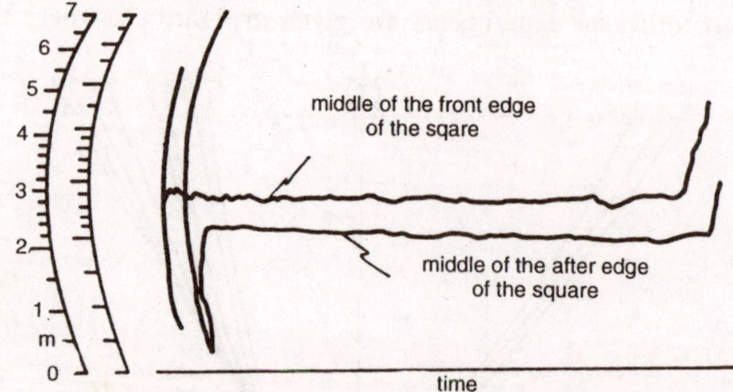

Figure 1.34 : Example of the records of the net height meter measuring two points of the net height simultaneously.

Measurements during trawling have been made with a trawl of the following dimensions under following conditions:

Headline 38.9 m. Footrope 53.6 m. Sweep lines 90 m.; Warp length 250 m. Water depth 70 m. Warp angle to horizontal 13 degrees,

Table 1.6

Time Ex. Mark	Mins. 0	1	2	3	4	5	6	7	8	9	10
	Angle between footrope and towing direction										
a	27.5°	27.5°	27.5°	28.5°	30.0°	28.0°	27.5°	27.0°	26.0°	25.0°	24.0°
b	32.5	34	34	34.5	33.5	32.5	30.5	29.5	27.5	27.5	22.5
c	34	34	34.5	35	35.5	36	34.5	32.5	32	31	28.5
d	43.5	43.5	43.5	43.5	43.5	43.5	42	43	44.5	43	43
e	46	45	45	45	45	45	46	46.5	47	48	49.5

N.B. Initial distance between the boats: 400 m.

Towing speed 2.5 knots, Bottom mud, Wind fair, 10 glass floats 26 cm. diameter on headline bosom, 19 glass floats 20 cm. diameter on each wing.

The height of the wings and the square is obviously unsatisfactory. The main reason for this is the fact that with these trawls the main pull acts on the headline. If, by changing the length relation between headline and footrope the pull would be shifted to the footrope, it

Table 1.7

Measuring Points	***Vertical distance from footrope***
Wing point 1	0.72 m.
Wing point 2	1.39 m.
Wing point 3	1.82 m.
Wing point 4	1.93 m.
Middle of headline bosom	2.03 m.
Middle of after edge of square	1.93 m.
Middle of front edge of throat	1.65 m.
Middle of front edge of codend	1.50 m.

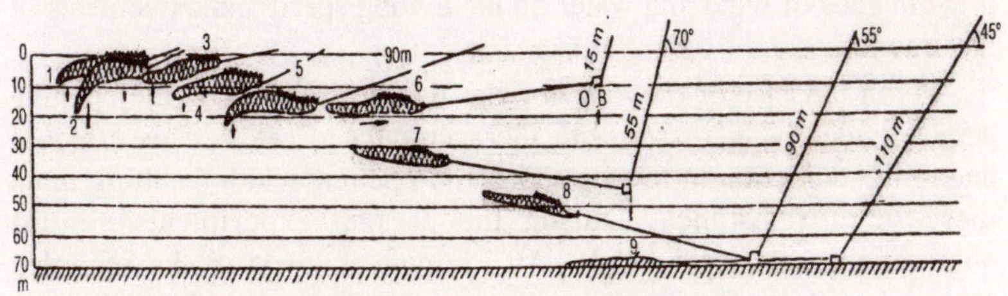

Figure 1.35 : Behaviour of the trawl during shooting.

may tend to cut into the bottom; This could be overcome by means of a suitable bobbin footrope. Furthermore, the suitability of the net construction. has to be checked including the flapper which in the present form tends to restrict the proper water flow into the codends.

When the trawler changes course the speed of the net and the horizontal opening decrease. Consequently the vertical net opening height increases because of the reduced resistance on the floats and the slack in lines and webbing.

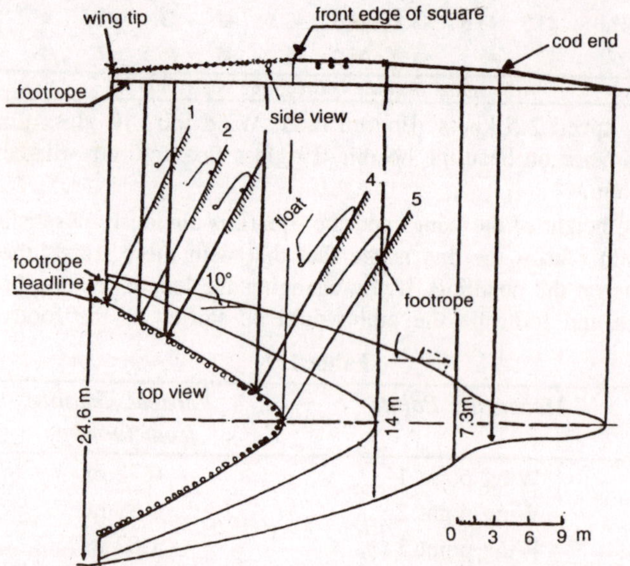

Figure 1.36 : Shape of the net as found by means of the net height meter.

The influence of wind direction was found to be very noticeable. The opening was much higher with head wind than with fair wind. Furthermore, with head wind the opening height oscillated considerably while with fair wind it was stable. This effect, of course, is caused by the influence of wind and water on the towing speed and movements of the trawler.

As in the case of trawling for flat fish in Bristol Bay, the opening height gradually increases with the accumulated catch. This effect is due to the reduction in towing speed and opening width resulting in an increase of resistance. During the present experiments similar observations have been made. An example is given of the recorded heights in figure. During this haul the catch amounted to 6,000 kg.

TABLE 1.8 : The Net Height when Towing with Head and Fair Wind

Measuring Point	*Average height with head wind*	*Average height with fair wind*
Middle of headline bosom	2.6 m.	1.8 m.
Middle of after edge of square	2.4 m.	2.1 m.
Quarter point	2.25 m.	1.8 m.

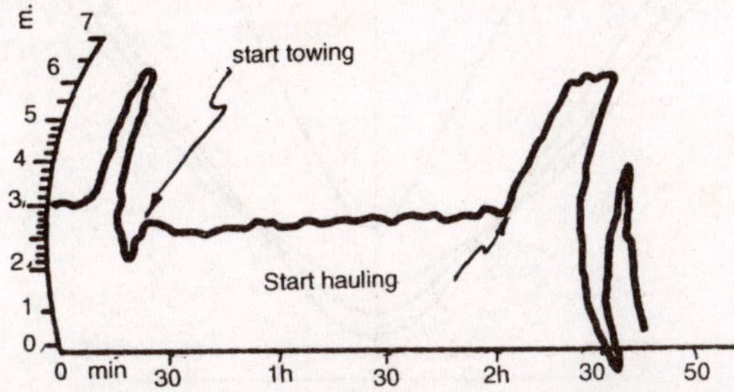

Figure 1.37 : Record of net height meter attached to the middle of the headline bosom, showing the influence of increasing amount of catch.

As the buoyancy of the glass floats is considered to be unsatisfactory at higher towing speeds, experiments were carried out with the addition of wing shaped floats.

It was found that the usual net covering of the glass floats is unfavourable, and plastic floats with a smooth surface give better results. The wing shaped floats are superior because of their hydrodynamic lifting power. In order to make the best use of this lifting power, sufficient surplus of webbing should be provided in the net mouth to allow for a high opening.

COMPARISON OF BOTH METHODS

Table and figure elsewhere in the chapter give a comparison of

Table 1.9

	With glass floats alone	*With glassfloats and wing-shaped floats in addition*
Height of net opening	1.60 m.	2.95 m.
Height of left quarter point	1.35 m.	2.20 m.
Towing speed	3.7 knots	3.7 knots

two-boat and one-boat trawling. In two-boat trawling the duration of a voyage is 20 to 30 days. In the East China Sea, otter trawling is excellent for catching hair tail or prawn and two-boat trawling for guchi *(Nibea argentata, Pseudosciaene manchurica, Nibea nibe,* etc.).

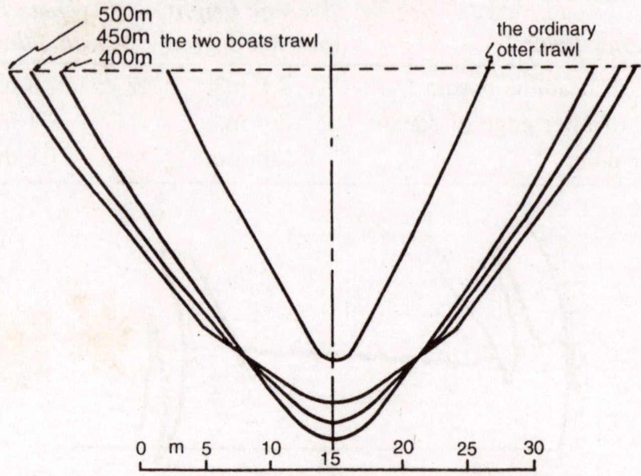

Figure 1.38 : Curvature of footrope and opening width of the one boat otter trawl in comparison with the two boat trawl at different distances between the towing boats.

	Two boat trawl	*Otter trawl*
Depth	54 m.	95 m.
Warp length	750 in.	300 in.
Sweepline length	-	90 m.
Footrope length	74 m.	53 in.
Towing speed	2 knots	3 knots
Angle between warps	—	12.5 degrees

THE USE OF ECHO-SOUNDING

Recently attempts have been made to accelerate the development and improvement of trawling gear by using measuring instruments. The problems connected with this new approach, such as, for instance, identifying those important characteristics of a trawl which should be measured, and the different ways of doing this without affecting the behaviour of the gear, need not be dealt with here. Instead, only one of the numerous measuring or observing methods will be discussed, the echo-sounding method.

To the author's knowledge, the first experimental observations of a trawl in action by means of an echosounder were carried out by Wood and Parrish. They used the sounder from a rather big boat which was towed by the trawling vessel and in this way operated over the gear.

The writer, as an employee of the German "Institut fur Netz- and Materialforschung", has since 1953 used a motor-driven rubber boat about 16 ft. long and 7 ft. in beam, with a battery-driven echo-sounder. The main advantage of this boat is its good manoeuvrability and the unlimited possibility of circulating freely. Numerous observations on trawling gear were carried out with this equipment, mainly to define the opening height of bottom trawls and their contact with the ground, but also on the behaviour of fish coming near such gear.

Later, the equipment proved of even higher value for studies directed towards the development of pelagic trawls. The following echograms were taken during pelagic trawl experiments carried out with the German Fisheries Research Vessel *Anton Dohrn* in June 1957, in the Baltic Sea. Three one-boat pelagic trawls were tested, all being equipped with hydrofoil otter boards, but with different nets and different riggings. As only the suitability of the method will be discussed, a description of the gear is not needed.

DEPTH OF THE NET AND OPENING HEIGHT

Two important characteristics of a pelagic trawl can be very easily, and also exactly, measured by echosounding: the actual depth of the net and the opening height. Figure elsewhere in the chapter shows the changes in these characteristics connected with the increasing length of the warps. With this gear, under the conditions in question, a lengthening of the warps by 25 m. caused an increase in depth of 6 to 8 m. Furthermore, with increasing warp length, the opening height decreased in favour of the opening width. With 100 m. warp length the opening height was 10 m. whereas with 175 m. and more it decreased to only 8.5 m.

POSITION OF THE LASTRICHES

Besides the headline and footrope, the lastriches also give good traces. They appear in all sections of fig. 2 and both are in the same or almost the same depth, except the last section, where a difference of nearly 1 .5 m. is shown. In this case, the length of the warps was not equal. It was observed that an inequality of about one fathom may cause the net to assume an oblique position, with the lastrich connected with the shorter warp being as much as 3 m. higher than the other lastrich.

POSITION OF THE OTTER BOARDS

Furthermore, the last section of figure elsewhere in the chapter shows that with this gear, under the given conditions, the position of the otter board was not level with the middle of the net opening, but it travelled about 2 m. deeper.

SLOPE OF THE LEGS

During another observation of the same gear, more care was taken to establish the slope of the legs. Coming from astern, the sounding boat was steered accurately over the legs to the otter board and then up the warp. Here the otter board is also about 2 m. below the middle of the net opening. Consequently, the legs have a down-going tendency, especially the lastrich one. A good trace is given by the heavy weight, which is fixed to the leg a short distance in front of the lower wing tip to keep it down.

Figure elsewhere in the book shows traces of another gear equipped with four legs to each wing, danleno and bridle. The record shows that the danleno travelled about 1 m. below the middle of the net opening. The bridle is not well recorded. The otter board obviously had the same depth as the middle of the net opening. The clear reproduction of the four legs, in this case, has a special interest. When the net was hauled in, it appeared that both lastrich legs of this side of the gear were broken, causing the net to be completely torn. The echogram proved that this damage had not occurred before hauling and the measured values could safely be relied upon. Likewise, the echosounding method can be successfully used to check the performance of a gear during trawling operations.

POSITION OF THE NET

Figure elsewhere in the chapter shows traces of the third gear which was equipped with two legs from each wing to the otter boards. Besides the depth of the different parts of the gear and the height of the net opening, the slope of the legs and the weight keeping down the lower wing can easily be recognised. In this case, the influence of the towing speed on the slope of the legs is obvious. At lower speed (left) they have a greater slope, and at higher speed (right) they are almost stretched. The position of the

otter board to the net opening is naturally influenced by the speed. At lower speed, the board is almost at the same depth as the middle of the opening, but at greater speed it travels about 2 m. higher. The most interesting feature of this record is the trace of the net (right). It shows that the net has no horizontal position at all. To control this, the sounding boat, coming from astern, was steered over the whole net from the codend to the opening and then followed one pair of legs to the otter board and up to the warp. It showed that, with this gear, under the given conditions, the codend travelled at almost the same depth as the footrope and, consequently, about 5 m. below the middle of the net opening. Furthermore, this record shows that in these cases the headline exceeded the height of the upper wingtip by about 1.5 m. (right, higher speed) to about 2.5 m. (left, lower speed) whilst the depth of the footrope did not seem to differ much in relation to that of the lower wingtips. The reason for this may lie in the fact that heavy weights were keeping down the lower wings, whilst the upper wings had no additional lifting device.

Although incomplete, these examples clearly indicate the value of the echo-sounding method for a quick test of the technical performance of certain characteristics of pelagic trawls. Such a test provides an objective background for estimating the probable catching ability which, as a second step has, of course, to be proved by real fishing.

DECREASE THE TOWING RESISTANCE OF TRAWL GEAR

For optimal economy in trawling, the towing resistance of the trawl gear should be as low as possible. It is well known that the commonly used plane otter boards are very unsatisfactory from a hydrodynamic point of view. Moreover, nets of manila or sisal, i.e. natural fibres, have to be made of thicker twine than those of synthetic fibres (e.g. nylon or Perlon). So, two simple ways are open to decrease the towing resistance of trawl gear:

1. Hydrofoil otter boards.
2. Nets made of thinner synthetic twine.

To test the effect of these changes in the German herring bottom trawl, measuring experiments were carried out with the Fisheries Research Vessel *Anton Dohrn,* during June 1957, in the Baltic Sea.

These experiments consisted of measurements of:

(a) Towing resistance of the complete gear.

(b) Towing resistance of that part of the gear behind the otter boards. The difference between (a) and (b) gives the share in total towing resistance of warps and otter boards.

(c) Towing speed.

(d) Distance between the otter boards.

(e) Height of the net opening.

(f) Angle of attack of the otter boards.

(g) Angle of attack of the kites.

Weather, course, depth, bottom conditions, propeller revolutions, "cutoff" and boiler pressure, were also taken into account.

Of the total of 27 tows, 17 had to be rejected because of changes in weather or bottom conditions, unsatisfactory conformity in the size of the gear opening or damage to the gear, such as broken lines or torn net. Of the remaining 10 tows, 5 were made with each of the two following types of gear:

1. Common German herring bottom trawl with 160 ft. ground rope, manila net and common otter boards. (Hereafter called "Common gear.")
2. German herring bottom trawl rigged in the same way with a net of the same construction and size but made of thin Perlon, and with "Siiberkriib" otter boards. (Hereafter called "Experimental gear.")

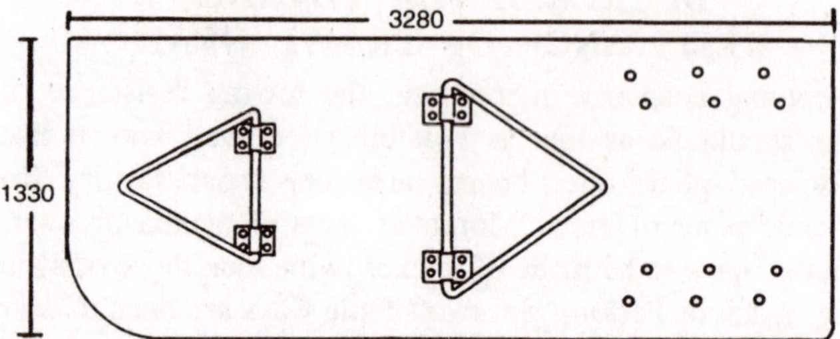

Figure 1.39 : Construction drawing of the common Otter-board.

Results

The measurements were made at a water depth of 80 to 100 m. on a rather hard clay bottom and with 225 fathoms of warps (22 mm. diam.).

The angle of attack of the otter boards was ascertained from the traces caused by the bottom friction on the iron shoe plates of the boards. These traces give good average values which were found to be as follows:

Common otter boards about 35 degrees "Suberkrub" otter boards about 12 to 15 degrees

These values are almost optimal for both types of boards.

According to the "*Gottinger Messungen*", and with regard to the influence of the water only, the following values are calculated for resistance and shearing force at 3-8 knots towing speed:

	Resistance	*Towing Shearing Force*
Common otter board	0.8 tons	1.1 tons
"Suberkrub" otter board	0-2 tons	1.1 tons

This shows a decrease of resistance with the "Suberkrub" type of boards of 1.2 tons (for both boards) or 75 per cent. compared with the common boards. But this calculation neglects the influence of the bottom friction, which should be higher for the common boards with their long lower edge. The measurements actually showed a difference of 1-6 tons. The 0.4 tons exceeding the calculated value is at least partly due to the lower bottom friction of the "Siiberkriib" boards under the existing bottom conditions.

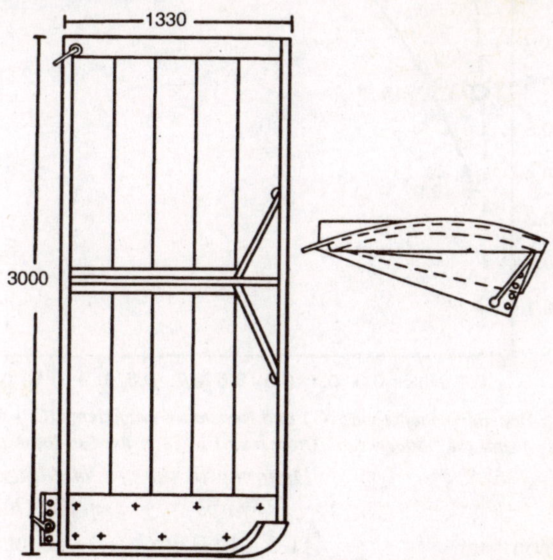

Figure 1.40 : Construction drawing of the "Siiberkriib" Otter board used in the experiments.

The size of the net opening should be as similar as possible for both types of gear. The width of the opening was not really measured. Instead, the distance between the two warps 1I m. behind their cross-over in the sliphooks was controlled. This, of course, does not give accurate values, but is, at least, a basis for comparing the opening width of equally rigged nets. This is an old fisherman's method for controlling the behaviour of the boards. The following values were obtained:

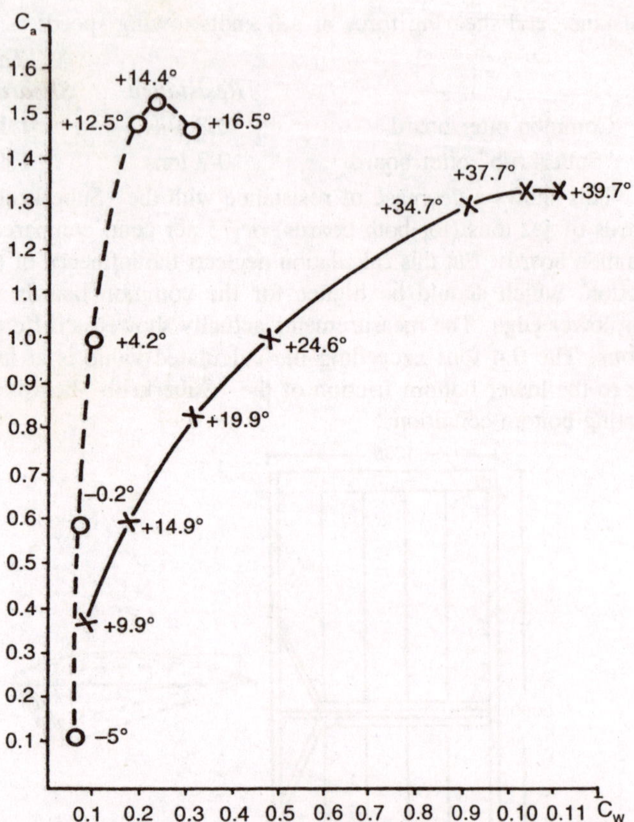

Figure 1.41 : Buoyancy coefficients (C_a) and Resistance-coeficients (C_y) for a common Otterboard (—) and the "Suberkrub" Otter board (- - - -) used in the experiments.

	Distance of the Warps	*Calculated Distance of the Otter Boards*
Common gear	11.5 - 12.0 cm.	about 48 m.
Experimental gear	13.5 - 14.0 cm.	about 55 m.

The slight difference indicates that the shearing power of the "Suberkrub" otter boards was too strong, at least for the light Perlon net. As the most simple way to decrease the shearing power, it was suggested that the angle of attack should be made smaller, but as this could lead to fouling the gear when shooting, it was not tried. It was not possible to reduce the size of the boards on board ship, so this difference was accepted as of no great importance.

The opening height was measured by means of an echo-sounder, installed in a motor driven rubber boat. The following values were obtained:

	Distance from the Bottom		
	Headline	*Ist Kite*	*2nd Kite*
Common gear	3.0-3.4 m.	7.0 m.	12.0 m.
Experimental gear	2.5-3.5 m.	6-0-7-5m.	10.0-12.0m.

Because of the greater width of the net opening, the false headlines of the experimental gear had to be slightly lengthened. The conformity thus obtained in the net openings was regarded as satisfactory.

The angle of attack of the kites was measured by the jelly-bottle method. The values lay between 28 degrees and 34 degrees, a favourable range.

Despite efforts to keep the same **towing speed** in all experiments, a deviation between 3.6 and 4.1 knots could not be avoided. The measured values of the towing resistance, therefore, had to be converted to an average speed of 3.8 knots assuming, as conventional, the resistance being proportional to the square of the speed. The speed was measured by means of a "Kempff"resistancelog.

The towing resistance was measured on both warps, close behind the sliphook (total resistance), as well as on both bridles, close behind the otter boards (resistance mainly caused by the net). As warps, bridles, danlenos, legs, kites and false headlines were the same for both gears, differences between the measured values are due to the otter boards and the net bag. Converted to a speed of 3.8 knots, the following average values were found (.the range of deviation is shown below in brackets).

This comparison shows a very remarkable difference, the experimental gear offering about 30 per cent. less resistance. Of this 30 per cent. about 24 per cent. was due to the hydrofoil shape of the boards and about 6 per cent. due to the lighter net.

This decrease of towing resistance means that, with the same engine power, the experimental gear could be towed about 0.7 knots faster (i.e. at 4.5 knots) than the common gear (3.8 knots), or its size could be increased by about 30 per cent. or, thirdly, a corresponding amount of fuel could be saved.

OTTER TRAWL GEAR AND TOWING POWER

It is a well-known fact that bigger boats use bigger trawls. But there exists no information about the common relation between size or

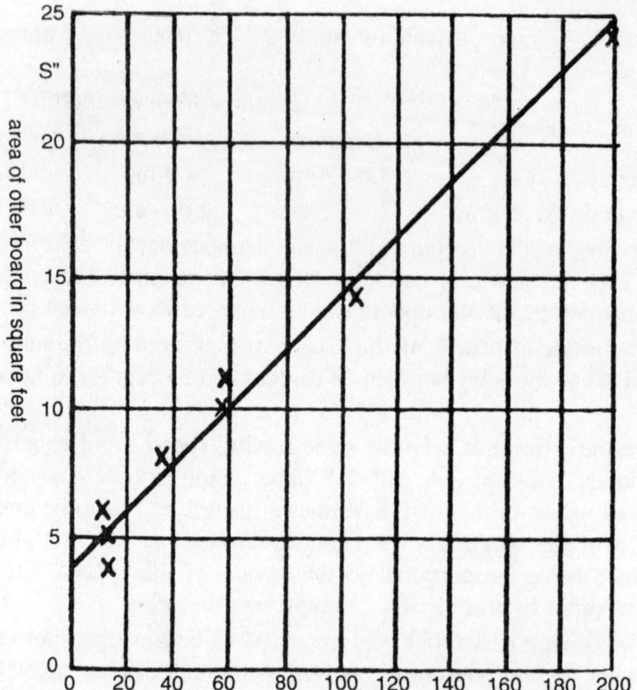

Figure 1.42 : Relation between h.p. of engine and area of one otter board.

engine power of the trawler and the size of the trawl gear and between the size of the trawl net and the size of the otter boards. Therefore, data has been collected mainly from trawlers in India and also from a few Japanese trawlers which are presented below.

ENGINE POWER AND SIZE OF OTTER BOARDS

Figure elsewhere in the chapter shows that in the common use the area of the otter boards is proportionate to the h.p. of the engine. If the area of one otter board is called S" and the h.p. of the engine P, the relation found can be expressed by the following equation:

$$S'' = 0.105\,P + 4$$

The ratio between the length and the width of the otter boards usually is 2 : 1 approximately. If B denotes the width, the length will consequently be 2B. Then B can be calculated by means of the following formula:

$$B = \sqrt{\frac{0.105\,P + 4}{2}}$$

SIZE RELATION BETWEEN OTTER BOARDS AND NETS

The hydrodynamic resistance of the net and the boards is proportionate to their area. Assuming that the otter boards of the trawls included in the collection data are being worked at approximately the same ratio of lift to drag, not only the lift needed for keeping the net mouth open but also the resistance of the otter boards should be *proportionate* to the size of the net.

The size of a trawl net is usually represented by the length of the headline. For purpose of comparison, therefore, the area of a trawl net can be represented by the square of the headline length.

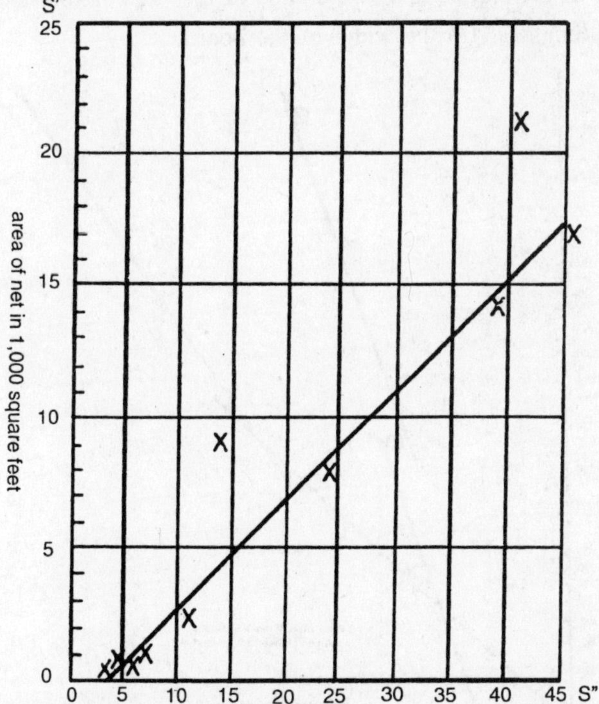

Figure 1.43 : Relation between area of one otter board and the area of the net.

The relation between the area of the otter boards and this expression of the area of the nets, is given in fig. 1.43.

If S" denotes the square of the headline length, the values found can be expressed by the following equation:

$$S' = 415\ S'' - 1000$$

THE ENGINE POWER (H.P.) AND THE SIZE OF THE NET

Using the equations given above, the relation found between the h.p. of the engine and the size of the net can be expressed by the following equation:

$$L=\sqrt{43.6\,P+660}$$

where L is the length of the headline and P the h.p. of the engine.

Weight of the Otter Boards

The weight of the otter boards was found to be proportionate to the h.p. of the engine and to the cube of the $\frac{a+b}{2}$ expression where a is the length and b the width of the board.

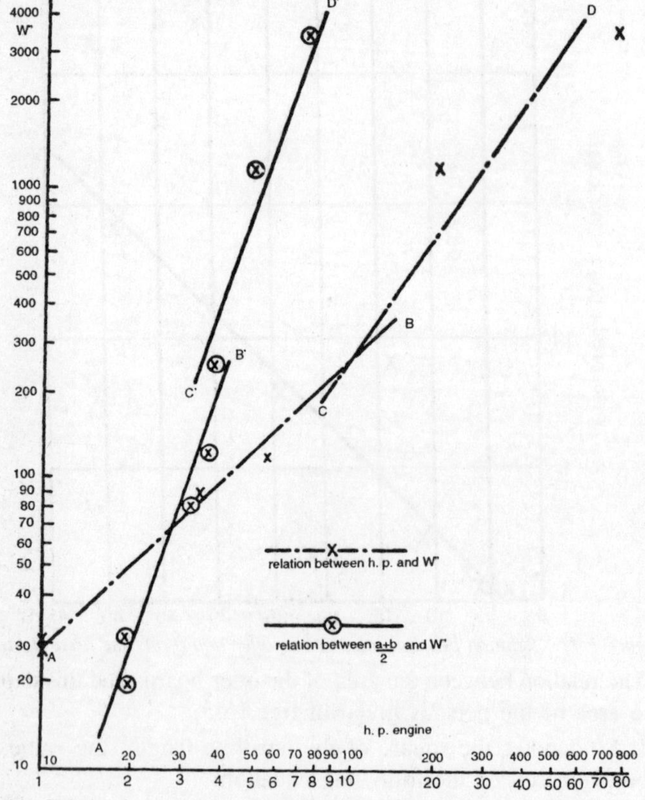

Figure 1.44 : Relation between h.p. of engine and weight of one otter board and relation of average length of otter board in feet and the weight of the board.

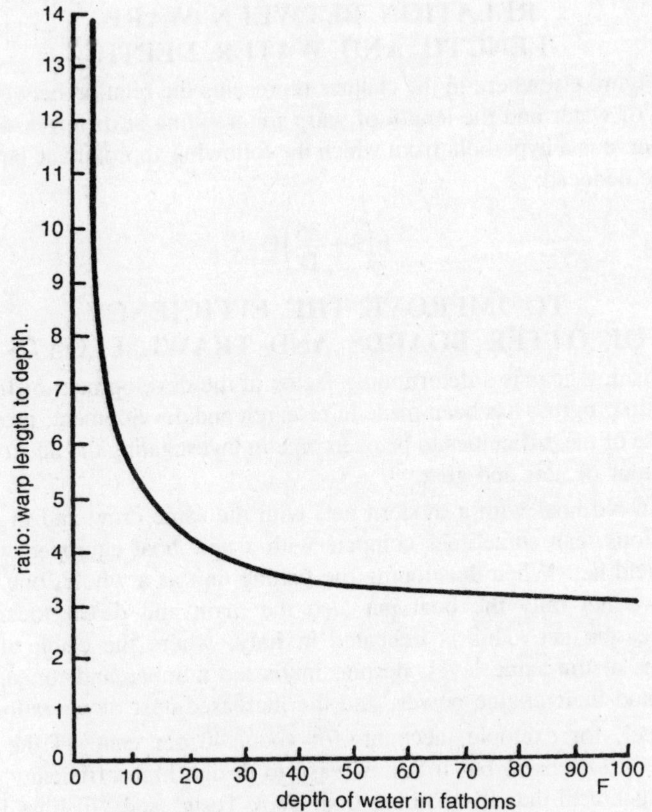

Figure 1.45 : Relation between the water depth and the ratio of warp length to water depth.

The findings shown graphically on logarithmic paper in figure elsewhere in the chapter can be expressed by the following equations:

Up to 100 h.p. W″ = 2.7 P

100 to 66 h.p. W″ = 6.5 P - 400

Up to 4 ft. average size of board W″ = $3.2\left(\frac{a+b}{2}\right)^3$

More than 4 ft. average size of board W″ = $7\left(\frac{a+b}{2}\right)^3$

where W″ is the weight of the board (tbs.), P the h.p. of the engine and $\frac{a+b}{2}$ the size of the board (ft.). 2.

RELATION BETWEEN WARP LENGTH AND WATER DEPTH

Figure elsewhere in the chapter represents the relation between the depth of water and the length of warp for trawling at different depths. The curve is a hyperbola from which the following approximate equation can he deduced:

$$F\left(3+\frac{25}{D}\right)D$$

TO IMPROVE THE EFFICIENCY OF OTTER BOARDS AND TRAWL FLOATS

Fishing gear is a determining factor in the development of fishing but little progress has been made in research and development, probably because of the difficulties to be overcome in investigating the underwater behaviour of nets and gear.

An old boat with a modern net, with the same crew and in equal conditions, can sometimes compete with a new boat equipped with a low yield net. When developing the fishing unit as a whole, one must improve not only the boat but also the main and direct means of capture, the net. This is indicated in Italy, where the catch of fish remains at the same level, despite increased number and tonnage of boats and their engine power, and the increased cost of operations.

Fuel, for example, accounts for about 40 per cent. of the total operating expenses, but it is now easy to design high efficiency hulls using technical data in "Fishing Boat Tank Tests" and "Fishing Boats of the World", both published by F.A.O., and so cut fuel costs. But trawl nets and gear are still made by rule of thumb, based on practical experience, and not on hydrodynamic laws. However, technicians in many countries have been studying how to increase the horizontal and vertical openings of the mouth of the trawl by the use of otter doors and floats.

OTTER BOARDS

The boards are rectangular sections which have badly finished surfaces because of the frame, reinforcements, nuts, etc.

Resistance to towing at a determined speed depends upon the area and angle of attack and is influenced by the finish of the surface and the density of the media. But, in the majority of cases, these factors are completely ignored. The dimensions, weight and angle of attack of the otter doors vary considerably not only in different countries but also

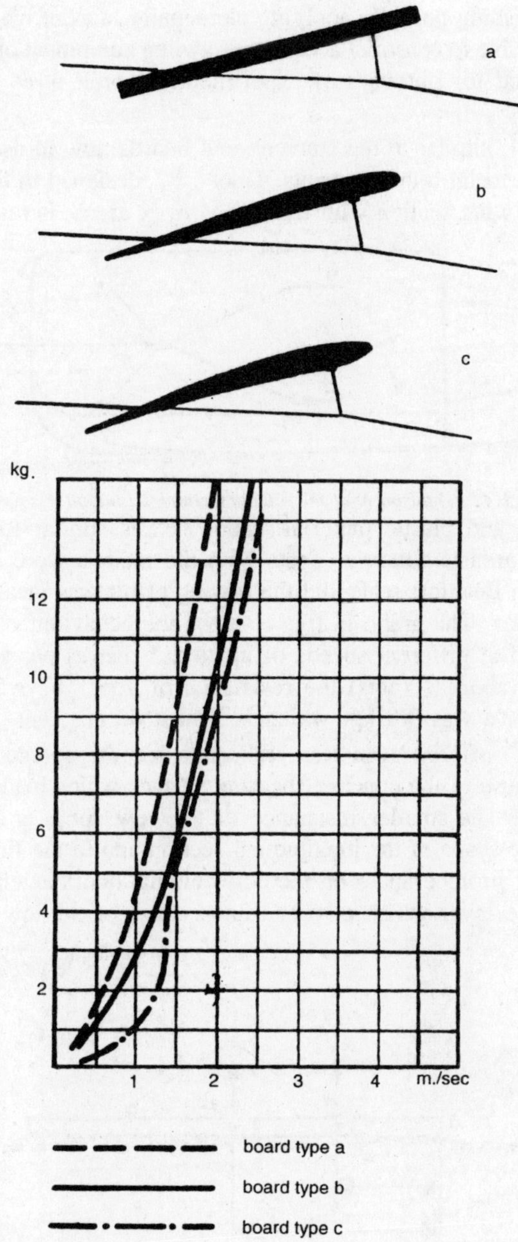

Figure 1.46 : Profiles of trawl boards and graph of performance.

in the same fishing port. By applying clementary laws of mechanics, it shouldbe possible to construct accessory trawling equipment of improved efficiency, and for purposes of experiment we built three models of otter doors.

Door "a", similar to the conventional boards now in use, is made of wood with metal reinforcements. Door .'b", designed in the style of an aeroplane wing section with flat and convex areas, is made of cast

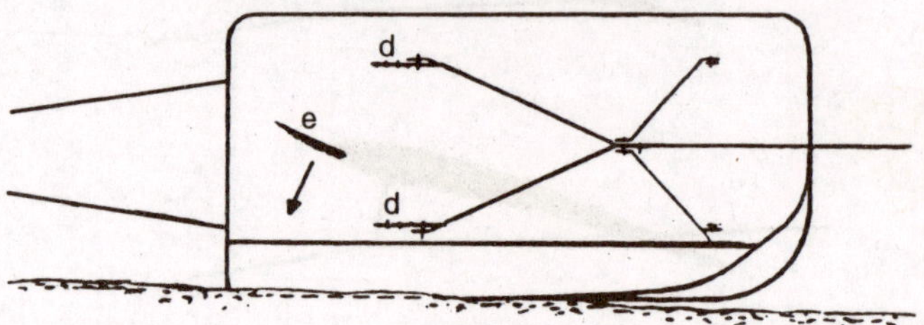

Figure 1.47 : Adjustment of the angle of attack on the otter boards.

or plate iron and plastic material. Door "c" is similar to "b", but the shearing area is concave. Tests with the models were made in a calm sea-zero Beaufort scale-and the towing effort was measured with a dynamometer. The graph in fig. 1 shows the behaviour of the three models towed at different speeds of up to 2.5 metres per second. At 1.54 metres (about 3 knots) the resistance of door "b" was 6.2 kg. while that of "a was 9.9 kg. which is almost 60 per cent. more.

Door "c" offered even less resistance but the spreading power was less because of the concave shearing surface which made the door less efficient. The smaller resistance of the new doors is due to the hydrodynamic shape of the longitudinal section and to the finish of the surfaces. The protuberances on the conventional doors extend through the turbulence layer which is very thin because of the low Reynolds

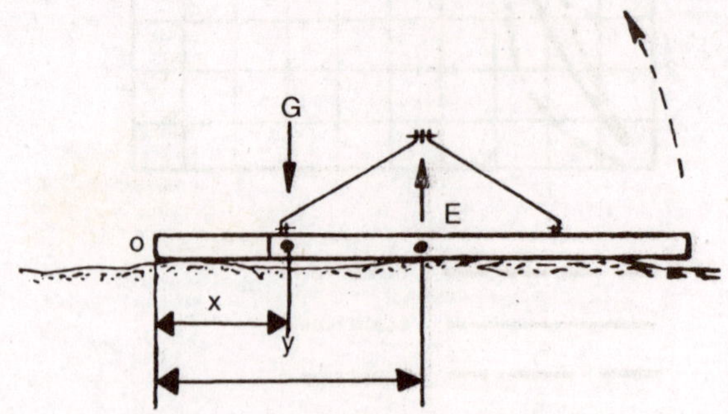

Figure 1.48 : Forces acting on the otter board.

number. The results of the tests and the graphs possibly contain some mistakes, but these can only be eliminated by making tank tests.

To vary the shearing power of the trawl board, the angle of attack may be changed by using different holes in an iron plate "d" for fixing the brackets, or by changing the points of attachment of the towing cables on the back of the boards.

We also studied the problem presented when the otter door falls flat on the bottom or gets wedged when turning on end. This is dangerous on muddy ground and often results in the loss of the net. The fins, are meant to act as stabilisers; they push the front up and the rear down and make the door rest on its after end. The doors may also fall flat because of a momentary interruption in the tension of the towing cables which may easily go slack in a head sea. If the doors fall inwards the damage is not serious, but when they tilt down outwards they get wedged in the bottom when the tension returns in the towing cables and soon the cables snap. An important characteristic of the new trawl doors is that they remain vertical in all circumstances, except on becoming entangled in reefs, etc.

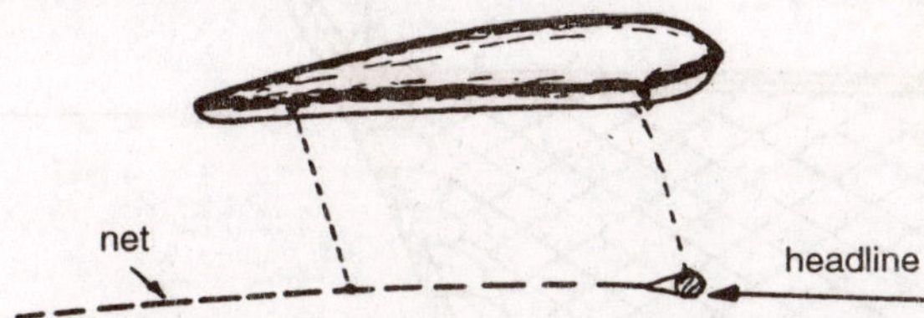

Figure 1.49 : Position of the plastic kite on the headline.

The centre of gravity "G" has been placed very low in the door while the centre of lift "E" has been placed very high so that the product of weight "G" times the distance "X" from the point 0 (tilting moment), is less than the product of the lift "E" times the distance "Y" from point 0 (lifting moment). During the experiments it was noticed that each time the submerged door fell flat on the bottom, it turned back to the vertical position again.

The new door models studied showed the following advantages:

(1) they offer smaller resistance;

(2) they do not fall flat;

(3) they are less liable to dig into the bottom;

(4) they should last longer.

During this study we were unable to tow the door models at the speed corresponding to mechanical similarity, because at a low speed they do not spread out well and advance with irregular movements.

Although we were unable to establish the resistance of full-scale doors, we got enough facts to establish the marked differences among the three models.

FLOATS

Glass balls are generally used to increase the height of the mouth of the trawl net, chiefly because of their good resistance to pressure and their low cost. New types of floats, metallic or made of other materials, are being used in many countries. They may be spherical, with reinforcement rings, or similar to kites. It is known that spheres

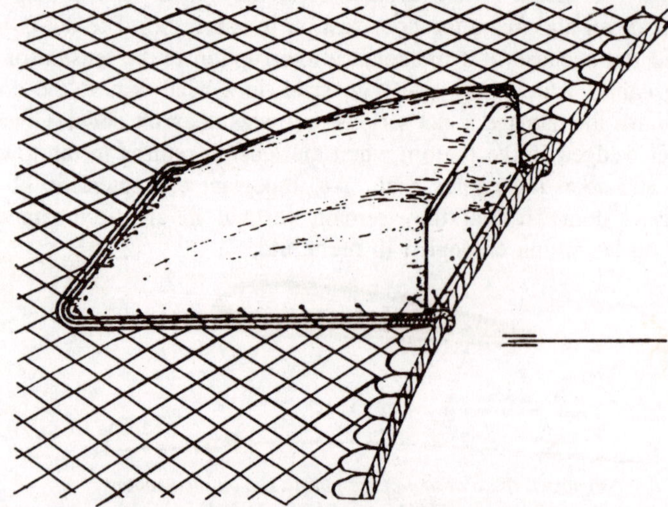

Figure 1.50 : Elevating hood.

have a high coefficient of resistance and with increasing speed they lower the headline. Kites would avoid this disadvantage and we have built a plastic float with a hydrodynamic section. In operation, we tied it to the headline with chains to avoid entangling. We found this device offered a very low resistance to towing compared with the spherical floats and had a fairly good lifting power. Reinforcements of the same material were used to resist the hydrostatic pressures. Kites are more expensive than the ball type floats, but the extra cost should be offset by increased catches.

LIFTING DEVICE FOR TRAWL NETS

The study of, the hydrodynamic floats revealed certain disadvantages while trawling, so a "lifting device" (similar to a prompter's box) was built. It was attached to the top of the net at the back of the headline to increase the opening of the mouth without offering excessive resistance.

The most important advantage of this lifting gear is that it is firmly attached to the trawl net, which makes efficient operation possible and facilitates handling.

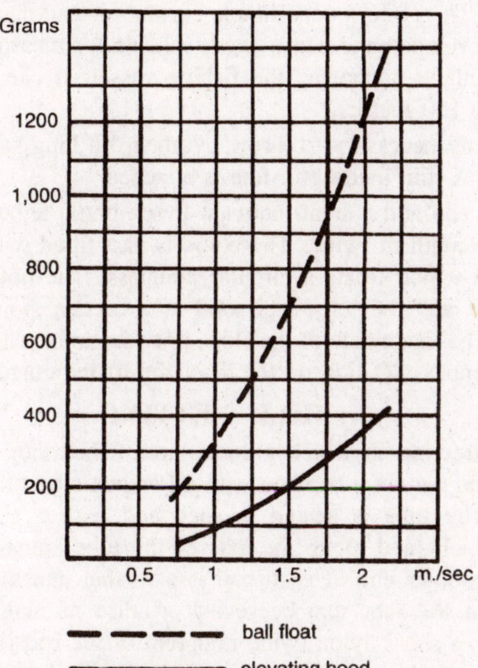

Figure 1.51 : Comparison between towing resistance of a ball float and the elevating hood.

The top of the lifting device slopes down towards the net to form an angle of attack in the direction of the advance. The walls are also of hydrodynamic profile and offer a minimum of resistance. As in the case of the kites, higher speeds increase the lifting power of the device and because of its smaller coefficient of resistance less towing power is needed than in the case of spherical floats.

The graphs compare tests of this lifting device with those of a ball float, and show the respective resistances for the same area and speed. The new lifting device withstands 60 to 80 kg./cm.^{2} hydrostatic pressure.

SIMPLE DEVICES FOR STUDYING THE GEOMETRY OF VARIOUS GEARS

THERE are fishing methods and gear of commercial importance such as bottom trawling, gillnetting and longlining, which are much affected by the strength and direction of underwater currents as they may influence the distribution and behaviour of the fish and the operation of the gear. Fishermen agree that, in certain cases, advance knowledge of the current prevailing at the fishing depth would be a great advantage. Therefore a very simple current-meter for use by fishermen has been developed and constructed, suitable for measuring underwater currents without anchoring the fishing vessel. It can be used in any depth up to 150 metres.

It is only necessary to throw overboard a long stick, weighted at both ends. A thin hauling-in line is attached

to one end and a small buoyant Pyrex bottle tethered at the other by a short length of twine. The bottle is part filled with a hot gelatine solution on which floats a circular compass. The bottle is canted by any current and the jelly sets solid at a certain slope, gripping the compass. The magnitude of the slope provides information on the speed and the compass reading on the direction of the current.

GILL NETTING

The direction in which gillnets have fished can easily be learnt by lashing to the net a bamboo with a Perspex tube full of hot gelatine solution. The tube is sealed at each end with a solid rubber ball through which (and along the axis of the tube) runs a slender brass rod. At the outer end of each ball is a washer and a travelling Wing nut, so that the tube can be sealed. A disc of magnetised ceramic hangs from a short nylon twine midpoint of the rod (inside the tube). The disc, which carries north and south marks, orientates itself resolutely into the magnetic meridian. It is easy to arrange for the gelatine solution to remain fluid long enough in cold water by enclosing the Perspex tube within a wider tube of polythene filled with hot water.

LONGLINING

Fishermen who longline for halibut off the Faroes and in Denmark Strait would like to know how the deep currents strike their lines. With this they could build up a body of knowledge connecting current strength and direction with tidal state, as represented by hours before or after high water at a port of reference. Then they would know when to fish safely near the edge of a bottom declivity at places where uprising water is held to produce good feeding conditions.

There is a very easy way in which the longline fishermen can collect such information. All that is necessary is to fix a small yardarm to his hand/anchor line just above bottom (one simple tie of a lanyard) and hang from it a bottle half filled with hot gelatine solution and half with hot coloured oil. A brass rod screwed into the underside of the bottle's cap runs down the axis of the bottle and carries a directional disc hung from a nylon thread at its free end. Any current will cant the bottle and a sloped frozen interface of known direction will give information about speed and direction of the bottom current when the lines are hauled. Sensitivity is increased by pushing a tight-fitting polythene "cuff" on the bottle, and trailing a tassel from it.

A very simple valve is used to avoid bottle breakage if the pre-heating is too severe, and for interior/exterior pressure equalisation when the hot liquids chill and contract deep in the sea.

GEOMETRY OF FISHING GEAR, PARTICULARLY TRAWLS

Until fully reliable telemetering devices are available for use from ordinary fishing vessels, it is perhaps of value to make observations by using some very elementary gelatine solution devices which present evidence of how towing cables sloped during a tow, their azimuth, the shape and height of a headline, and so on. The slope of a towing cable can be studied by fixing to it (by means of stretched "garters" cut from a motorcar tyre inner tube) cylinders of clear Perspex filled with gelatine solution and each containing a rolling inclinometer. This is contrived from a schoolboy's celluloid protractor mounted on a rod running through the cylinder and weighted so as to remain always pendulous. A weighted pointer registers the slope of the cylinder. The correct slope is registered whatever the aspect of the cylinder on the cable-whether below, above or to either side. The jelly sets and grips the pointer at its protractor reading, thus making a record of cable slope.

In the same way, the shape taken up by a trawl headline during the tow can be easily ascertained. We used six cylinders fixed at intervals along the headline of a very big trawl to establish the slope at each of the points. To measure the opening height, a similar tube was fixed with one end to the middle of the headline, and a length of rope and plummet (longer than the expected headline height) attached to its lower free end. During the tow the line pulls tight and draws the tube into alignment. The liquid interface in the bottle becomes "frozen" so that, on hauling, one learns the headline height by multiplying a known length (tube + rope) into the cosine of the angle of slope realised.

A more ambitious version of the simple Perspex cylinder has a ring aircraft compass in its pendulum so that the direction of the obliquity as well as the slope of a net or cable is learnt. By using several such devices, it has been possible to measure the angle of divergence of a pair of trawl warps and something of the distribution of direction along them. It would doubtless be easy to glean such information of value by using a number of these simple instruments.

A RESEARCH ON SETNETS

Characteristics and Possibilities for Improvement

The immobility which is characteristic for setnets has made it very difficult to make any notable improvements to this gear, which is based on the principle of inducing the fish to enter the trap after leading them to it by obstructing their normal route. After more than thirty years' development of this typical Japanese type of gear, there is still scope for improvement. More thorough research is needed on the behaviour of the fish, and on the physical characteristics of the net construction before really effective improvement can be expected. Bad weather and high tides have for many years caused severe damage and loss of gear. Efforts are being made to prevent this loss of material and labour by improving the construction of the gear, by reduction of the diameter of twine and ropes, and by reducing the tension on the webbing by decreasing the buoyancy of floats and the weight of sinkers.

The most effective way of preventing loss and damage is to completely submerge the usual type of floating setnet so that it becomes a submerged or bottom setnet.

The fish caught in the submerged setnet are mostly of the demersal kind, but many species of fish which normally swim in the upper layers are also caught. In fact, many so-called pelagic species are frequently found near the sea-bottom, especially when they come near to the coast.

In fact it is even possible to catch such pelagic fish as salmon, sea-trout, sardine and mackerel with the submerged setnet.

Some advantages of the submerged setnet over the floating type are: economy in material and labour: less chance of damage by tide and swell, because of reduced resistance in the water. The reduced resistance at the same time allows the net to be used in stronger currents. With the floating type of net, the trapped fish often escape from the bag over the floatline which becomes submerged in bad weather or strong tides; this is avoided in the improved type of net, where the fish are completely enclosed.

When constructing a floating setnet the buoyancy on the floatline is made as great as possible to prevent the net being submerged under the influence of the tide current. On the other hand, the submerging of the net can also be avoided by decreasing its resistance to water currents and waves. With the setnet for yellow-tail the buoyancy of the floatline is decreased so that the floats are allowed to submerge in heavy weather or strong tides; this is because the fish usually stay well below the surface. However, a certain amount of fish does escape over the floatline whenever it submerges.

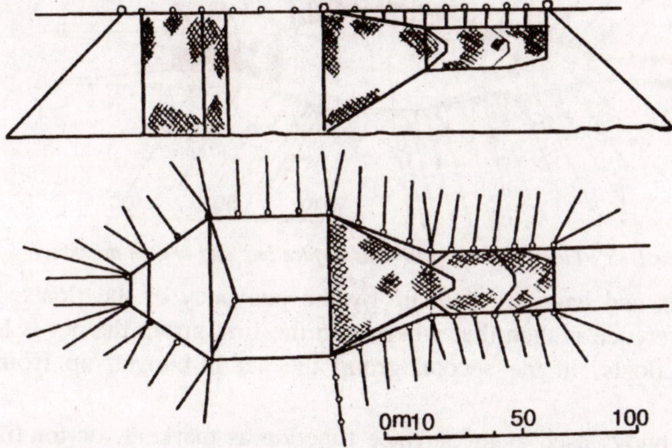

Figure 1.52 : Surface setnet with submerged bag for Yellow-Tail.

With the double-trapping net, one of the recent new types of setnets, the trapped fish is induced to enter a box-net by passing through a funnel. Once inside the box-net the fish cannot escape as they are completely enclosed. This retaining feature of the double-trapping net has been incorporated in the submerged setnet.

CLASSIFICATION OF TYPES

The construction of setnets can be divided in two groups. In the first group there are two kinds, the first of which has the ante-chamber and the front shoulder of the funnel buoyed up so that the floatline is at the surface, while the rear end of the funnel and the entrapping bag are completely submerged.

Another way is to have the whole net submerged while the floatlines are held up partly by submerged floats and partly by surface buoys.

In both cases the webbing is suspended from the floatline and held down by sinkers so the nets could be called surface- and midwater setnets.

In the second group, the sinkers keep the net on the sea-bottom and the webbing forming the walls of the net, as well as the top parts

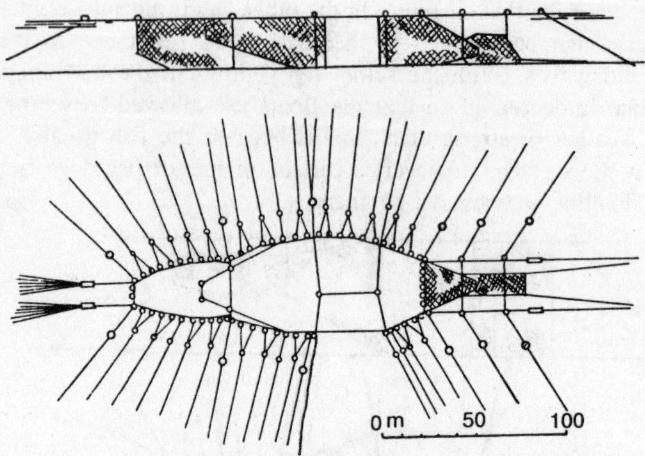

Figure 1.53 : Floating setnet with one surface bag and one on the bottom.

of funnels and bags are held up by the buoyancy of the floats. The main difference is then that whereas in the first group the net is hung from its floats, in the second group the net is buoyed up from its sinkers.

The buoys used at the surface function as markers, or for lifting the submerged net, and have nothing to do with the floating or buoyancy of the net.

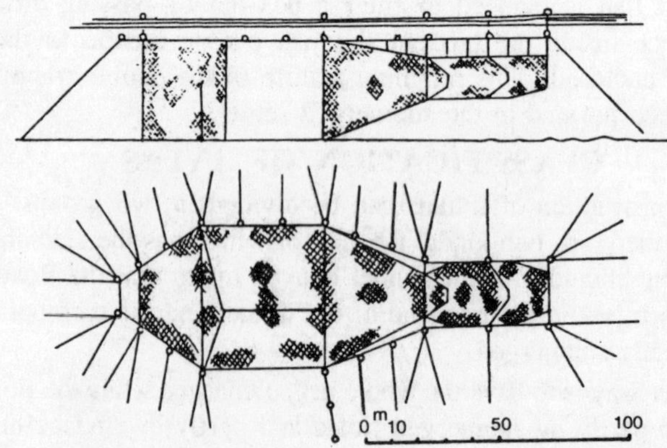

Figure 1.54 : Submerged setnet with one bag in midwater.

This group can be called bottom setnets and is of different construction. In figure elsewhere in the chapter, the net has two wings and one bag; the fish is led straight to the entrapping bag, through the funnel. In other constructions, as in figures elsewhere in the chapter, the net has two funnels and two bags but only one leader. The relations between the buoyancy, sinker weight and the fixing power of the nets classified according to the above grouping.

THE EFFECT OF DIRECTION AND SPEED OF CURRENT ON THE SHAPE OF SETNETS

The influence of speed and direction of current on the webbing structures of several types of setnets was investigated by the author, resistance of the webbing and the holding power of the sandbags were examined and measured; some merits and defects in the construction being ascertained.

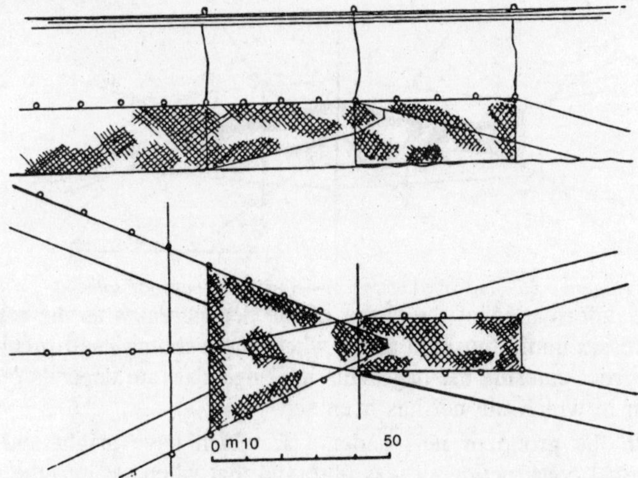

Figure 1.55 : Submerged bottom setnet with one bag on the bottom.

Almost all the above mentioned nets, though there is a slight difference, usually take a forward-bent position against the flow of tide. The bagnet, funnel and the ante-chamber all bend inwards on the up-tide side, while on the lee-tide side they tend to balloon outwards. The deformation is more side pronounced on the up-tide side than on the lee-tide of the tide.

The lifting movement due to current influence of the ante-chamber floor section of the nets in Group 1, was also examined and very little lifting was observed in the case of the nets classed under 1-a, which are left and right symmetrically constructed nets. However, with the increase of the current velocity the bag on the uptide side was pushed down to the sea bottom. In other nets, a lifting movement was generally caused by a current velocity of 13 to 38 cm./sec. ¼ to ¾ knot). The lifting occurred in both tidal directions; when the tide was directed towards the ante-chamber as well as when it was directed towards the bag.

With the 1-b category net, the current directed towards the bag had more effect than when directed in the antechamber. In both cases, the floor of the net lifts completely at a current rate of from 40 to 45 cm./sec.).

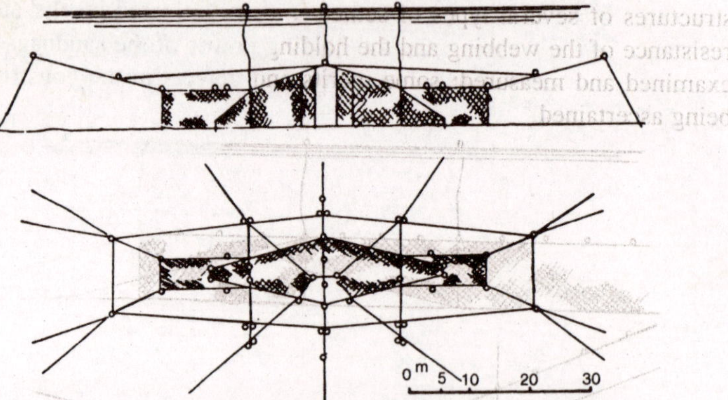

Figure 1.56 : Improved bottom setnet with two bags for salmon.

The deformation of the shape of the net increases as the rate of tide increases until a position arises where the webbing itself precludes the fish from entering the bag. This limiting tidal rate depends on the direction in which the net has been set.

With the group of nets under 1-a, which have aright and left symmetrical construction, it was observed that when set into the tide, the limiting current speed was 38.6 to 51.4 cm./sec.), while in the opposite direction the net remained fixed beyond this rate of tide.

With the revised A type fitted with single bagnet, the current velocity needed to lift .the enclosing walls was found to be 25.7 to 50 cm./sec. (about 1- to 1), while the bag end was lifted already at 19 to 38 cm./sec. (g to knot). Both these nets are of the single bag type.

With the revised B type (1-b) the current speed needed to lift the enclosing walls was found to be 50 cm./sec. (about 1.0 knot) and for the bag end 45 cm./sec. (knot) respectively.

In the second group the 2-a category nets showed little deformation when the current entered at the antechamber, but when the current set in the opposite direction, they were easily deformed. In the first case the bagnet began to be lifted off the sea-bottom at 38 to 51-4 cm./sec.), the entrance of the net was closed and the height curtailed to ¾ of the original.

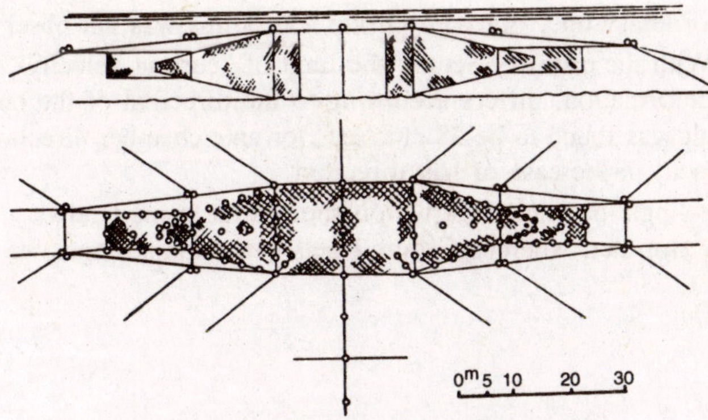

Figure 1.57 : Improved submerged salmon and trout setnet with two bags in midwater.

When the current pressure was acting directly on the bagnet, the entrance of the bag was already closed at 26 cm./sec. (about I knot); at 38 cm./sec. Q knot) the net was near to collapsing. In case of the 2-b category nets, strong deformation was observed over the whole net when set in up-tide direction, while in lee-tide deformation was less, and even at a current velocity of 26 cm./sec. (about I knot), the height of the floatline and the form of the net remained the same, only being bent forward; at 38 cm./sec. there was some decrease in the height of the net; and at 51 cm./sec. (1.0 knot) the height decreased to about one half. Except where small sinkers were used, only slight lifting up of the net bottom was observed. Fish were still being caught even at a current rate of 38.6 to 51.4 cm./sec. (4 to 1.0 knot) when set in up-tide direction, and at 51 cm./sec. in the lee-tide direction.

DISCUSSION OF RESULTS

By comparing the above mentioned net-constructions with the ordinary setnet, the following differences were observed.

1. With the ordinary type net, during low tidal rates the huge surplus buoyancy prevents the floatline from being submerged; whereas with the nets under category I even a small current velocity sinks the floatline.

2. With the ordinary type of net, the ante-chamber starts to be lifted from the sea-bottom at 13 to 37 cm./sec. (about 4 to knot); the funnel at 26 to 32 cm./sec.; and therefore the net is easier to be lifted when the current is directed towards the ante-chamber than when directed towards the bag. With Group I nets, which have only one bag, the net-bottom of the ante-chamber lifted at the same current velocity as that

for the ordinary one. But with Group 2 nets this was not observed.

3. With the ordinary Setnet, the limit of "current velocity" which causes deformation, differs according to the direction of the current. This limit was found to be 38 cm./sec. for ante-chamber direction and 26 cm./sec. in the case of bag direction.

The single-bag net construction showed the same tendency as the ordinary trap net; whereas Group 2 nets can stand stronger lee-tides.

2

FISHING RODS

Before World War II, there was only one quality material from which fine fishing rods were made - split bamboo. The names of fine rod producers such as Thomas, Payne, Leonard, Edwards, Winston, Granger, Orvis and others still are hallowed by connoisseurs of fine tackle. If you can acquire one of these at a fair price, you probably have located a prize.

A very few of the historic split-bamboo rod manufacturers still make rods just as fine as they always did, but their high prices put them more in the "pride of possession" category than in the realm of fiscal common sense.

About the time of World War II a change occurred in the rod business. The enamel-rich Tonkin cane from which the best bamboo rods were made came from China. In anticipation of a dwindling supply, a few of the historic rod makers hoarded as much as they could acquire. While there seems to be no shortage now, this is partly due to the advent and popularity of new space-age materials for fishing rods; in particular, fiberglass, graphite and boron (or boron-graphite).

Capable makers of fine split-bamboo rods still fill the demand for them - a demand influenced by pride of manufacture and nostalgic reverence for the art more than for efficiency. Labor costs of hand craftsmanship have escalated to the point that a fine split-bamboo rod now costs several times what it did a generation or so ago. In addition to high price, split-bamboo is

heavier and more perishable than more modern materials but, as far as diminutive wands are concerned, the pride and pleasure of owning and using them cannot be disputed. In general, however, the newer materials are less costly, more practical and far more popular.

The earliest of these is fiberglass. Since the method of manufacture into rods is somewhat similar to those of other substances, let's discuss it rather thoroughly because knowledge of procedure is important to sensible rod selection and to fishing uses.

When I handled my first fiberglass rod many years ago I didn't think much of it. But, year by year, the quality builders increased their skill to the extent that today many of the fussiest and most affluent anglers would as happily use a quality fiberglass rod as the most expensive one made of split bamboo. Many, in fact, now prefer fiberglass to anything else.

Why? Not particularly because fiberglass rods are cheaper, or because they can stand abuse, or because it hurts less in the pocketbook when loss or accidents happen. Mainly because modern technology can plan graduated tapers for superb actions much more precisely than ever could be done with the mitering plane, some glue and a file.

In either case, the hidden ingredient in fishing rods is quality, and quality rests on the reputation of the manufacturer. Since the great majority of rods today are made of fiberglass and since most of us don't realize the vast amount of scientific knowledge that goes into them, let's see how topquality fiberglass rods are made.

PREPARING THE GLASS CLOTH

A tubular glass rod basically is a bonding of cloth-like woven glass fibers that are impregnated with a resin base and fabricated in such a way as to form a tube that is wide in diameter at the butt and gradually and scientifically tapers to a much smaller tip. The kind of glass cloth and the method of converting it into a tube are very important to quality.

The cloth first is woven to the exacting specifications of the manufacturer. These specifications can vary in diameter of thread, the count of threads per inch running in either direction, the width of the bolts of material, the number of imperfections allowed, and so on.

Added to all this is the type of resin-based material into which the finished cloth is dip-treated, and the method by which this is done. Since this is a plastic-based liquid resin, the process must be rigidly controlled to produce correct color, stiffness, final curing temperature and other factors.

All this may sound a bit complicated, but it indicates the great pains that top manufacturers take to produce quality fishing rods. Manufacturers of cheap rods (even cheap rods sold at too-high prices) find many ways to cut quality, thus producing equipment which looks much better than it really is.

After dipping, the cloth is run through drying towers and the viscous plastic resin that is left in the cloth after it has passed through squeegee-type rollers is allowed to dry and partially cure. At this time, when the cloth is rolled into bolts, it is separated by a layer of thin plastic sheeting to prevent the semi-tacky layers of cloth from sticking together.

Until used, these bolts are stored in humidity- and temperature-controlled rooms because, over a period of time, an aging process otherwise would take place which would make the cloth hard and unusable. The wide cylindrical rolls of bolts of cloth are hung for processing at the ends of large stainless-steel tables. Here, extreme care must be taken to keep dirt and other contaminants away from the cloth, because these would show in the finished rod sections.

BUILDING THE ROD TUBES

This cloth is unrolled and drawn over the tables in a small pile of multiple layers. A stainless-steel template, carefully engineered for each type of rod tube, is laid over the layers, which then are cut to the shape of the pattern. One side of the template is not straight. When cloth cut to this shape is rolled into a rod tube, the wall thickness of the tube will vary in diameter according to the number of thicknesses (or wraps) planned when the template was made.

The predetermined thickness of the rod tube helps to build proper action into the rod, so this step is very important. Hundreds of templates are made, each for a particular design of rod section, but there is only one which is correctly engineered for a specific rod tube.

When cut, these pieces of glass cloth are sent to another part of the factory where the straight edge of each piece is tacked with a hot iron to a steel rod called a mandrel. This is placed between high pressure rollers to wind the cloth evenly and under extreme pressure around the steel mandrel. This pressure also helps to determine the density and thickness of the finished rod tube.

Mandrels are designed with straight or stepped tapers. The combination of the design of the taper of the mandrel and the pattern of the glass cloth provides the predetermined action required in the finished rod tube, also called a "section," or a "blank."

This tube of glass cloth, compressed around the mandrel, then is machinewrapped with strips of cellophane, applied under tension. It then is hung vertically in a curing oven controlled so that the temperature will rise gradually, level off, rise again, and so on until the final curing heat is reached.

These curing steps serve to evaporate volatile materials from the plastic binders to assure that no bubbles or trapped air remains in the synthetic walls of the rod blanks.

After this curing, the mandrel is pulled from the glass, thus leaving a hollow tube covered with cellophane. The cellophane is lightly slit and is blasted away from the glass tube under extreme pressure.

ASSEMBLY AND INSPECTION

Final processes of polishing and trimming to desired length complete the manufacture of the rod sections, or tubes. Identical bundles of these then go to the wrapping rooms where ferrules are applied and bonded with epoxy glue to the ends of the blanks. There is little or no grinding in fitting to prevent any possible weakening of the tubes.

Pre-assembled handles, or grips, on wooden dowels or aircraft aluminum tubes are fastened to the butt sections in the same manner. A catalytic finish is applied to the glass blanks by dipping and slowly withdrawing them from the liquid to prevent any runs or imperfections.

Line guides are positioned exactly by use of templates and the guides then are wound in place with nylon thread. The wraps then are treated with color preservative and lacquer, and the finished rods go to groups of inspectors for final examination and marking identification. Inspections go on through all of these various processes, and blanks or material containing even the slightest imperfections are discarded.

All this may seem rather complicated, but anglers should understand the vast amount of care, knowledge, equipment and skill necessary to produce rods that they will be proud to own. Of course, manufacturing processes vary somewhat with different manufacturers, so it should be stated that the method of rod construction we have been discussing is the one used by the Wright & McGill Company of Denver, Colorado, in the manufacture of its famous Eagle Claw fiberglass rods. We have not discussed several manufacturing refinements used by this company because they are more or less secret.

GRAPHITE RODS

Early in the 1970s, the fishing-tackle industry borrowed from the discoveries of the aeronautics industry and began to incorporate graphite fibers (man-made carbon fibers) into fishing rods. This was because the newly discovered graphite fibers excelled glass fibers mainly in providing greater tensile strength and stiffness with lighter weight and smaller diameters. Thus, graphite rods, as compared to similar ones of fiberglass, provided lighter, stronger and more powerful qualities with exceptional resistance to fatigue.

If graphite fibers are substituted wholly or in part for fiberglass ones, the manufacturing process for making graphite rods is similar to that for making fiberglass rods. An oversimplification, perhaps, but this is not a technical book.

My own preference is to use fly rod and fly for big fish such as Atlantic salmon around the north Atlantic, rainbow trout in Alaska, brown trout (sewin) in Wales, or perhaps any of several saltwater species which can test tackle and skill to the utmost. Outwitting smaller trout may be an art, but fighting much bigger battlers is a contest; a sportier one if the odds are about equal. For these I usually use a 9-foot rod. My first, before fiberglass days, was of split-bamboo weighing over 7 ounces -an arm-tiring weapon to handle all day. Then came a fiberglass one weighing 6 ounces, followed by a graphite of 4 ounces, which still is a favorite. Of these three the graphite has more power, less weight and better casting ability than any of the others. Such is progress!

I asked a world-famous angler his opinion of graphite rods versus fiberglass. He replied, "I think most of us agree that glass rods, properly made, are pretty good, and their reasonable cost makes them ideally suited for the novice fly fisherman or for a person who is only casually involved in fly fishing. For certain heavy-duty applications, such as saltwater fishing or freshwater bass fishing, glass rods may be even more practical than graphite. I am infatuated with graphite and believe it to be the best currently available material for making fine fly rods. It is lightweight and, probably most important of all, is so responsive that it will help almost anyone present a fly line easier and more effectively. I should emphasize that just because a rod is made of graphite doesn't necessarily make it a good rod. It must be designed and manufactured properly. There are a lot of poor rods on the market labeled `Graphite' because there are no standards that require identification of the graphite fiber content of a 'Graphite' rod. I think it important that one who is contemplating the purchase of a graphite rod should be very conscious of the reputation of the manufacturer, and that he should only consider rods made by firms who are knowledgeable about the design requirements for producing fly rods."

BORON (AND BORON-GRAPHITE) RODS

Coincident with the introduction of graphite, another high-temperature space-age product was introduced into fishing-rod manufacture. This is elemental boron made into fibers as will be described. It features outstanding lightness coupled with hardness, stiffness and resistance to stretching, and is said to exceed the strength of steel with less than the weight of aluminum.

Boron filaments are not all boron. They are a "plating" of boron on fine tungsten wire caused by its exposure to a gaseous mixture of boron trichloride and hydrogen under very high electrical heat. These tiny boron-coated fibers then are enclosed in tape to prevent them from rubbing against each other. The tapes then are combined into sheets which are cut to size and rolled into rod blanks. Manufacturers have different ways of doing this.

For reasons of cost, but mainly for efficiency, boron fibers are "diluted" with other fibers such as graphite to provide fishing rods with actions deemed most suitable to each manufacturer. Thus, rods called "boron" actually are a combination of boron and graphite and perhaps other filaments compressed with a binder (such as epoxy resin) to form the "cloth" from which fishing rods are made. The percentage of boron combined with other substances varies widely and depends in part on whether the result is to be used for fly rods, spinning rods or whatever. Manufacturers set their own standards. Thus, it is possible to call a rod "boron" even if it contains only a trace of that element. For this reason, we can't take rods called "boron rods" purely at face value, but should trust the integrity of the manufacturer. Famous makers have their reputations to protect and won't stint on quality. Some others may be suspect.

Boron rods currently cost more than ones of fiberglass or graphite, and even this brief account should make the reasons obvious. Are they worth the difference? They seem a bit lighter, and some experts consider the rods of some manufacturers to provide superior action, but every angler must decide for himself.

3

Fishing Knots

More than twenty reliable and well-tested fishing knots are brought together in this chapter. Some are relatively new and unknown. Others are old standbys for which, in many cases, new information is provided which makes them stronger or easier to tie. You only need to learn a few, but you may want to refer to others from time to time, or to select from the alterna-tives described. Since tackle can be no stronger than the knots which join its parts together, knowing how to tie appropriate knots correctly is of major importance.

LEADER KNOTS

Which describes how to make leaders, we noted that the Basic Blood Knot is preferred for joining strands of similar diameters. Those who have trouble with it may prefer the Surgeon's Knot, which fol-lows. Use the Double Strand Blood Knot when joining strands of very unequal diameters. The Perfection Loop Knot often is used on the end of the leader so it can be joined to the line, although other connections are given which many anglers think are superior. Favorite methods, for tying in droppers also are included in the next section.

Basic Blood Knot

This joins two lengths of monofilament together in making leaders and in mending monofilament lines. Since beginners often are bothered by the ends pulling out when drawing the knot together, start by lapping the two pieces to be joined so the ends above the crossing are a good 5 inches long and hold them between the right thumb and forefinger. Twist the leftward-pointing end around the other strand five times and place the end between the strands at point X, holding it there. Transfer

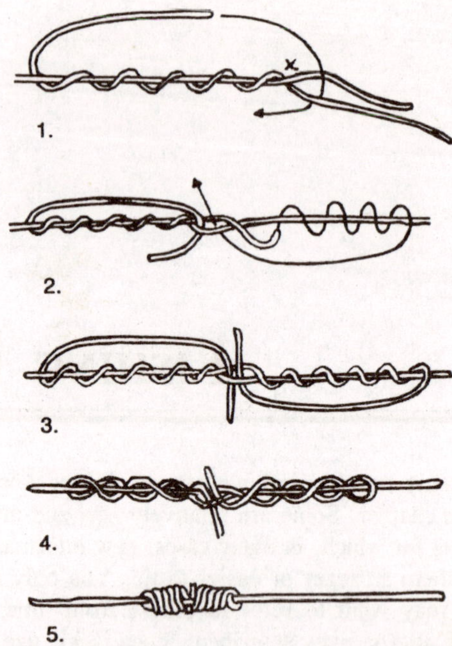

the knot to the left thumb and forefinger and wind the other end around the other strand five turns in the *opposite direction*. Pass the end through the knot beside the other end, but in the opposite direction from it. The knot now can be released. Now pull on both strands of monofilament, being sure while doing so that the ends don't pull out. The knot will gather. Now pull the knot as tight as possible, to test it. Clip the ends close to the knot.

Five turns in each direction are correct. Six improves the strength very little, and are more trouble. Only four decreases strength by about 15 per-cent, and only three by 25 percent. When tied as above, the knot has nearly 100 percent of the unknotted line strength. Tightening a knot creates fric-tion, which induces heat, which can weaken the knot. For this reason, expert anglers wet the knot between the lips before pulling it tight, to decrease friction.

Double-Strand Blood Knot

The Blood Knot is used for tying together monofilaments of the same or nearly the same diameters. If monofilaments are of greatly unequal diame-ters, such as when tying in a shock leader or a bumper line, this knot won't be strong enough. In such cases the Double-Strand Blood Knot should be used.

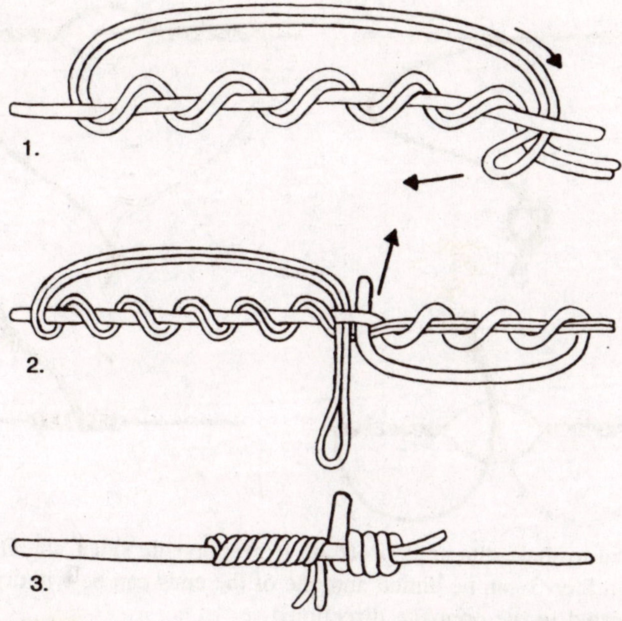

This is tied in the same manner as the Basic Blood Knot except that the end of the smaller monofilament is doubled. The doubled end is used as if it were a single one. Wrap the doubled end around the single strand at least five times. Do it six or seven times if the diameters vary very greatly. The single end is wrapped around the doubled end only three times. Cut off both single and doubled ends about ¼ inch from the knot.

Basic Blood Knot (alternate method)

Beginners often have trouble tying the Basic Blood Knot because one of the ends pulls out when the knot is being drawn together. A gillie on the River Spey, in Scotland, showed me this alternate method, which pre-vents this and thus may be easier. Do it in these 4 simple steps, as shown in the sketches (one leader is shown black, for clarity): Tie the ends to be joined together by using a simple Overhand Knot. Cross the monofilament above this knot to make a fairly wide loop. Wind the two looped strands around each other five times. Pass the joined ends through the loop thus formed. Pull each strand to draw the knot tight, wetting the partly-tightened knot between the lips before pulling it completely tight. Clip off excess ends closely.

(In this case the two ends protrude from the same side of the Blood Knot, which seems to make little or no difference in strength. If it is

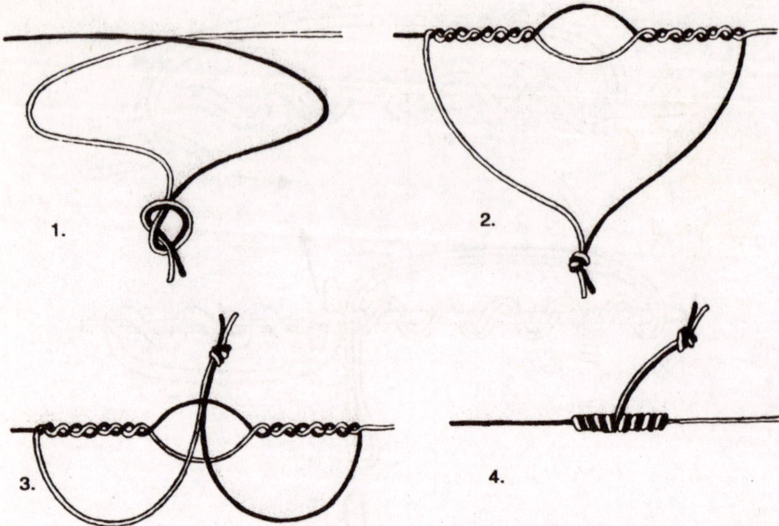

de-sired to make the ends protrude from opposite sides, the Overhand Knot in Step 3 can be untied and one of the ends can be withdrawn and rein-serted in the opposite direction.)

Surgeon's Knot

This also is used to join two strands of unequal diameter. It is easier to tie than the Blood Knot, but not used as extensively. A popular use is for tying a shock tippet to a leader point.

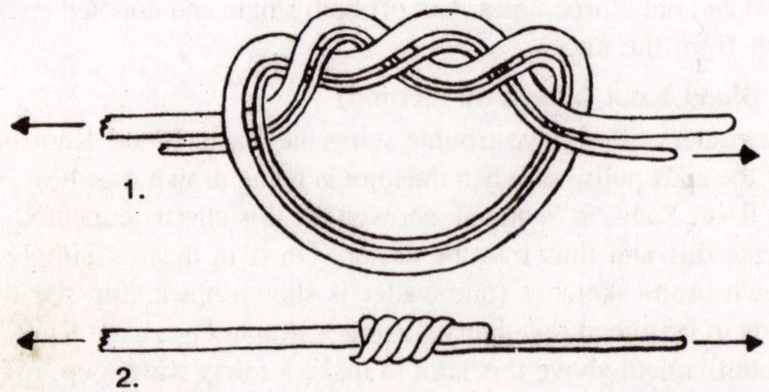

Lay the two strands together, with tips in opposite directions, with ample overlap. Treating the two as a single strand, tie a loose overhand knot in the two lines, and then repeat the overhand knot by pulling the two strands all the way through the loop again. Holding both strands at both ends, pull the knot tight. Clip the ends short.

Perfection Loop Knot

This knot is used to make the loop at the butt end of the leader. It provides a detachable leader which is tied to the line by using the Tucked Sheet Bend or the Jam Knot. When a loop is spliced into the end of the fly line, the two loops can be joined by threading the fly-line loop through the leader loop and pulling the leader end through.

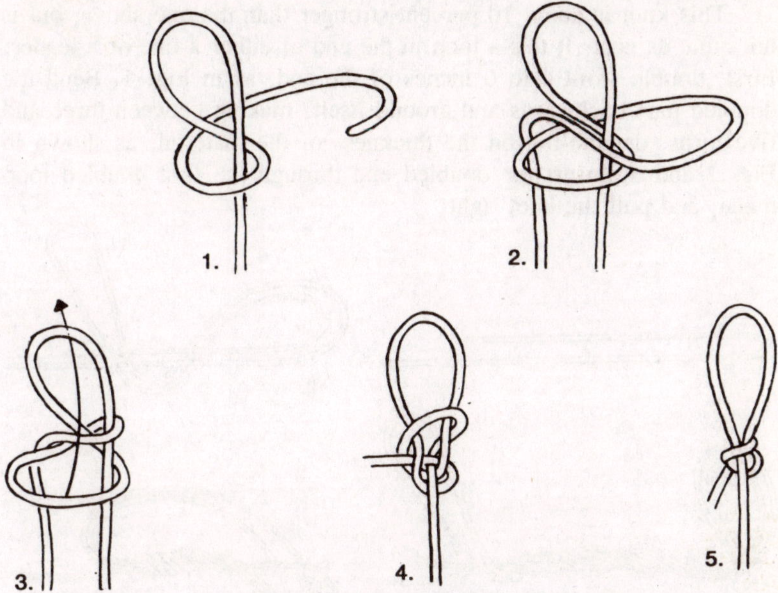

To tie the Perfection Loop Knot, hold the end of the monofilament between left thumb and forefinger so that about 6 inches of it extend and point upward. Holding the end with right thumb and forefinger, throw a small loop to the left so it crosses behind the standing end. Holding this loop between thumb and forefinger, bring the end toward you and pass it around the loop, clockwise, also grasping this between left thumb and fore-finger. You now are holding two loops, the second one in front of the first. Now take the short extending end of the monofilament and pass it between the two loops, also holding it between left thumb and forefinger.

Now grasp the front loop and work it through the rear loop, at the same time pulling it out a little. Still holding this tightly so the knot won't slip, pull on the lower extension of the monofilament, thus closing the smaller loop. If the remaining loop starts to twist, keep it from twisting and pull downward on the lower extension and upward on the remaining loop until the knot is tight. The knot can now be released from the fingers. Put a pencil or something similar through the loop and pull the loop as tight as possible. The short end of the monofilament will now extend at a right angle from the loop, and it can be clipped off closely.

Improved End Loop Knot

This knot is about 10 percent stronger than the one above, but is not quite as neat. It ties a loop in the end of either a line or a leader. First, double from 4 to 6 inches of the end, as in Fig. 1. Bend the doubled part backwards and around itself, making between three and five turns, depend-ing on the thickness of the material, as shown in Fig. 2 and 3. Insert the doubled end through the first doubled loop made, and pull the knot tight.

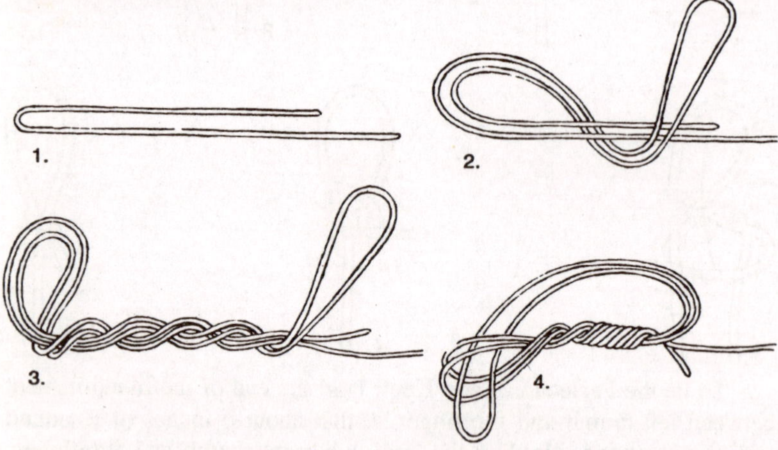

Making a Dropper

A dropper is a length of monofilament extending from a leader or mono-filament line. One can be tied in by making a Blood Knot, qne end of which is left as long as is needed for the dropper. A fly, a hook or a lure is tied to the end of the dropper.

Another way to tie on a dropper is to make a Perfection Loop Knot in the end of a piece of monofilament of the desired length and strength. If the leader lacks a Blood Knot where the dropper is to be tied in, cut the leader and tie one there. Lay the Perfection Loop under the leader above the Blood Knot and put the dropper's end over the leader and through the loop. Pull the connection tight and slip it against the Blood Knot. Tie a fly or hook on the end of the dropper.

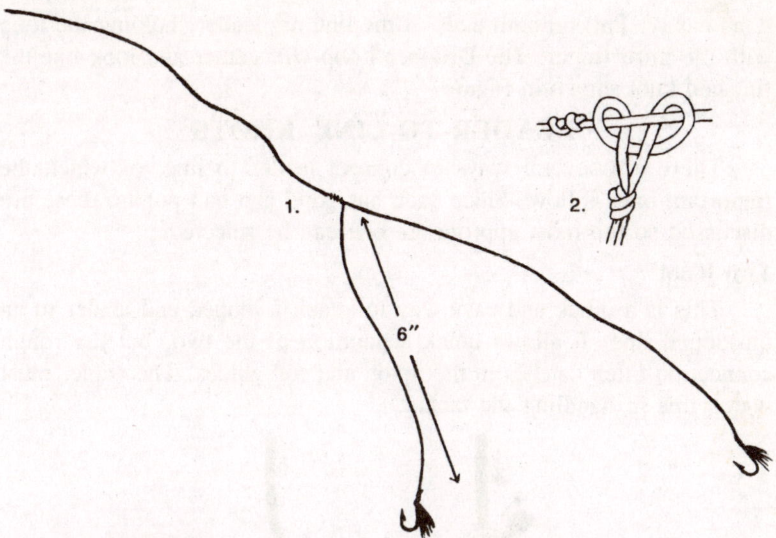

Dropper Loop Knot

This is used to put a loop in a leader or monofilament line. A dropper can be attached to the loop for removal when desired. Usually this drop-per, which is short, has a Perfection Loop Knot on one end, and a fly, hook or other lure attached to the other end. Put the Perfection Loop around the Dropper Loop and put the dropper end through the Dropper Loop,.pulling the connection tight so both loops are connected in the same way.

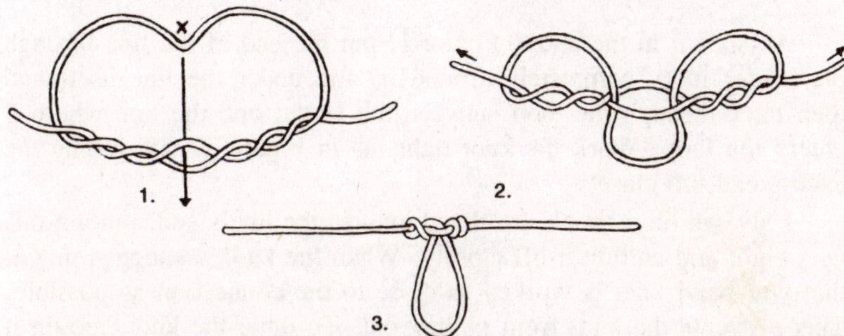

To make the Dropper Loop Knot make a loop in the line or leader and wrap the end overhand seven or nine times through the loop. Pinch a small loop at the point marked X in Figure and push it between the middle of the turns, as shown by the arrow. A pencil or pin, inserted in the middle turn, helps to keep the strands separated so this can be done easily. Pull on both ends of the line or. leader, holding the loop with the third finger. The Dropper Loop will gather and look like the finished knot shown in Figure.

LEADER-TO-LINE KNOTS

There are several ways to connect leader to line, of which the important ones follow. Since each has good and bad points, these are discussed so the most appropriate one can be selected.

Jam Knot

This is a quick and easy way to attach a looped end leader to an un-looped line. It allows quick separation of the two, but the rough connection often catches in the tiptop and rod guides. The angler must watch this in handling the tackle.

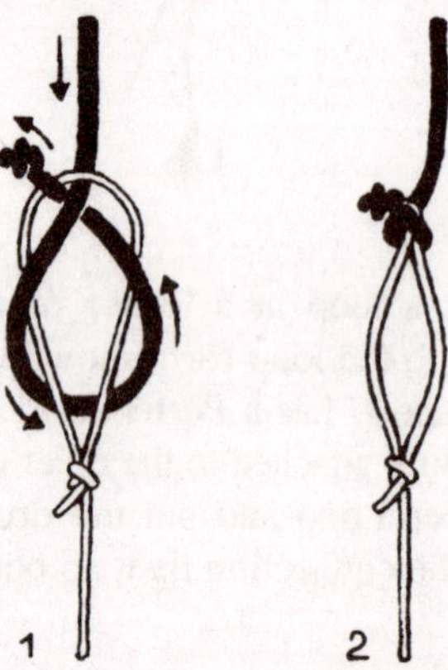

As shown in the above Figure 1, put the end of the line through the leader loop, completely around it, and under the line itself and then back through the loop between the leader and the line where it enters the loop. Work the knot tight, as in Figure 2, not cutting the excess end too closely.

I always tie a simple overhand knot in the line's end, pulling this very tight and cutting it off closely. When the knot is snugged down, the over-hand knot is worked as close to the connection as possible. This prevents the knot from pulling out. To untie the knot, loosen it by pushing the end of the line against it.

Tucked Sheet Bend

This has the same purpose as the Jam Knot. The line's end is passed through the leader loop, around it, then between line and loop, and down through the crossing that was made in the line. This is a

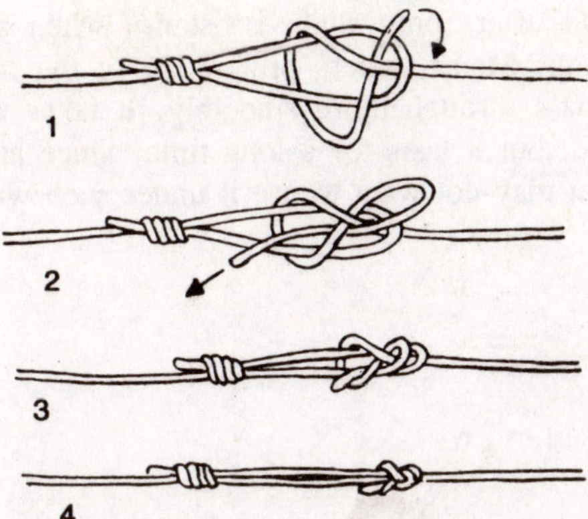

simple figure. Be sure that the leader loop doesn't slip over the line before the knot is pulled tight. Since the clipped end of the line points toward the leader's tip, this is a neater knot than the previous one.

Making a Spliced Line-Loop

With a jackknife, scrape off the line's coating ³/s inch from the end, and fray the core with a pin. About an inch above this, carefully scrape another 3/8-inch segment of coating off, trying not to damage the threads of the core. Lay the two scraped parts together to make a

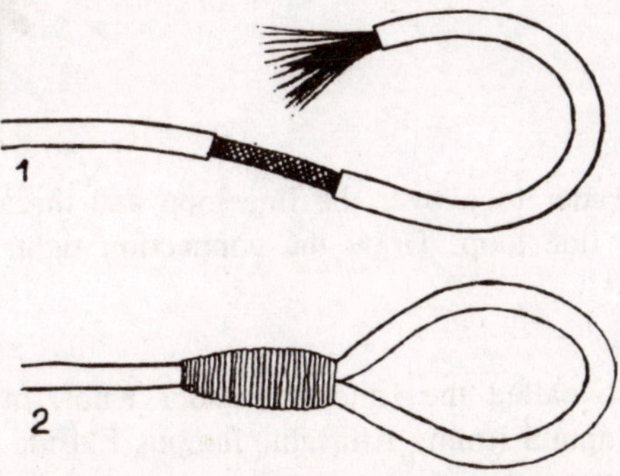

loop about 1/2 inch long. Lacquer tie two scraped parts and whip them (wind them) with fine thread to make a smooth binding. This can be secured by two or three half-hitches, but preferably by a whip finish. Apply two or three coats of lacquer to the joining to make it smooth. (Spar varnish, fly-dressing lacquer or Pliobond Cement can be used.)

Joining Looped Leader to Looped Line

Some anglers prefer this connection to any other because it is very secure and offers only slight resistance when passing through guides and tiptop. Many anglers prefer one of the following knots, thinking they pass through more smoothly. It takes time to make a good line splice, but it lasts for a long time. Since algae can collect in the knot, you may not want to use it under such water conditions.

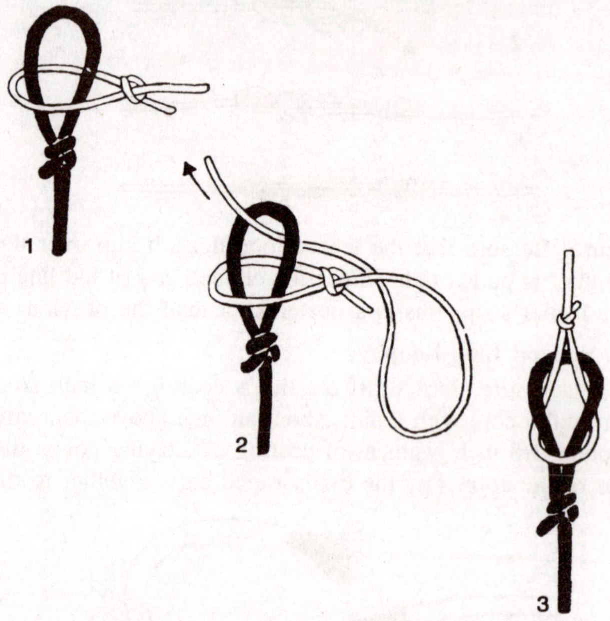

Slide the leader loop over the line loop and thread the leader's tip through the line loop. Draw the connection tight so both loops join.

Key Loop Knot

This is also called the Line to Leader Knot, or the Albright Special, after Captain Jimmy Albright, famous Florida fishing guide.

Double the line's end and lay a similar reverse loop of leader over it, with the line's end through the leader's loop, as shown in Figure 1. Make five or six turns of the leader's end around the line loop and then run the leader's end through the line loop. Pull the turns tight and work them down to the end of the line loop, then pulling them very snug. Cut off excess ends fairly closely and varnish the connection. (Two or three applications of Pliobond Cement, instead of varnish, are recommended.)

This is a jam knot which pulls even tighter under tension. It is not quite as strong as the Blood Knot, but is very good also for joining two

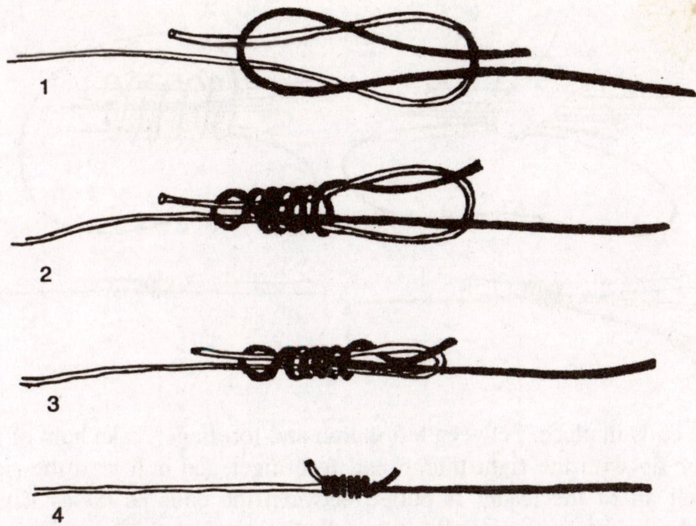

strands of monofilament of widely different diameters. It is recommended for joining stranded wire to leader material.

Leader Whip Knot

The Nail Knot is very good for joining line to leader, and will be dis-cussed later for use in joining a fly line to its backing. For joining a level leader section to a fly line I prefer the Leader Whip Knot. This is a new knot and my favorite for the purpose. It forms a semipermanent connection of 2 or 3 feet of leader-butt to the line to which a tapered leader can be tied by a Blood Knot if desired.

As shown in Figure 1, hold the end of the line between left thumb and forefinger so it extends about an inch. Also hold the leader butt (leader extending to the right) so about an inch of it is between thumb and fore-finger. Insert a heavy embroidery needle between leader and line, its point beside the line end. Take hold of the forward end of the leader and also put it between left thumb and forefinger, holding its tip so that it points to the right and lies beside the line end and the needle. (The rest of the leader hangs in a loop below the fingers. The loop is shown much smaller in the drawing than it actually is. You now have four items grasped between left thumb and forefinger: the line end, the needle, the leader butt and the leader tip.)

Grasp the part of the leader butt extending from thumb and forefinger and wind it clockwise to the left around the four items, making seven tight, close coils just in front of the thumbnail, as in Fig. 2. Holding

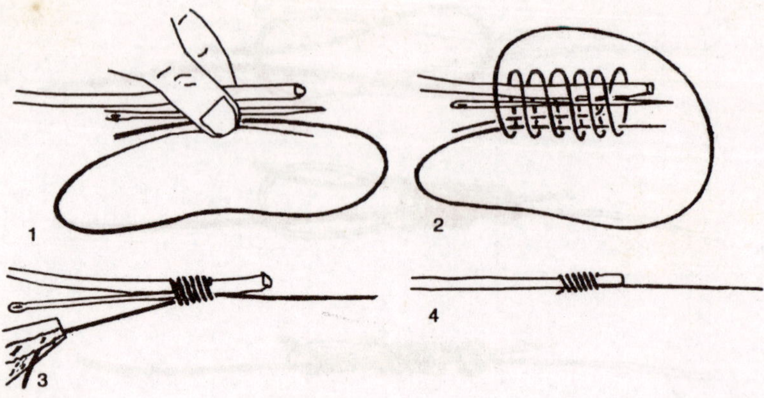

these coils in place, between left thumb and forefinger, take hold of the leader tip with the right thumb and forefinger and pull it to the right so that all of the leader is pulled between the coils as far as it will go. (At this point be sure all seven coils lie closely together.) Now you can let go of the knot, pull out the needle, and grasp the rearward protruding end of the leader butt with a pair of pliers, as shown in Figure 3. By also holding the leader which extends from the knot and pulling in both directions, the coils will bite into the line to form a rigid connection.

Now snip off the excess butt end of the leader and the excess tip end of the line, and test the connection to be sure it is tight, as in Figure 4.

The rest is optional. The connection can be coated with Pliobond Ce-ment to make it smooth. When partly dry, mold it around the connection for greatest smoothness. When very dry, add another coat or two, if you wish. In addition, you can leave about 1/4 inch of leader butt and line end and whip them and the knot for added smoothness, also coating the whipped knot with cement or varnish.

Pin Knot

This is very much like the previous knot but the leader comes out of the line's end instead of beside it, providing a slightly smoother connection.

Using a medium-sized common pin or needle, push it into the end of the line and out the side so that ¼ inch of the line is strung on the pin. Pliers are needed for this and for holding the pin during the following operation.

Hold the pin's point with the pliers and apply heat from a match or cigarette lighter to the head until the part of the line nearest the heat begins to smoke *very slightly*. Push the line toward the pinhead and, hold-ing the pin with pliers at this end, apply heat to the point until that end of the line which is strung on the pin also begins to smoke very slightly. (Avoid overheating so as not to impair the strength of the line. All that is needed is only enough heat to prevent the hole made by the pin from closing up when the pin is withdrawn.)

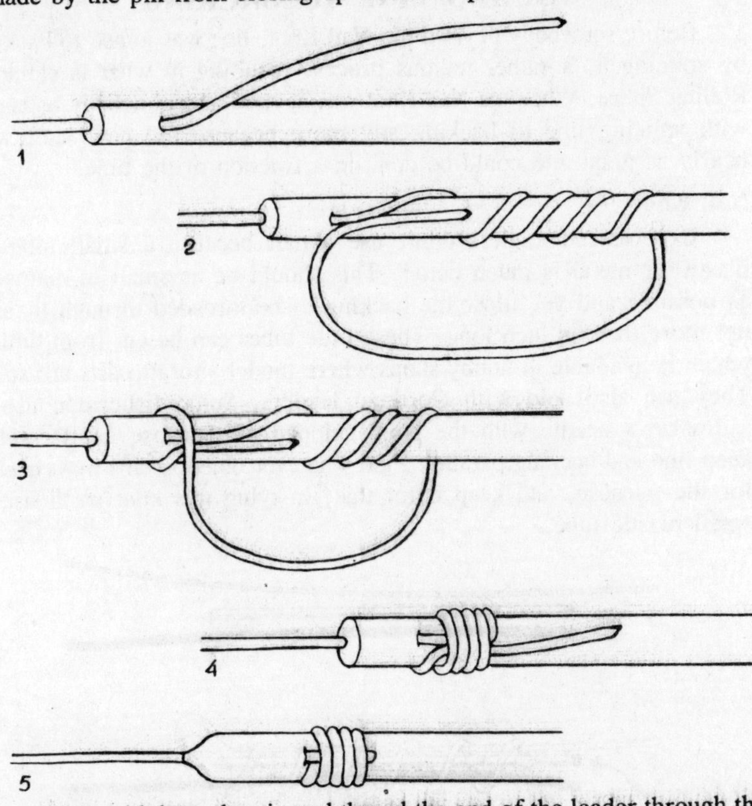

Now withdraw the pin and thread the end of the leader through the hole in the line's end. It helps to do this if the end of the leader material is shaved to a point. Pull an ample amount through because ex-cess can be recovered. Make five turns of this leader end around the line end fairly loosely and an inch or so up from where the leader comes out of the line. Place the leader end beside the line end, pointing away from it. Now, holding the loop thus formed, wind the leader back around the line, starting the winding so the spot where the leader comes out of the line can be seen. Wind tightly in close coils until all five coils are used up. Holding these coils in place, pull on the leader where it comes out of the line to take out all slack and to make a snug connec-tion. Pull on line and leader to be sure the coils cut into the line slightly. Snip off the excess leader tip and varnish the connection to com-plete the knot. Quite obviously, this is merely a five-turn whip finish.

JOINING LINE TO BACKING

Before somebody devised the Nail Knot, line was joined to backing by splicing it, a rather tedious process resulting in what is called a Rolling Splice. When the Nail Knot was devised almost nobody bothered with splicing line to backing any more because the new knot was nearly as good and could be done in a fraction of the time.

Nail Knot

Experienced anglers don't use a nail because a small tube of plastic or metal is much better. This should be as small in diameter as possible and yet allow the backing to be threaded through it, and not more than an inch long. These little tubes can be cut from tubing which is available in hobby shops where model aircraft parts are sold. They are also sold with Cortland leaders. Some fishermen use a sailmaker's needle with the point snipped off because its flat sides keep line and backing parallel. Find whatever object seems most useful for the purpose, and keep it for that. In tying this knot we'll use a small plastic tube.

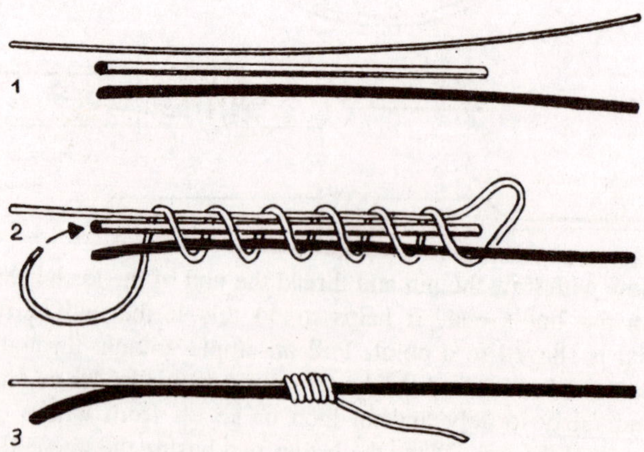

With sufficient backing on the reel, pull out a comfortable length and lay the tube against it, allowing about 6 inches of the end to work with.

Lay the end of the fly line beside the tube, as in Figure 1. Hold the three between left thumb and forefinger. Taking the protruding end of backing in the right hand, wind it in close and tight coils backward toward left thumb and forefinger and almost touching them. Wind at least five coils or as many as six or seven. With too many it is more difficult to draw the knot tight.

Holding these coils so they won't overlap, bring the end of the backing to the left and push it through the tube to the right. Still holding the coils tightly, carefully pull out the tube, pulling the backing out as much as possible while this is being done. With the tube pulled out the coils will appear soft, but don't let go of them. Pull the backing from both ends alter-nately, allowing thumb and forefinger to let the coils gather. When no more backing can be pulled from either end, let go of the coils and adjust them if necessary. Then pull hard again on both ends of backing to make the coils bite into the line. (If you use pliers to do this, be sure they don't injure the backing behind the knot. They can be used without harm on the excess end, because this will be removed.)

To be sure the knot is tight, now pull cautiously and then hard on fly line and backing. If the knot has gathered tightly the fly line won't slip, and further pulling will tighten it more. Clip off excess ends, as in Fig. 4, and coat the knot with two or three applications of Pliobond Cement or its equivalent. Let the knot dry thoroughly before winding the fly line on the reel.

KNOTS FOR TYING HOOKS AND LURES WITH STRAIGHT EYES

The knots in this section are used mainly for joining lures to lines because such lures (unlike most flies), as well as swivels, have straight (ringed) eyes; that is, eyes with no bend in them. Don't use these knots on hooks or flies with upturned or downturned eyes.

Improved Clinch Knot

This is a basic knot everyone should learn for tying lines or leaders to hooks or lures with *straight* eyes, and for tying on swivels, snap swivels, etc. The knot is dependable, easy, and retains nearly 100 percent of line strength.

Stick the end of the monofilament line through the eye and make *at least five* twists around the line. Then push the end through the first loop (nearest the eye) and then back through the big loop. Holding the lure or swivel and the line, pull the knot tight so it looks like the

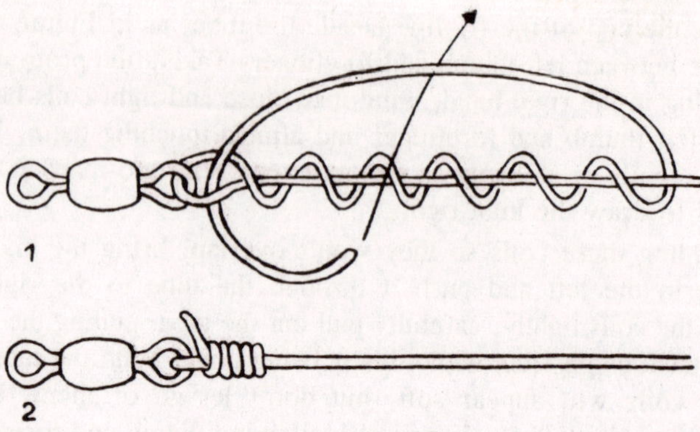

knot in Figure 2. Cut off the end closely. Five turns around the line are essential for proper strength. If more than six turns are made, the knot can't be pulled tight enough. The knot must be pulled tight so the mono-filament won't slip and cut itself. When using heavy monofilament, wind a handkerchief around your hand to protect it while pulling the knot tight. Three or four turns are enough for extremely heavy monofilament, such as 60-pound test and stronger.

The regular Clinch Knot is the same except that the line end is not passed back through the big loop. The Improved Clinch Knot is usually considered better, but some anglers prefer the regular one.

Loose Loop Clinch Knot

This variation of the Improved Clinch Knot is useful for making a

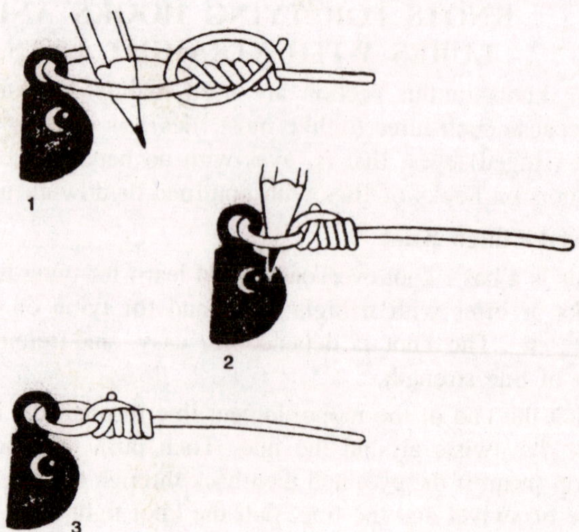

loose connection to a lure such as a jig to give it more action. Make an Improved Clinch Knot, but do not draw it tight. Insert something such as a pencil or pen into the loop and *then* pull the knot tight. Pull out the pencil and clip off the excess monofilament. This knot should not slip, but if it does, it probably would be due to the pull of a fish, which then would make little difference because it merely would revert to an Improved Clinch Knot.

Duncan Loop Knot

This is an alternate to the previous knot and also makes a loose-loop connection. Try both and take your choice.

Thread the tippet through the hook's eye, pulling out about 5 inches. Turn the end of the tippet toward the hook's eye to make a loop.

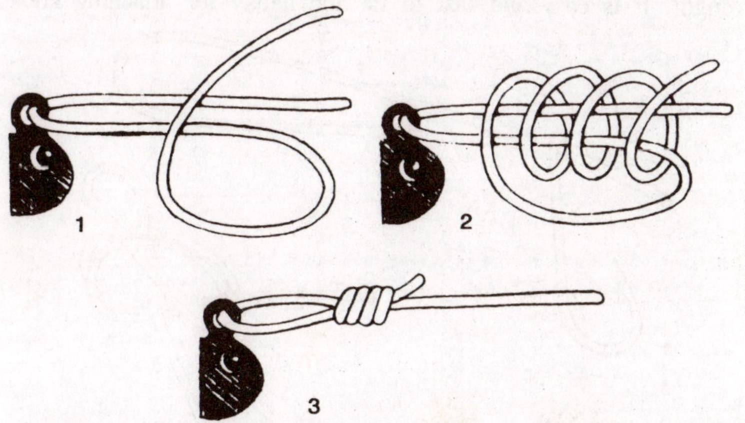

Wrap four or five turns around the two strands of tippet, going into and out of the hook's eye, doing this *inside* the loop you just made.

Pull the tippet end slowly until the knot is almost tight. Slide the knot to make as big a loop as you want and pull again on the tippet end to tighten the knot firmly. Cut off the excess end. This loop will close on the pull of a strong fish, but it can be opened again by sliding the knot up the leader.

Lark's Head Knot

This isn't quite as strong as the previous knot, but is easier to tie. Being a looped knot, it is handy for attaching and detaching such things as swivels, eyed sinkers and short lures. Of course the loop can be made as long as desired, but the doubled part of the line makes it more obvious to fish.

Double the line for a loop as big as desired and tie a simple

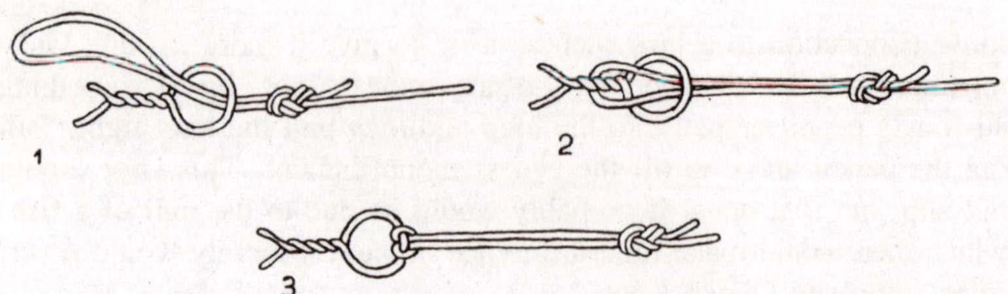

overhand knot in it. Clip off the excess end. Pass the loop through the ring or swivel and over (around) the lur. Draw the loop back to the ring until it upsets the loop around it.

Palomar Knot

This one is a basic knot providing almost 100 percent of line strength. It is easy and fast to tie and handy for attaching hooks,

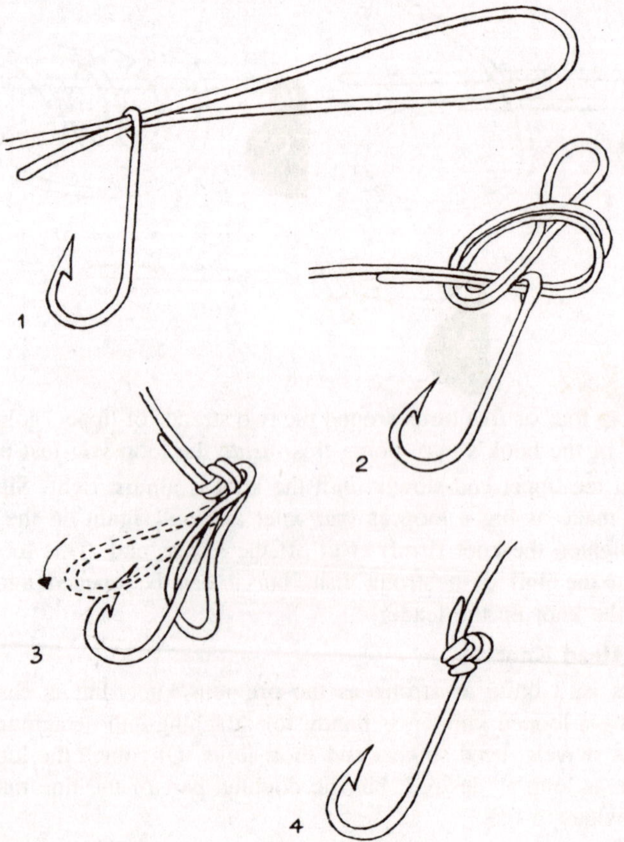

swivels, eyed sinkers and small lures to lines or leaders. Put the line or leader through the eye and then return it through the eye to make a loop 3 or 4 inches long. Holding line and eye be-tween thumb and forefinger, use the other hand to bring the loop back over the doubled line, making an overhand knot around the eye. Without tightening the knot, put the remaining loop over the hook or what-ever is being tied on. Then pull on the line to draw the knot to the top of the eye. Pull either the line or the knot end to tighten the knot and cut off the excess end fairly short.

KNOTS FOR HOOKS WITH TURNED EYES

The knots in this section are for flies or hooks which have *upturned* or *down turned eyes.* The idea is to give the leader a direct line of pull to the fly or the hook, which can't be done properly when using the knots in the foregoing section.

Double Turle Knot

While this knot has only about 60 percent of the strength of the line or tippet it is tied on, it is used very successfully by fly fishermen everywhere.

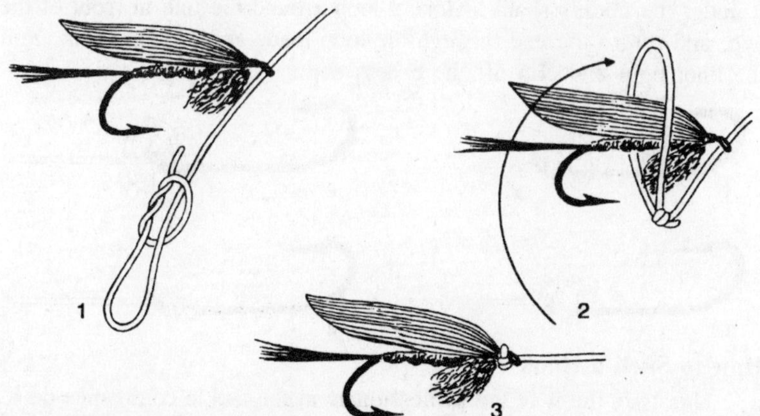

String the fly on the leader and let it slide down, out of the way for the moment. Double the end of the line or leader and make a noose or slip knot in it, but put the end through the knot *twice,* as shown in Figure 1. Pull the knot tight and open the noose or slip knot enough so the fly can be passed through it. Holding the noose in one hand, raise the leader so the fly slides down it into the noose. Gather the noose around the neck of the fly *(around the back of the eye* of the fly) so the knot is set against the back of the downturned or upturned eye to give the leader a direct line of pull when the leader pulls the knot tight.

Single Turle Knot

This is the same as the Double Turle Knot except that the end is put through the knot only once. Anglers have found that this knot sometimes pulls out, which evidently was why the Double Turle Knot was developed. I use the Single Turle Knot with this improvement: Tie a common over-hand knot in the end of the line or leader; make it very tight, and cut off the excess closely. Now tie the Single Turle Knot but work the noose up to the common overhand knot so the two knots are joined into what ap-pears almost as a single knot. This never has been known to pull out, even when handling very big and very active fish.

Return Knot

The purpose is similar to the Turle Knots', and is somewhat stronger. However, it is better for use with hooks that will be baited than for arti-ficial flies because the hackle of the fly gets in the way when tying the knot. Run the line or leader through the eye and pass it under the hook's shank. Make a loop around the line in front of the eye, and return the end through the loop made around the shank. Pull the knot tight and clip off the excess end.

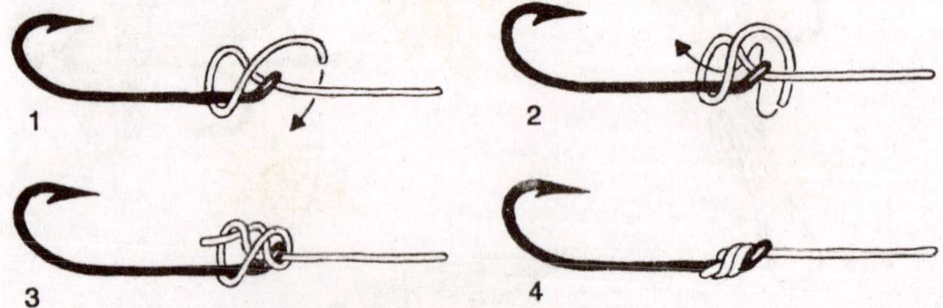

How to Snell a Hook

This isn't the way the professionals in the tackle companies do it, but it is good enough. After spending hours watching the pros at work, I still don't know exactly how they do it, which proves that the hand is quicker than the eye. But a snelled hook is a snelled hook, and here are two easy ways to do it.

Run 4 or 5 inches of leader material through the hook's eye and bend it in a loop under the shank, with the end of the material laying along the length of the shank. Holding the loop, wind the forward part around the shank back of the eye, making as many tight and even coils as desired. (The whole loop will wind around the hook each time.) When enough windings have been made, hold them tightly and pull on the line to remove the remainder of the loop and to make the snelling tight. Clip off the excess end, and varnish the snelling, if desired.

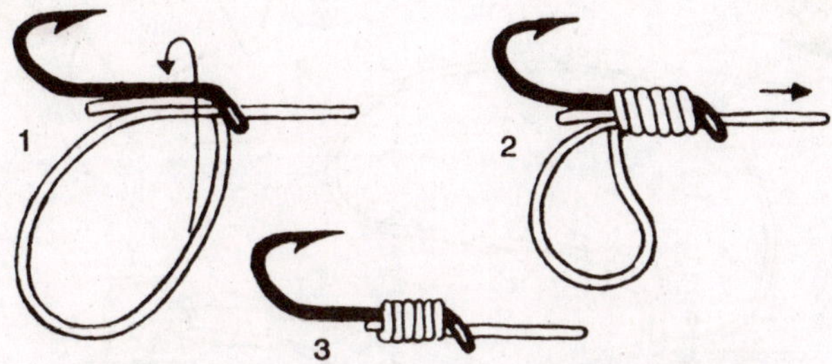

A double or triple gang of hooks can be made by running a much longer amount of leader material through the eye to leave the free end about 6 inches long, or more. After snelling the leading hook, there should be enough material to snell the trailing hook in the same way. Be sure the trailing hook is properly aligned with the leading hook, and is at the desired distance from it, before snelling it in the same manner.

A Quick Snell Knot

A quicker way to snell a hook is similar to tying a Clinch Knot in a leader. As in Figure 1, run the line or leader through the eye and then wind it around the shank five (or more) times. Push the line or leader tip under the material just back of the hook's eye (between the material and the shank). Pull the material tight from the line end and clip off the excess fairly closely. (It may be necessary to push the coils toward the eye to make them lay together.)

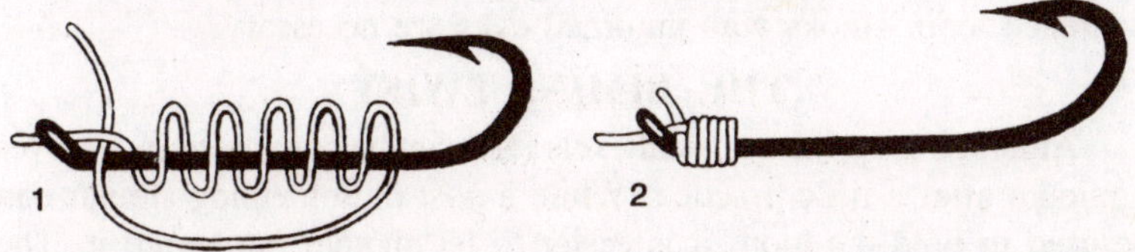

The quick snell is more suitable for use with monofilament of 10-pound test or over. If the end which is clipped off is not clipped too closely, it can be used as a bait holder.

The Steelheader's Looped Snell

This method of snelling provides an adjustable loop for holding

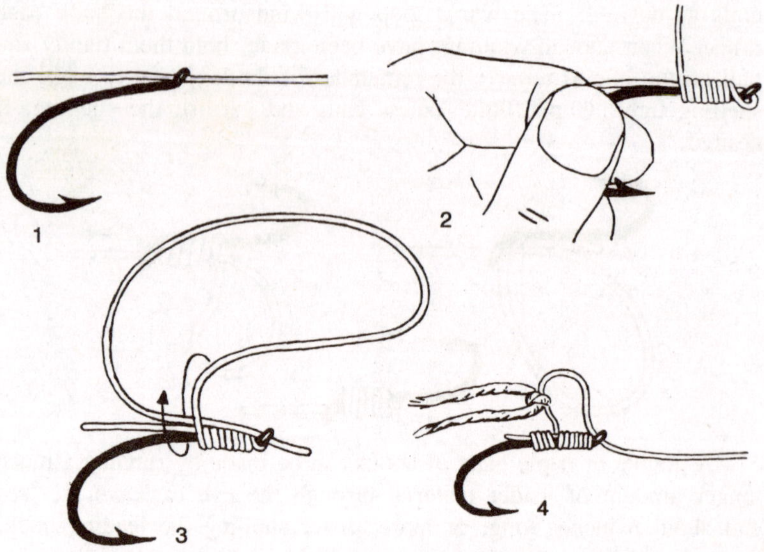

at-tractors such as bits of yarn, foam plastic or, more particularly, a cluster of salmon eggs. First, slip the leader material through the eye and hold it near the bend of the hook. Take hold of the forward end and make about eight tight turns down the hook shank. Pass the forward end of the leader through the hook's eye, leaving a fairly large loop over the shank. Grasp the end of the loop that was used to wind the coils, and make four more tight, close coils with it. Pull the rest of the leader through the eye to tighten the connection.

The leader can be pushed back through the eye to provide a loop which will hold an attractor or bait. The rig is used primarily on steelhead rivers for drifting a cluster of salmon eggs. These are held in place by pulling the leader forward, thus binding them in the tightened loop. Hooks with upturned eyes are necessary.

THE BIMINI TWIST

Although it appears difficult, this knot can be completed in not over a minute after a little practice. While a post or something similar can be used to hold the loop, it is easier to let another person do it. The "Bimini Twist," or "Bimini Roll," is the formerly preferred way to double a line and retain nearly 100 percent of the strength of the unknotted one. It is used both in fresh and saltwater fishing to attach leaders, make shock tippets and to strengthen terminal tackle.

The best description of tying this knot was prepared by the famous angler Stu Apte for the booklet "How to Use Du Pont Stren Monofilament Fishing Lines," and is reproduced here with permission.

Spider Hitch

A new end loop which serves the same purpose as the Bimini Twist has been developed and is recommended by DuPont, the makers of Stren monofilament. Their tests prove that it is from 98 to 100 percent as strong as the unknotted line strength, and it is very easy to tie.

First, make a long loop in the line and hold the ends between thumb and first finger, with the first joint of the thumb extending beyond the finger. Use your other hand to twist a smaller reverse loop in the doubled line. Slide the fingers up the line to hold the loop securely, with most of it extending beyond the tip of the thumb. Wind the line from *right* to *left* around *both* the thumb and the loop, taking five turns. Then pass what remains of the large loop through the small one. Pull the large loop to make the five turns unwind off the thumb. Use a fast, steady pull-not a quick jerk.

HOW TO TIE LINE OR BACKING TO A REEL SPOOL

Run the line or backing around the arbor of any type of reel spool and tie an Overhand Knot, jammed tight and clipped closely, to the line's end. Tie another Overhand Knot around the line going into the reel spool. Pull the second Overhand Knot tight while working it close to the first one, thus forming a noose which can't pull out because of the first knot. Pull on the line to run the knot down snugly to the reel's arbor. If the knot is tied against the direction of reeling it will jam tight and will not slide when line is wound onto the reel.

Successful fishermen take pride in tying knots carefully, and in select-ing the right one for the job. This chapter describes knots for several purposes, including some variations and alternatives. Of course we don't need to learn them all, but it's handy to know they are available for use when needed.

While overhand (or wind) knots occasionally are used as "stoppers" to insure against other knots pulling out, even careful casters accidentally get them in monofilament lines or leaders occasionally. To avoid losing the "big one" it bears repeating that, since they decrease line or leader strength approximately by one-half, the line should be checked periodically to be sure none are there. If one is found it should be picked out immediately, perhaps using a pin carried for this purpose.

Fly fishermen get them most often on windy days, or when casting into the wind. Usually they are caused in casting by tipping the fly rod forward before pushing it ahead. The remedy in this case is to tip the rod ahead only at the conclusion of the forward cast.

4

Baits and Rigs

While it may not be intentional, fishermen often become amateur naturalists. In baitfishing, part of the fun is to relax while watching the line or bobber for signs of action. Meanwhile we observe the swimming of baitfish, and learn from them the kinds of action that should be given to streamer flies and bucktails to make them appear lifelike. We see pollywogs in dead water in the spring and perhaps bring a few home. We see salamanders emerging from under wet debris or dead leaves and we learn that they, too, can be used for bait. If we are very observant we see many kinds of underwater bugs and sometimes can watch their transformation from nymphs into flies.

THE FIRST STEP IN FISHING

When we were very young, we probably were fortunate enough to know an older fisherman who taught us the fundamentals of angling. Our instructor cut and trimmed a moderately whippy green branch from a tall bush and tied to its end a short length of line to which he then attached a yard or so of light monofilament and a small hook. Perhaps he put a split shot or two on the leader to sink the bait. We tossed this into the water from a dock, a boat, or from a rock jutting into a pool, and soon landed our first fish. It mattered little that it was a chub or some other undesirable species because it was our first one, and we probably were hooked on fishing from then on.

I exposed my own kids to fishing in this elementary manner and believe the fun of this first step should not be neglected. If interest is aroused, as it often is, the youngster later can be rewarded with an efficient fishing outfit, which he will treasure all the more because it is his graduation present from the "first step in angling."

Many adults never progress from baitfishing to using artificials with a fly rod. They realize that baitfishing can be a science in itself, and that even Izaak Walton evidently never went any farther. An important step in baitfishing is to observe the development of nymphs into flying insects, not only because this is one of the wonders of nature, but also because the larger nymphs provide very good baits in themselves.

AQUATIC INSECTS

When the water of a stream becomes warmer in late spring and we turn over a few near-surface rocks, we find clinging to them various kinds of buglike creatures known as nymphs. We may sit by a stream in late spring or summer and watch one climb from the water onto a stone or branch where it dries in the sun. Suddenly its body splits and it wriggles out of its drab covering, crawling away from the shuck as a large, shining and colorful insect. The insect quivers and seems to enlarge. The top of its body expands to form gauzelike wings. As body fluid is pumped into the wings, they extend until suddenly it becomes a handsome fly -a dragonfly, an alderfly, a caddisfly, a cranefly or one of many other varieties. Its wings pulsate more rapidly, and it quickly flies away.

These flies make this transition from nymphs to flying insects on land near streamside or on the water. When some of them do it on the water anglers say that a hatch is occurring, and they know that fish feed on them avidly. More will be said about this in Chapter 21.

The flies that escape the feeding fish quickly mate and the females skim the water to lay their eggs in pond or stream. Many of the eggs will hatch, eventually to become larvae, which is the elementary stage of the nymph we saw emerging from the water.

Hellgrammites

The elementary stage of the dobsonfly is called a hellgrammite and, because it is quite large, it is one of the very best baits for fish. Hellgrammites live under rocks or amid other debris in or near the water for a year or so, gradually developing until instinct tells them to come up into the air on a warm and sunny day to hatch into flies. About 2 inches long, hellgrammites are tasty tidbits for big fish, so fishermen often make a habit of turning over rocks and debris in shallow water to search for them. Their pincers can bite, but the nip is harmless.

During the summer many hellgrammites leave the water to crawl under logs, boards and flat stones near the bank before changing into flies. There they may be discovered in their small hollowed-out homes. They can be kept for several days in damp leaves or moss if not exposed to the sun, but ones taken from water will live longer than those found on land. Hellgrammites should be hooked through the tail, or under the collar, so they will be free to wriggle. Fish them *close* to the bottom; if they get *on* the bottom they will crawl under rocks where fish can't find them and they will be difficult to dislodge. They are excellent baits for all kinds of fish. A good way to use them is to drift them close to the bottom under a bobber.

How to Gather Nymphs

Fishermen who use live nymphs for bait often gather them with an old window screen. Have a friend help you. One person steps into rocky, shallow-flowing water and holds the screen against the bottom with its top above water and tilted downstream at about a 45-degree angle. The other person, working a few feet above the screen, turns over or dislodges rocks, thus freeing the nymphs and washing many of them downstream, where they are caught by the screen. A square yard or so of netting can be tacked on opposite sides to broomsticks so it can be rolled up for easy carrying. When unrolled, and the sticks spread apart, this serves the same purpose as a screen.

When you see several nymphs on the mesh of the screen, raise it from the water and pick off the large ones. Put these into a box containing dampmoss or leaves. They make excellent baits when impaled on small, lightwire hooks. Some kinds of nymphs, such as

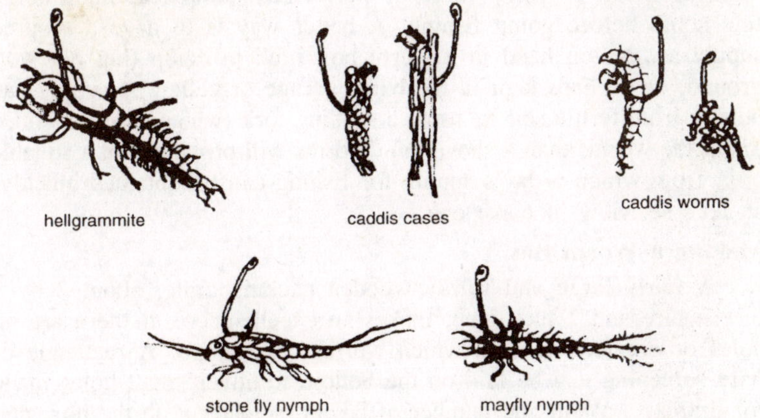

caddis worms, live in protective cases that resemble inch-long rough cylinders of bark, or an encrusted layer of tiny bits of gravel held together by their mucous. The nymphs in their cases can be impaled on light-wire hooks, or the nymphs can be removed and one or more can be put on the same hook, as shown. Stonefly and mayfly nymphs are hairy, buglike creatures, the former being identified by two tails and the latter by three, forked like a prong. Dragonfly nymphs, sometimes known as "perch bugs," are fat and somewhat resemble crickets.

WORMS: HOW TO KEEP AND RIG THEM

The common earthworm, which comprises more than a thousand varieties, is found almost everywhere, is easy to keep on hand, and is a favorite food for most species of fish. Since it is the most popular bait from coast to coast, too many fishermen think all one has to do is to put it on a hook and drop it into the water to get results. Fishermen who get the best results know there's a lot more to it than that, so let's see what successful ones know that the others don't.

Freshly dug worms are dark (because of the earth in them), soft and slimy. They should be prepared for fishing by scouring, which means leaving them in sandy, mossy soil or in a combination of sand and leaf mold for a few days. In such a preparation they expel the dark earth from their tracts and become a brighter shade of red. Their bodies toughen and are relatively free of slime. Thus they become more lively, easier to handle, and more attractive to fish.

The quick way to partially cleanse and scour them is to keep them in a can of damp moss or pulverized damp leaf mold for a few hours before going fishing. A better way is to have a scoured supply always on hand in a worm box sunk in damp (but not wet) ground, or perhaps kept in the barn, garage or cellar. A supply can be dug from fertile soil by using a spading fork (which is less inclined to cut the worms than a shovel is). Worms will propagate in a suitable box, from which a day's supply for fishing can be obtained quickly, so let's see how to make one.

Making a Worm Box

A fairly large and sturdy wooden packing crate, about 2 by 3 feet square and 2 feet deep, makes an excellent box if there are no holes or crevices through which worms can escape. A rectangle of wire screening can be laid on the bottom to cover small holes made for drainage. While the number of layers of material in the box, and their composition, can be varied, there should be a minimum of soil so the worms can scour themselves in alternate layers of sand and leaf mold or moss. Before filling, sink the box in cool, damp ground in a shady place, leaving a few inches of box above ground. The bottom layer should be of sand and the top layer of leaf mold or moss. Commercial bedding preparations, available at sporting goods stores, also can be used.

When the box has been prepared the worms are dropped on top so the lively ones can burrow in. The others should be picked out and discarded.

A quart or so of food also should be placed on top. This can be coffee grounds, vegetable cuttings, bread crumbs, cornmeal, chicken mash, etc., or a combination of such materials. Enough water should be added to dampen the contents. The box can be covered with canvas or burlap, held in place by boards. When a supply of worms is removed for fishing, the layers will be disturbed, but that is unimportant. If the box remains outdoors in freezing weather, leaves or other debris should be piled over it for protection. If small worms eventually are found it means that the box is functioning efficiently because a new crop is coming along.

Carrying Worms

A flat tobacco can makes a good container because it is handy to carry in the pocket. Air holes should be punched into the cover, and the can should be partially filled with damp moss or leaf mold. A difficulty with this is that the worms burrow to the bottom, so the contents may need to be dumped out to find a few. An alternative is to obtain a small round can which comes with a plastic cover; cocktail peanuts come in such cans. Cut out the bottom and put another plastic cover from a similar can on it. The worms can be seen through one cover or the other, and easily removed. To keep them lively they should be kept where it is shady and cool, out of the sun. Special cans for carrying worms can be purchased in tackle stores. These strap to the belt and keep the worms fresh.

Hooking Worms

Use small worms for small fish and large worms for large ones. If the worms are too big for the fish, the fish will steal the bait.

For trout, a single worm usually is best, hooked through the collar. Trout like worms that wriggle naturally, and they rarely will take them when they are impaled on the hook more than once or twice. I like to have the worm cover the barb because often fish feel

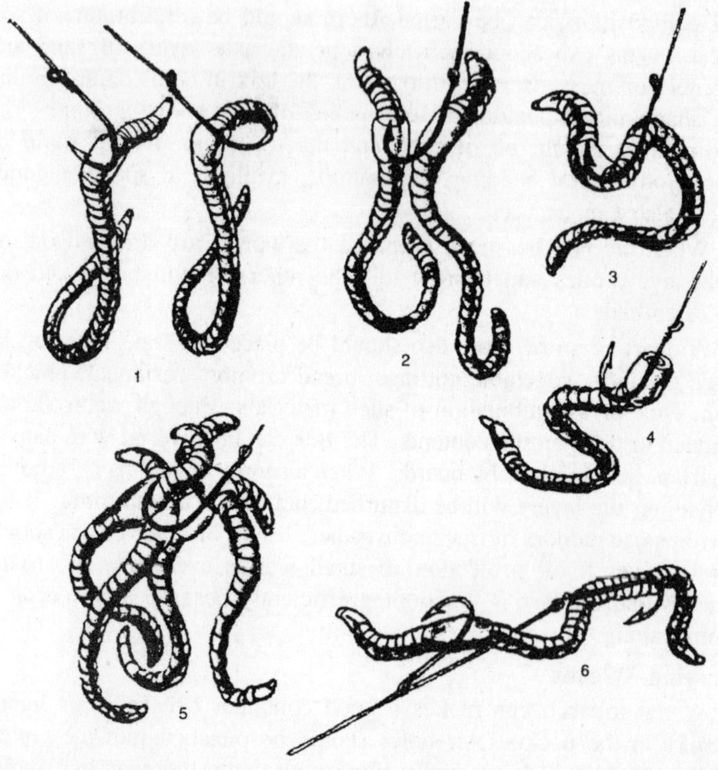

the barb and won't swallow the bait; also because this prevents the hook from catching in weeds.

When fishing for large fish, two small worms can be used, or a single large worm can be strung up the hook's shank.

Panfish are notorious bait stealers. If this happens, put half a worm, or a small end of one, on a small, light-wire hook. In quiet water, let the worm sink slowly to the bottom under its own weight.

Large bass, catfish and some other species prefer several worms on a No. 2 or larger hook, strung up the shank and down onto the barb.

In places where more than one hook is allowed, a tandem rig of two or three hooks keep the worm more securely on the hook for casting. Such a rig also is used for slow trolling, usually with a spinner or two ahead of it. Worm rigs also are popular for bobber fishing, and there are some tricks to this which will be described later in this chapter.

Most important rules in worm fishing are that the worm(s) should be bright and lively; hooked so it can wriggle naturally; be allowed to drift with the current (instead of being towed through it); and fished where the fish are -which usually means on the bottom or very close to it. To get down there we may need to use a sinker, or one or two split shot. If lead is necessary, the smallest amount that will do the job is the best, and it should be placed well up on the monofilament, rather than too close to the hook. Many beginners use hooks that are too large, or too heavy. Smaller ones of light wire usually are better.

Catching Nightcrawlers

Beginners use nightcrawlers too often, evidently on the theory that, if a smaller worm will do well, a bigger one will do better. Nightcrawlers often are so big as to prevent hooking fish of reasonable size that could be caught readily with a worm or two. Favor nightcrawlers for big fish, or for smaller ones with bigger mouths, such as bullheads, large bass and catfish. They are hooked as worms are, and are fished in the same ways. The best time to use them is during a rise of water after a rain.

There are several ways of obtaining nightcrawlers (and worms) in addition to digging them with a spading fork. Some fishermen shock them to the surface with electrodes pushed into the ground; others by producing underground vibrations made by raking a driven stake with a board as one would use a bow on a violin, or by throwing solutions of one thing or another on the ground so the liquid will drain into their burrows and drive them up. The simplest way is to gather them on the lawn after dark just after a rain, or after heavily watering the lawn with a garden hose.

Some sort of illumination usually is needed to see the nightcrawlers but, since they are sensitive to light, it should not be shown on them directly. A piece of red balloon rubber stretched over a flashlight lens with a rubber band usually provides enough light to see the quarry without frightening it.

There's a trick to gathering nightcrawlers. While they lie on the surface in the grass, their tails are inside the neck of the burrow, so they can slip into it instantly. The trick is to crawl slowly and quietly over the lawn until you spot a crawler and grab its tail end. The crawler will hold on to its grasp in the burrow. Hold onto the rest of it until it relaxes enough to be pulled out. Some crawlers may be entirely out of their burrows, but it's hard to tell for sure. They can be dropped into a container they can't crawl out of, and then can be kept in the worm box with the worms.

MINNOWS

While minnows are considered the second most popular bait, many fishermen would give them first place if they weren't so expensive to buy, so hard to catch, or so difficult to keep for several hours in a lively condition. While all baitfish technically are not minnows, the term is commonly accepted. Those obtained at bait shops usually aren't as lively and won't stay alive as long as the ones fishermen catch themselves. Gathering enough for a day's fishing usually isn't very difficult.

Minnows often school in large numbers in quiet backwaters of streams and lakes, and they congregate around docks when scraps of bread and other foods made from flour are dropped to them. There they can be caught with a minnow drop net or a minnow trap baited with such scraps. Some fisher men carry a wide-mouthed thin-meshed net for the purpose. They set it in a foot or two of water on the bottom in a place where minnows are present and they throw a few scraps of bread into it; the long handle extending to shore. When things quiet down the minnows will return for the bread, and when enough are over the net it is scooped up.

The baitfish then are transferred to a pail or a bait bucket. When fishing from shore this can be left in the water. Lacking such a container, you can scoop a small depression in the sand or gravel close to the water and keep the bait in this shallow-water corral. When fishing from a boat, you can tether the bait bucket overboard. The important thing is to keep the water cool and well oxygenated so the bait will stay as lively as possible. One way to keep it oxygenated is by shaking the container occasionally, or by changing the water from time to time. Oxygen tablets or capsules and other mechanical aids also help, and some bait buckets come rigged with them. If ice is available -perhaps a couple of refrigerator trays wrapped in newspapers-a cube or two dropped in the container helps keep the bait alive when the water begins to become warm.

Minnow fishermen who live lakeside usually make a livebox in which to store their bait. This is a mesh-sided box about a foot deep and somewhat wider and longer, with a trapdoor on top. It is tethered in water deep enough to avoid wave action, and it will keep the bait almost indefinitely if a little food is added every few days.

Rigging Minnows

Different sizes of minnows should be used for various species of fish. The smallest ones that can be hooked properly (under 2 inches long) are best for panfish. These also are suitable for average-sized trout, but they could be slightly longer for the bigger ones; even as large as 4 or 5 inches for lake trout and very big browns and rainbows or steelhead. Sizes for bass vary depending on the probable size of the bass, the average being a length of 3 or 4 inches. Extremely large species such as pike and muskellunge might call for baitfish from 5 to 10 inches long.

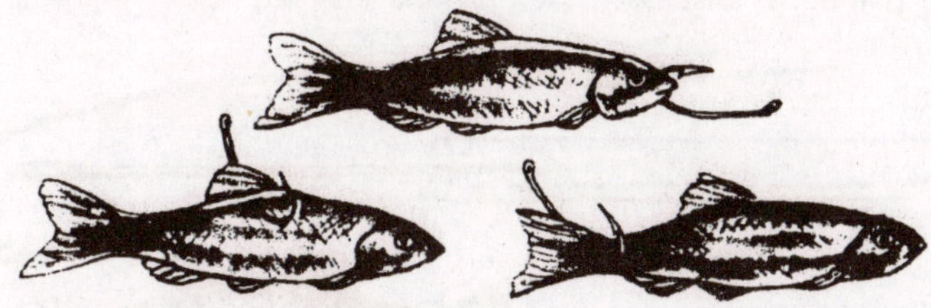

minnows rigged for stillfishing.

minnows rigged for casting.

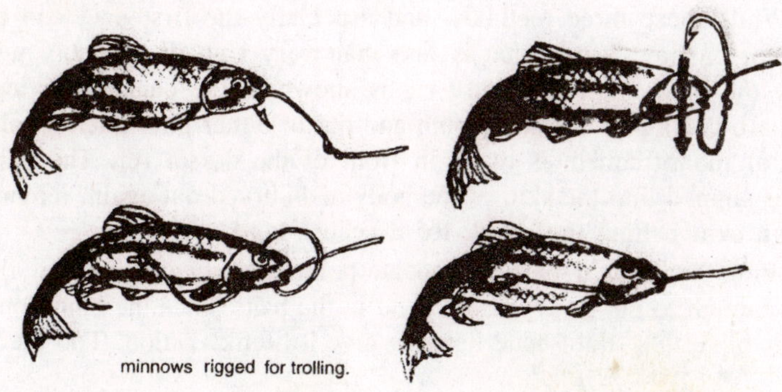

minnows rigged for trolling.

Three ways to rig minnows for stillfishing are shown on the next page. Since they must be active, they must not be hooked through the vital organs. The barb can be passed through the tip of the lower jaw and out of the upper one, as shown in the top illustration. A popular way is to push the barb under the dorsal fin just above the backbone, as in the second sketch. A third way is to put the hook through the rear of the body near the tail. Of these three methods the second one seems to allow minnows to swim most actively and, because the barb is in the midsection, it seems to hook more fish. These methods also are popular for fishing through the ice, and they are excellent for bobber fishing as discussed later in this chapter.

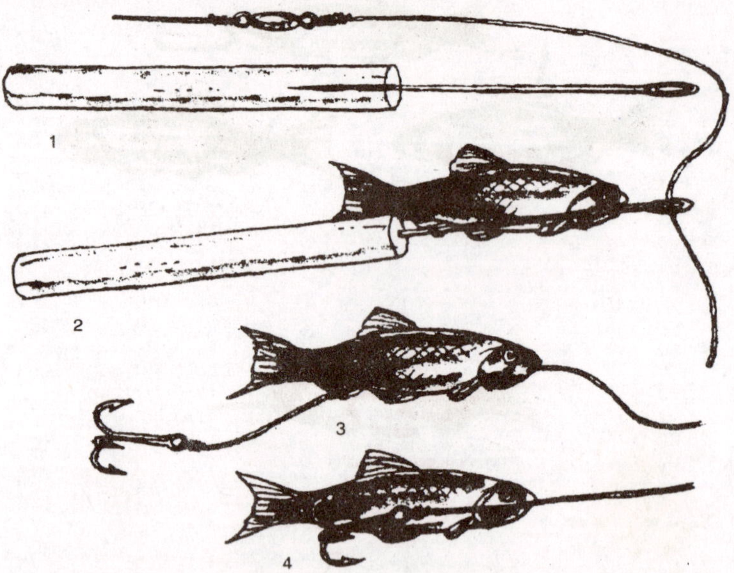

While these three methods, and especially the first one, can be used for casting, a cast that is less than very smooth probably will throw the bait. A more secure rig is shown in the center drawing. The barb is slipped into the mouth and out of either gill. Then a half-hitch of monofilament is made in front of the dorsal fin. The barb then is slipped into the skin of the body and curved out again, leaving enough skin behind the shank for a secure hold.

With the hook's barb and bend pointing outward, tighten the trace enough to put a very slight bend in the bait. Since the bait won't remain alive this slight bend helps to give it lifelike action. Too much bend would make it spin, which is undesirable. A favorite method is to cast up and across stream, retrieving fast enough to keep the minnow moving actively as the current carries it down. Since the hook is exposed one should strike immediately when a fish hits the bait.

A rig for trolling, as shown, must put a curve in the minnow so itt will rotate slowly as the boat moves along. First pass the hook down through the lower lip and then through the whole head from top to bottom just back of the minnow's eyes. Pull some extra line (or leader) through so this can be done easily. Then hook the minnow through the rear of the side of the body, as in the third sketch. Holding the bait in the palm of the hand so it will be curved slightly, pull the leader back through the head and then back through the lip to tighten the rig. Now test the bait in the water at proper trolling speed (about as fast as a man can walk) to be sure it revolves slowly. The bait can be made to revolve faster by tightening the rig to give it more curve, or by loosening it a bit to give it less.

Since this rig will put a twist in the line, it is advisable to tie in a ballbearing swivel or one or two link swivels between line and leader 4 or 5 feet ahead of the bait. If the line should become twisted it can be untwisted by rigging a bait curved in the opposite direction (with the hook coming out of the other side). By trolling this for about the same length of time the twist should be removed. If lead is necessary to fish the bait deeper, it should be put on the line just above the swivel(s), but not between swivels and bait.

If minnows are brought to a lake or stream from which they weren't caught, *never* dump the unused ones in the water. Many good fishing waters have been ruined by people introducing alien minnows which multiply and consume food which gamefish need. The result is fewer and smaller gamefish. It is better to take the unused minnows back where they came from, or to bury them far up on shore.

Almost everyone knows the black or dark-brown cricket which sometimes gets into houses and is supposed to bring good luck. The big black ones are best for fishing. Since they don't like dewy grass or rainy weather, look for them in cellars of old buildings; outdoors under boards, logs, stones or other refuse; in hay, wheat or corn stacks; and especially under strips of tarpaper. They can be caught by hand or in a fine-meshed net. One can be made by bending coathanger wire into a circle the size of the top of a stretched nylon stocking. Sew the stocking top to the wire, tie a knot a foot or so

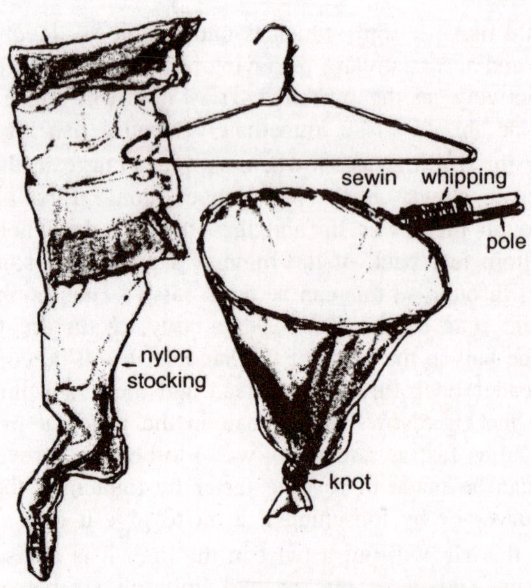

down on the stocking, and cut the rest off. Lash this to a wooden pole several feet long, and you're in business! Such a net also is good for catching grasshoppers because both kinds of insects have small spines on their feet which catch in the m esh. Ways to keep them for fishing uses are described in the next section. They are biggest in late summer and fall and can be caught more easily on cold mornings when they are partially dormant. Since their bodies are soft they must be hooked and handled carefully. Use finewire hooks of about size and hook them as shown. Other information about them is the same as for grasshoppers.

GRASSHOPPERS AND CRICKETS

When grasshoppers, locusts (cicada) and similar insects are buzzing over fields in summer, many of them land in the water and are favorite foods for fish. In early morning or during cool nights they are rather dormant and can be picked up easily. During the daytime a good method is to spread a woolly blanket on the grass and to chase the hoppers onto it. Their spurred feet catch in the wool, so they can be picked off easily. The nylon stocking net previously mentioned also can be used.

Some fishermen drop the insects into empty soda pop bottles from which they can't escape when the bottles are kept upright. It's easy to shake the hoppers from the bottle, one by one, as needed.

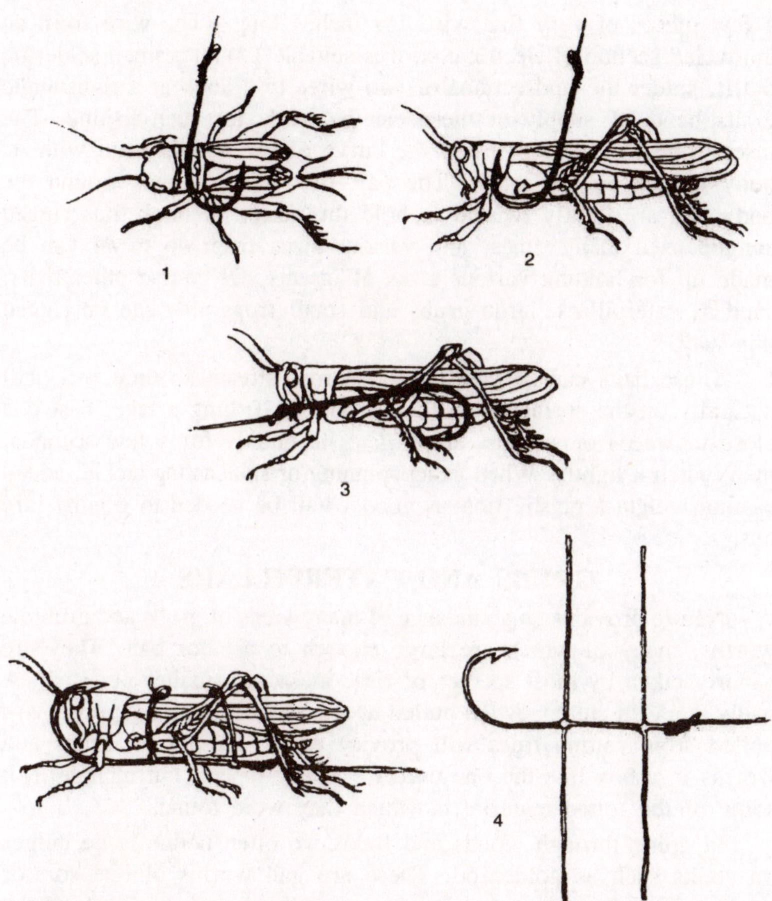

Another way is to put them in plastic boxes, preferably the ones with sliding tops. A little grass keeps them from jumping out. A better way is to put part of a nylon stocking in the box. The spurs on the insects' feet catch in the mesh and prevent them from jumping out. A few holes should be punched in the box for ventilation.

The illustration shows four ways to hook such insects, using light-wire hooks. Figure 1 shows the hook pushed down through a cricket's collar, with the barb then pulled up into the body. Figure 2 shows a similar hookup in reverse. In Figure 3 the hook is pushed into the insect's throat just below the head and then is turned downward to emerge from the body near the tail.

Many fishermen like to keep the insects alive on the hook. Cut a few pieces of very fine wire 1½ inches long. (The wire from an unraveled section of electrical cord is suitable.) With a small soldering outfit, solder the midsections of two wires to a hook at a right angle to its bend. A supply of these can be made for such fishing. The insect's head is placed inside the curve of the hook's bend with its body resting on the shank. The pair of wires are bent around the body and are lightly twisted to hold the insect. A hook thus rigged can be used many times, and various sizes from #6 to #4 can be made up for holding various sizes of insects. Of course other baits, such as caterpillars, large grubs and small frogs also can be rigged this way.

These baits can be cast up and across stream so they will drift naturally on the surface with the current. In fishing a lake, cast one close to weeds or stumps, allow it to lie quietly for a few seconds, then twitch it lightly. When using spinning or spincasting tackle, added casting weight-a plastic float is good - will be needed to get the lure out.

GRUBS AND CATERPILLARS

Nature provides an abundance of many kinds of grubs and grublike worms, many of which are large enough to use for bait. They are eagerly taken by most species of fish, including smaller pondfish. A badly decayed stump, when pulled apart, should yield a supply. Bark pulled from rotting trees will provide more. Keep such grubs and worms in a box like the one used for grasshoppers, but include in it some of the rotted material in which they were found.

In going through woods and fields we often notice large bulges on stalks such as goldenrod. These are gall worms of one sort or another and they are good baits, especially during the part of the season when it is hard to find others. Cut the stalk above and below the gall, or swelling, and take a quantity along. When needed, the stalk can be split to free the grub.

The best caterpillars are the kinds with no hairy fuzz, or with only a small amount of it. All these make good baits but, since their insides are very soft, they should be hooked through the head, or the Figure 4 rig should be

used. Cocoons contain caterpillar larvae, and they need not be taken from the cocoons until time to be used. Nests of tent caterpillars are common and often contain a day's supply. Cut the cobwebby nest from small growth or scrape it from large branches or trees. Take the whole thing along. If one collects a variety of grubs and caterpillars it is interesting to see which sorts the various species of fish like most, so the best can be selected for use on another trip. All these vary so much in kind from one part of the country to another that it is useless to try to mention them more specifically. Trees and bushes along streams and beside lakes drop them into the water in season and fish often collect there to gather them, so deep spots in such places can provide excellent fishing.

Even the more repulsive kinds of grubs and insects provide good baits, if one wants to use them. These include cockroaches, maggots, large mealworms and so forth. There seems to be little reason to bother with them, because the others are at least as good.

CRAWFISH

Looking like miniature lobsters, crawfish (also called crayfish or crawdads) are widely found in warm freshwater ponds, lakes and streams. They live there under rocks and logs on the bottom and often are in large colonies where they burrow into the peaty soil of undercut banks. They can be dug from the latter places and make excellent bait for bass, trout and other gamefish.

When you see crawfish in the water it is easy to catch them. Tie a rotting fish to a cord and throw it among them. When several crawfish collect to feast on the fish, raise it slowly by the cord and put a net under it as soon as possible to catch those that fall off. Some may hang on until they can be captured without a net, but using one adds to the catch.

Another way is to use a minnow drop net, baited similarly, and traps also are available for the purpose. Sometimes crawfish can be caught by hand, but one has to be quick, and not mind being pinched.

The best way to hook a crawfish is to embed all but the barb of the hook in the flesh of the tail just under the shell. Another way is to push the hook from under the tail out through the back. Since crawfish like to crawl under hiding places on the bottom, many fishermen break the claws off to prevent this. Some break off the body and use the tail part only, hooked as above. This is an excellent bait for catfish and other bottom feeders.

Similar to lobsters, crawfish shed their shells periodically as they grow bigger. Softshell crawfish, or "shedders," make the best bait. Being nocturnal creatures, most of them are out after dark.

Another way to fish crawfish is to drift the bait into riffles, runs and deep pools on a moderately slack line. A split shot or two may

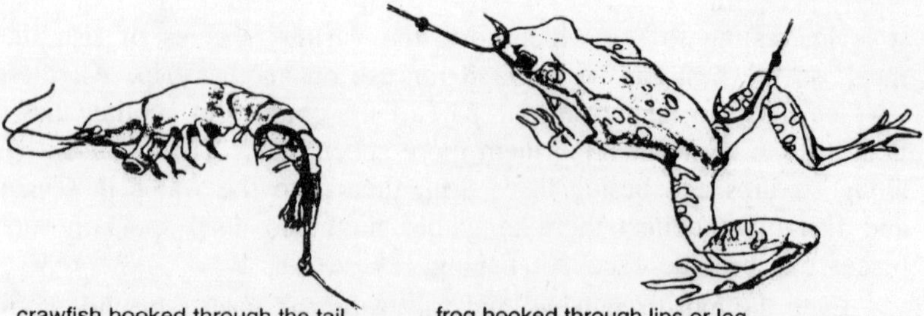

crawfish hooked through the tail. frog hooked through lips or leg.

be necessary to get it down into the deeper places. The larger crawfish are heavy enough to be cast out to reasonable distances with spinning or spincasting tackle. Longer casts can be made by tying in a plastic-bubble float partly or wholly filled with water (or another type of float) 2 or 3 feet above the bait. Fish will mouth the bait before swallowing it, so let them do this for a short time before striking. Usually, the time to strike is when they start to swim away with it.

FROGS AND TOADS

Small frogs are one of the world's best baits for bass and big brown trout, as well as for other trout and many other species. Bigger ones also take the bigger bass, members of the pike family (which includes muskellunge), and many other gamefish. Some states have restrictions on frog hunting, so the fish and game laws should be checked.

Frogs can be caught with long-handled nets, and by other means, but my favorite method is this:

On a pleasant night paddle in a canoe around the edge of a weed-lined section of a pond or lake. The person in the bow does the hunting and directs the paddler in the stern. The bow man is equipped with a flashlight and an old fly rod. On the rod's short and fairly strong leader is a hook of size 4 or 6 or so to which is attached a very small piece of red flannel or red balloon rubber. If there's a breeze we may need a few split shot on the leader to prevent it from blowing excessively. By shining the light along and near the shore, frogs are spotted. Lower the red bait on the long rod close to their heads, and they usually will grab for it. Hoist each one around so the man in the stern can remove it from the hook and put it in a damp burlap bag, or some other suitable container. (Of course this can be done by one man from shore, but it's more fun with two in a boat!)

The frogs can be kept in the bag overnight if it is laid in shallow water, or they can be transferred to a wire screened livebox set partly in the water and with a board or some rocks in it so the frogs can emerge to rest and breathe. If the big ones aren't needed for fishing why not saute them to make a delicious dish of frogs' legs?

I once took a big sack filled with frogs into the kitchen of a cottage where I was a guest so the frogs would be safe in the sink overnight. The cord tying the bag came loose and all the frogs escaped and hopped all over the place. The hostess discovered the frogs when she got up in the dark and stepped on one in her bare feet. Eventually calm was restored, but it became quite plain that it is much easier to catch frogs in the wild than to recatch them in a cottage. Everyone except the hostess thought all this was very funny, and it was a long time before I was invited back there again!

Frogs should be kept as lively as possible for maximum action while being fished. Two of the good ways to hook them are shown. The better way is to press the hook's barb upward through both lips. Another is to put the hook through a thigh, as shown. Others are to use one or another of the various harness arrangements to be found in tackle stores.

Spinning, spincasting or plugcasting tackle all provide excellent ways to fish with frogs, but spinning seems to be superior for the smaller ones. Cast to open spots amid lily pads or in the weeds and allow the frog to sit on the water for a minute or so before forcing him to swim in slow, short spurts. If a weedless hook is used, the frog can be cast onto the pads and then pulled off, or made to swim slowly between them. The commotion thus caused should bring a hard strike from a good bass. The same method can be used for members of the pike family. If big brown trout are in a stream, cast the frog into a run and let him drift downstream on a slack line. This often brings action.

Frogs can be fished deep, also, but some lead a foot or two above them on the line is necessary to get them to the bottom. A dead frog lying on the bottom is an excellent catfish bait, especially at night.

I remember having used toads for fishing only once, but without much success. Since toads are so helpful in eating bugs in the garden, perhaps they should be left alone in favor of better baits. Also, the smaller ones, which are best for bait, are hard to catch. A small, long-handled net usually will do it.

SALAMANDERS AND TADPOLES

Salamanders are the little "lizards" of various sizes and colors so often found when wet leaves and debris are disturbed along a stream or beside a lake. They can be caught by hand and kept in a can containing some damp moss or crumbled damp leaves. They are hooked either through both lips, through the base of thc tail, or through a thigh. Grab the salamander near its head; otherwise you may pull off its tail. If one gets away without its tail it will grow another one. Salamanders are excellent baits for all kinds of sportfish, but they are especially good for trout in the spring.

Tadpoles, often called pollywogs, are embryonic frogs which are found swarming in dead-water parts of streams and lakes in the spring. When they reach sizes big enough for bait they can be scooped out with a finely meshed net. They are hooked through the base of the tail, and often are used for bobber fishing. While they may be good baits on occasion, the ones mentioned previously seem to be much better.

STRIP BAITS

After catching a few fish it's not unusual to run out of bait. But actually you haven't. You can cut strip bait from the fish. Clean one of the fish and cut off the lower fins and some of the belly. The drawing below shows three ways to cut and rig strip baits. Cut them in a long V and hook them as in Figure 1. Or trim oval strips to a point and hook them as in Figure 2. Use the fins also; their fluttering motion adds to the attraction of the bait. Figure 3 shows both pectoral fins and the anal fin trimmed into one strip, but single fins can be used similarly. These should be given a mild bucktailing action to make them look alive, but, since they are actual baits, they often will take fish when being drifted, especially along the bottom. Such strips bring strikes by trolling when they are attached to streamer flies and bucktails. When plugs don't get results the addition of a strip bait on the tail hook can make a big difference.

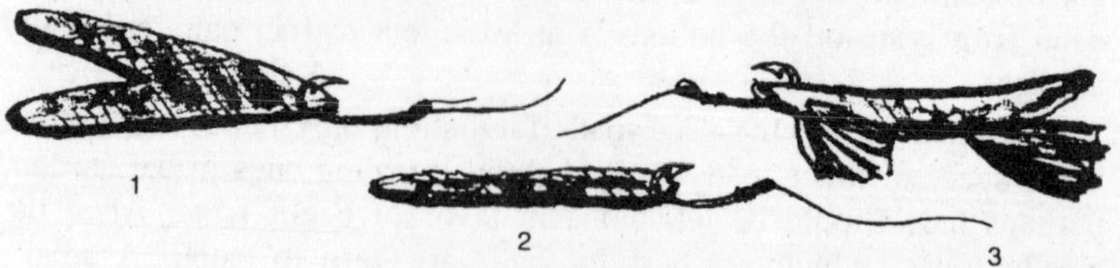

Lacking an actual fish belly, similar strips can be cut from such things as discarded white kid gloves and balloon rubber, but of course action must be given to them. They are very similar to the pork strips or pork rind baits.

HOW TO RIG
HOOKS WITH FISH ROE

Many kinds of fish, especially steelhead and other trout, are taken on salmon and steelhead roe and with the roe of other species. Preserved eggs can be purchased in jars in tackle stores, but the roe from fresh fish is better. When a fish is nearly ready for spawning the eggs may be large enough so a single one can be hidden in a very small hook.

Single salmon eggs are excellent bait for catching all species of trout in all but very large streams, especially those having spawning migrations. The best hook is a type made for the purpose: short-shanked, gold-plated (for less visibility), with a turned-down eye (so the eye will lock into the core of the egg more securely). Select the largest size-usually between 8 and 14-that can be concealed in the egg. Fasten this to the lightest practical leader tippet. Bait the hook as shown below.

To rig egg strips use a regular or shorter light-wire hook tied with a sliding snell. Push the snell down the shank to form a loop. Loop the roe around the hook so the part where the two ends of the strip cross is under the snell. Slide the snell forward and pull the loop to tighten the connection.

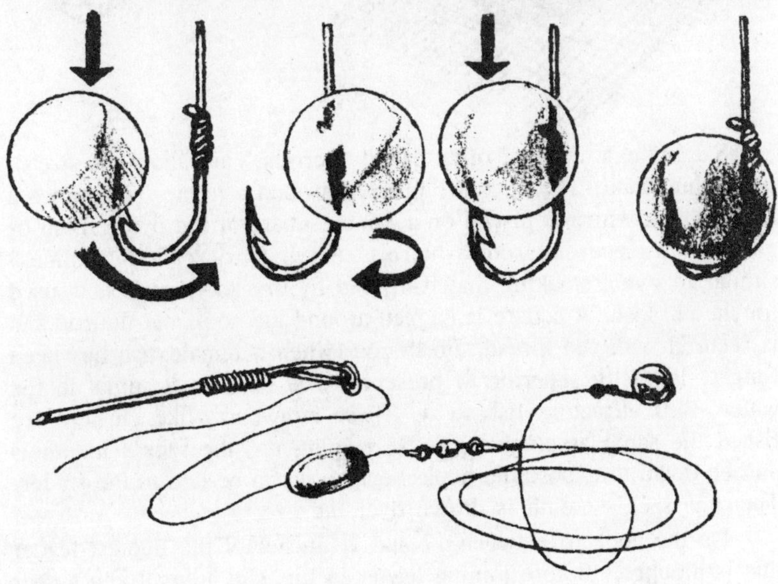

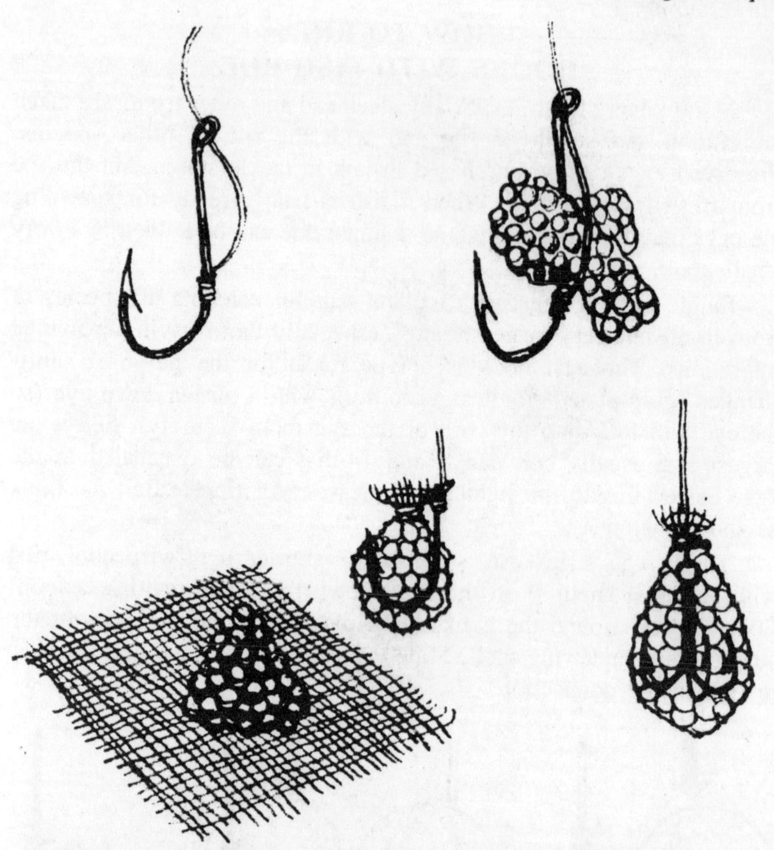

Salmon egg clusters, often called "berries," are about the size of one's thumbnail. The eggs are pressed around a treble (sometimes a single) hook which is placed on a 3-inch square of red nylon mesh or gauze or a piece cut from a hairnet. A red fabric called "moline," similar to nylon stocking material, sold by dry-goods stores, is used for the purpose. A square is bagged around the hook and the roe and is secured with red thread. Fresh roe (when a female fish has been caught) is vastly superior to preserved roe because it milks in the water, thus attracting fish to it. These strawberry-like clusters are fished the same as single eggs. In rigging up, the sack's mouth is pushed down to expose the hook's eye so it can be tied to the leader. Then the sack's mouth is drawn over the eye.

Tie the hook to between 12 and 20 inches of the lightest leader that is practical. Before joining leader to line slip a small egg sinker onto the line. This should be only heavy enough to bounce the bait along the bottom, hitting it every foot or two. Tie the smallest swivel between line and leader that will prevent the sinker from sliding over it. The idea is to let the fish take the lure and to move off without feeling much drag from the sinker, which will be lying on the bottom with the line running through it.

DOUGHBALLS AND STINKBAITS

Doughballs are popular baits for such bottom feeders as carp, bullheads and catfish. There are many ways to make them, and some fishermen think that the worse they smell the better they work. A small piece of the dough is pinched off the big chunk (if the balls aren't made separately); worked into a ball, and put on the hook. When fishing for small fish a very small ball covering only the barb of the hook may be enough. In such cases, however, it seems better to use a smaller hook so the ball can conceal barb and bend, even also perhaps covering the shank. The size of hook used is consistent with the size of the fish. Usually the baited hook is allowed to lie on the bottom until a fish picks it up and starts to go off with it. If no strikes are obtained in one location, cast to another.

Addicts of this sort of fishing all have their favorite recipes, but beginners can try these examples and eventually settle on favorites of their own.

Peel and grate a large potato, or two or three smaller ones. Add two tablespoons of cornmeal, about half a teaspoon of salt, and enough flour and shredded cotton to make a stiff batter. Pinch off bits and roll them into balls between a half inch and an inch in diameter, depending on the probable size of the fish and the size of the hook to be used. Cook these in boiling water until they float. Dip them out onto absorbent paper to dry, and carry them in a bait box or plastic bag. As an alternative, add a grated onion to this, or two tablespoons of syrup, sugar or molasses, or both. Carp, especially, seem to like them sweet.

The use of shredded cotton as a binder helps to keep the bait on the hook. Cornstarch or white of egg can be substituted, but none of these is necessary if the dough is stiff enough. If doughballs are made well in advance they can be kept in a refrigerator.

Another recipe is to mix equal parts of bread, oatmeal, cereal and sugar with enough water to make a stiff dough. This can be carried as one ball and pieces can be pinched off and rolled into small balls as needed.

A simple recipe is to mix flour and any kind of cheese into a firm ball and to use it as above. One that may be tastier (to the fish) is to mix equal parts of hamburger steak and cheese with enough flour and water to make a stiff paste.

Plastic sponges in such colors as red, pink, yellow and white can be soaked with various preparations after being cut into cubes of whatever size is preferred. One recipe calls for working cheese spread into them; the smeilier the better. Another is to soak the sponge cubes in blood and to let them dry. Liver of any kind can be cut up for use. Entrails of small animals (especially chicken) are used as bait, as well as the soft meat of clams or mussels. Almost anything goes!

Fishermen who don't want to bother to make their own dough and stinkbaits, or who have wives who think they should draw the line somewhere, can buy them already prepared in tackle stores. However, addicts of this sort of fishing usually prefer their own formulas, of which these are only a few of the milder examples. Because lady anglers may read this and then refuse to cook the catch afterwards, it seems prudent to go no farther!

These baits of course are for certain varieties of bottom fish, and each variety has its preferences. Thus it may be helpful here to discuss some of these bottom feeders to see which baits they prefer and how best to catch them.

Carp

Carp are bottom feeders which frequent muddy places in ponds, lakes and sluggish rivers. Too often they thrive in such places to the detriment of more desirable fish because someone allowed Junior to empty his goldfish bowl there, thus doing serious and perhaps permanent damage to the fishery. Ponds in parks often contain carp for this reason, and they may provide easily accessible sport to young people.

Carp have small mouths and feed largely by smell, so small baits on small hooks are necessary, except for the very big ones. They should be fished on a loose line, preferably without a sinker, because they are wary and won't take a bait if they think something's wrong with it. A good time to fish for them is on dark, rainy days, although they will bite at other times. When one is hooked he should be drawn away from the others to avoid disturbing the school.

While carp eat snails, algae and various forms of vegetation, they also take small hooks baited with worms, peas, kernels of corn (several of both strung on the hook), partly cooked bits of potato, doughballs (as mentioned above), pellets of bread, small pieces of fish, shrimp, balls of moss scraped from underwater rocks, and even marshmallows. The doughballs can be sweetened. Other live baits, such as minnows, are not productive, and smelly baits are unnecessary.

Catfish and Bullheads

Bullheads are a species of catfish which inhabit the muddy bottoms of ponds, lakes, slow rivers and streams, canals and ditches, preferably in areas where there is protective vegetation. While they feed mainly at night, they will bite during the day. Both bullheads and catfish will take a wide variety of baits, including live minnows, frogs, a gob of worms, shrimp, crawfish tails, cut fresh bait (especially mossbunkers cut in pieces, because they are oily), pieces of rotted fish, stinkbaits and doughballs, chicken and animal entrails, liver of any kind, grapes, kernels of corn strung on a hook, nightcrawlers. Their diet includes the entire gamut from overripe cheese to laundry soap! Hooks of sizes 4 or 6, on line testing about 10 pounds, are good choices for bullheads, but the larger catfish may need something more substantial. Hooks often are of the baitholder type, with barbs on their shanks.

Catfish are found in both fresh and salt water. In fresh water they live in such places as protected holes in rivers, near brushpiles, fallen trees, and undercut banks. Fishing in tidal rivers is best when the tide is running, rather than at the high or the low. Bigger baits attract the bigger fish.

In addition to usual rod and reel fishing, catfish often are taken on trotlines, where such are allowed. These are long, heavy cords, perhaps extending across a stream, which are anchored at both ends, such as by tying them to brush. Middle sections may need to be kept near the surface by using plastic bottles as buoys. Short lines with hooks attached hang down from the main one, the hooks being baited with a choice of the stuff which has been described. Another way to catch catfish (where it is allowed) is by "jugging." Short lines with baited hooks are attached to handles of corked plastic jugs. Several of these are put overboard and allowed to drift in a slow current, being watched and guided by people in a boat. When a big catfish takes such a bait it may pull the jug under from time to time, or may tow it for a considerable distance.

Chubs

These are primarily river fish that often are taken by fly fishermen when fishing for trout. They like water between 2 and 6 feet deep with from slow to moderate flow, so they sometimes occupy the same habitat as smallmouth bass and trout. They like a gravel bottom and often collect near jetties, wharves and piers, under overhanging brush, or in beds of weeds. Tackle used can be the same as for smallmouth bass and trout. Chubs eat vegetable matter, small shellfish, insects, minnows, caterpillars, grubs, bread-balls, soft cheese, doughballs, etc. They also often take small spinners and dry or wet flies. Young people on their first fishing trips can have a lot of fun catching chubs, and older fishermen sometimes enjoy them, too, except when they interfere with the catching of species which are considered more desirable. Chubs are not fussy about water temperature and they feed more or less constantly. While lacking the challenge provided by more palatable species such as trout and bass, they can provide good sport.

Suckers

Although they're rated low as food fish, suckers can furnish fishing fun when not much else is going on. Neophyte fishermen often see several big ones lying clearly exposed in sandy pools, slow-water eddies, or backwaters and excitedly fish for them under the misapprehension that they are trout. Of course, they are not where any respectable trout would be, and a closer look at their snouts dispels the illusion.

Since these coarse fish don't fight much, any sort of light tackle will do. Their small mouths call for small hooks, which can be baited with a piece of mussel or one or more small worms. The bait is dropped to lie near them. Wait until one starts to go away with the bait before striking.

BAITFISHING WITH BOBBERS

While many kinds and sizes of floats or bobbers are available, it is important to select the right one for the kind of fishing that is to be done, because just any one won't do. When a bait is suspended from a bobber floating on the surface, the bobber must float just enough to prevent it from being pulled under by the weight of the bait. When a fish takes the bait he should feel almost no resistance. If he tugs at the bait, and the bait seems to tug back because of too much resistance from the bobber, the fish usually will leave the bait, unless his first approach happens to hook him. Since fish sometimes mouth the bait before taking it, this usually isn't the case.

British anglers have this very important bit of information reduced to sort of a science. They carry their bobbers (which are more like pencils, and which they call "controllers") in sets of various sizes or degrees of buoyancy, and they usually carry spares of more than one of each kind. In rigging a bait they put on a light bobber to see if it will hold the bait up. If it does, they try a smaller one, and if it doesn't they try a larger one until they find one that barely remains on the surface to support the pull of the bait, or the combination of it and the pull of the current on it. The result of this rather precise bit of experimenting is that, when the fish takes the bait, he feels almost no resistance at all. Feeling almost no resistance almost guarantees hooking him.

Other than that, it makes little or no difference what color or type of bobber one uses, but I like to use one which comes in only two sizes which cover all requirements. This is a colorless plastic ball float with a rim around it to which line and leader can be attached. A removable plug permits the float to be filled with just enough water or mineral oil. When entirely filled it is about the same specific gravity as water, so it will drift down in the current and yet provide an ample weight to cast tiny baits long distances with spinning or spincasting tackle. Since this float is colorless it resembles a bubble on the water, and thus attracts no attention from fish.

The main purpose of a bobber is to hold the bait off the bottom so it won't catch and so it will drift a bait naturally downstream without excess speed or drag. The leader (a piece of the spinning line) should be long enough to keep the bait as close to the bottom as possible without snagging it. Thus, with such a bobber, we have a suitable weight to cast a small lure long distances and we also have a means of keeping the bait off the bottom so it can cover a lot of ground as it drifts down in the current.

A second consideration is that the size of hook should be appropriate to the fish we are seeking. Most people use hooks that are too large, and probably too heavy in the iron. If in doubt, select the smallest and lightest hook that will do the job.

The illustration at left shows three ways to rig a plastic bobber. In Figure 1, one side of the bobber is affixed to the spinning line and the other one to the leader. The bobber is filled with just enough water to allow it to hold up the bait. (An eyedropper is handy for this.) The leader is just long enough to keep the bait off the bottom. Two or three split may be needed to get the bait down, but use them only if necessary. Cast upstream, take in slack and then let out or take in enough line to guide the float where you want it to go. If the bobber is suddenly pulled under, strike hard.

In Figure 2, a sliding float rig is shown. This rig permits the fish to mouth the bait without feeling tension on the line. String the float (through one eye only) on the line and tie on a barrel swivel. Tie a couple of feet of line to the swivel and tie the hook to the end.

The rig shown in Figure 3 is for holding the bait off the bottom. Pinch a small lead on the line 2 or 3 feet behind the float, which may be partly filled with water. The lead holds bottom and the float seeks to rise, holding the bait off bottom. With this rig the bait can be cast out and left alone for a while. If there's no action in that spot the rod can be raised, thus raising the sinker to allow the rig to drift downstream a few feet more.

SINKERS AND OTHER ACCESSORIES

Sinkers come in many shapes and sizes from which specific ones are chosen for various reasons. Some of the more important types will be described and illustrated.

The old familiar *split shot* is available in a range of about ten sizes between #1 (.38 inch diameter) and "B" (.17 inch diameter), one or more being pinched on line or leader with pliers. These can be attached lightly to a short dropper so they will pull off if snagged. A more recent improvement is the pinch-on type designed for application without pliers.

Strip lead, also used where small amounts of weight are needed, is furnished in folders like matchbooks; the strips being torn off like matches and spirally wound around line or leader. If this isn't done neatly it can gather bits of weeds.

Dipsey sinkers are rounded in a teardrop design to help prevent them from catching on rocks. They are popular for fishing bait on the bottom in fresh water. They come in a range of sizes between 1/8 and 8 ounces.

Egg sinkers are oval, with a hole in the middle, allowing them to slide on the line. They sink a bait while allowing it to roll in a current, and also are used where there is little or no current action. An egg sinker can be cut in two parts for use as a bullet-shaped weight for plastic worm fishing.

Keel sinkers come in a great variety of shapes and sizes, all designed to prevent line twist (usually in trolling) and to provide weight. They normally are connected between line and leader. They can be made (or bought) for light lure or bait fishing by cutting lead sheeting into heart-shaped pieces which are crimped around line or leader to form small keels.

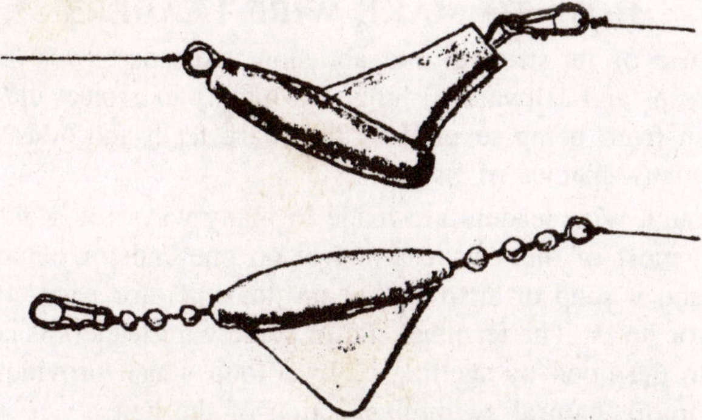

Diamond sinkers are of elongated diamond design with rings on both ends to provide a streamlined shape for minimum drag when trolling. A hook is added to one end for jigging.

Drails are more or less boomerang shaped, with rings at both ends, thus simulating keels. They usually are heavy to fish baits deep. Some have a clothespin-type snap at one end so the line can be clipped to it for deep trolling, allowing the line to be pulled free on a strike so the fish can be played without weight. They also are used in surf fishing with bait when casting from shore.

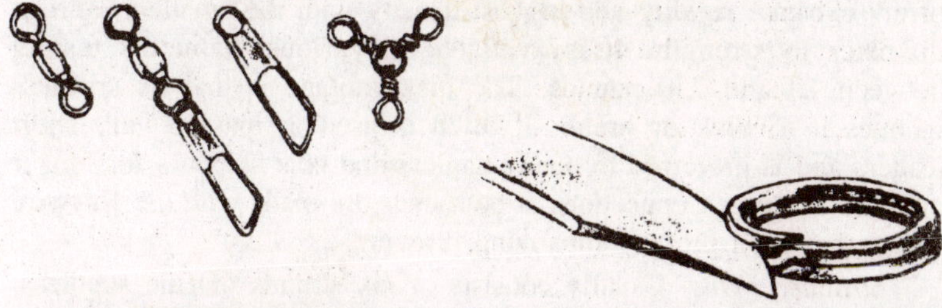

Several different types and sizes of *safety snaps, swivels* and *split rings* are shown in the illustration on page 120. Snaps and swivels are used only when necessary, and then only in the smallest sizes suitable for the tackle. Most anglers agree that black or bronzed finishes are superior to bright ones because they are less noticeable, although bright ones are commonly used in salt water, and in fresh water when the fish being sought are not presumed to be tackle-shy.

Split rings make handy connections and are easy to apply when one knows how. Slip a jackknife blade between the loops to raise one of the ends. Clip the hook eye or connecting loop around the raised end; withdraw the blade, and turn the ring until the connection(s) snap inside it.

HOW TO MAKE WIRE LEADERS

Because of its strength and abrasion resistance, wire has many uses in fresh- and saltwater fishing, but mainly to protect the terminal connection from being severed by the sharp teeth and other abrasive parts of many species of fish.

Although wire leaders are made in many ways for a diversity of purposes, most of them have a swivel on one end for connection to the line and a snap or snap swivel on the other for easy attachment of hooks or lures. The terminal end of some wire leaders is connected directly to the hook by snelling or by a loop which provides a loose hook for more natural swimming action of the bait.

Wire leaders vary in length from about 4 inches to 6 feet, the 4- to 12-inch ones are used in casting. The end of the leader which is connected to the line can be fitted with a ring, or with a swivel, which diminishes twisting, when trolling or retrieving the lure.

Since each of several kinds of wire has its advantages and disadvantages for making leaders (or for trolling), it will be helpful to discuss them and how they should be connected to the tackle.

Solid Wire. Sometimes called "piano wire," but it isn't; true piano wire is carbon steel while that used for fishing is stainless steel. It offers extreme rigidity and highest density with the smallest ratio of diameter to strength. It is available in various diameters testing between 10 and 250 pounds. The disadvantage is that its stiffness inclines it to kink or break. It often is used in making very short leaders and is preferred by many anglers and boat skippers for longer ones for trolling. Connections at both ends are made with the Haywire Twist shown in the accompanying drawing.

Stranded Wire. Usually consists of six strands of fine stainless-steel wire wound around a seventh wire which acts as a core, thus providing a Sevenstrand (its trade name) miniature cable, as shown. The stranded wire is given a bronze camouflage finish by heat treating, which also provides more uniform strength and greater flexibility. This wire is very unobtrusive in the water and is amazingly fine for its strength which, in its many sizes, is from 18 to 600 pounds.

Nylon-coated Stranded Wire. Called Sevalon, this is the same as above except that it is covered by a thin coating of transparent nylon. This gives it a smooth surface with much less tendency to curl and kink. The nylon coating makes the wire more durable but it adds slightly to its diametera disadvantage so minor that its qualities far outweigh it.

Knots and Fastenings for Stranded Wire

Making loops in coated or uncoated stranded wire for looping it to the line or for attaching hooks, snaps, snap swivels or lures is easily and quickly done. An inexpensive crimping tool and several sizes of soft metal (usually brass) swaging sleeves are sold for the purpose. A sleeve is slipped onto the wire, the hook or other connection is threaded on, and the wire's end is pushed into the sleeve, whereupon wires and sleeve are crimped into a permanent connection. A stronger loop can be made by doubling the wire through the eye.

When a wire leader is needed that is so long that it must pass through rod guides, or when the crimping tool and metal sleeves are not available, the wire leader can be joined to the monofilament line by the Key Loop Knot. Other knots for joining stranded wire and monofilament, such as the Figure-8 and Surgeon's Knot, are either unsafe or impractical.

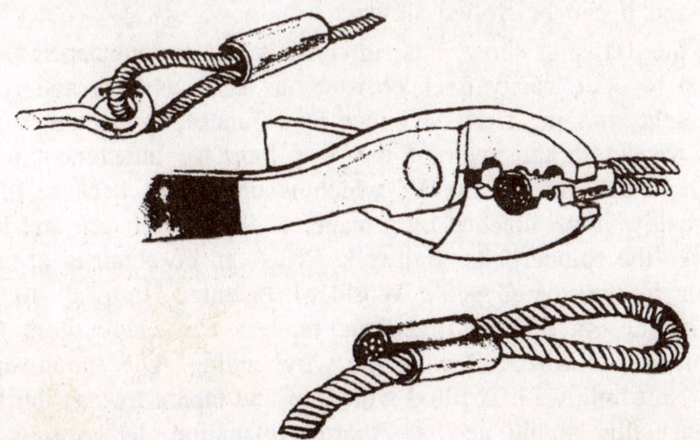

How to make a proper crimp: 1. Slide sleeve on stranded wire and make a loop by also threading end through sleeve, but don't allow end to protrude. 2. Hold loop horizontally, being sure wires are not crossed inside sleeve. Apply crimping tool to sleeve and crimp tightly in one or more places, depending on length of sleeve. 3. Proper crimp; wire should be flush with end of sleeve.

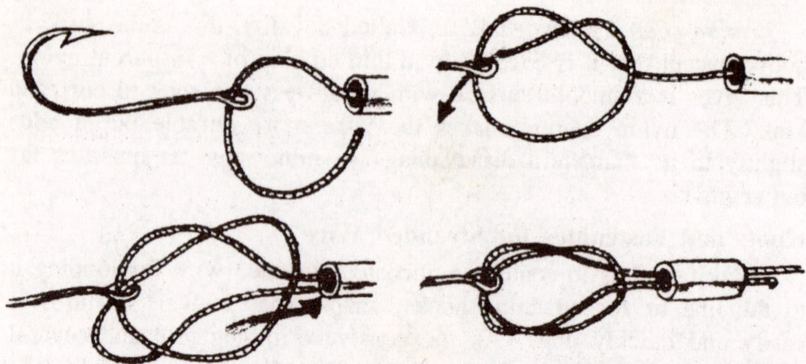

To make a stronger loop with nylon-coated wire: 1. Thread wire through sleeve and eye of hook, swivel or lure. 2. Make an overhand knot. 3. Pass wire through eye a second time. 4. Make a second overhand knot; pass wire end through sleeve and crimp twice.

THE BAITWALKER RIG

When fish aren't taking foods on or near the surface they will be on or close to bottom; almost never in middle depths. A very productive way for hooking bottom huggers is with the baitwalker rig. This can be used with many kinds of natural baits or artificial lures. It can be handled with spinning, spincasting or baitcasting tackle, and it can be trolled slowly.

As the drawings show, this rig consists of a spoon-shaped weight designed to slide easily over obstructions such as logs and rocks. This weight, ranging from ¼ ounce to 6 ounces, is affixed to a V-shaped wire with the apex of the V a loop for attachment to the line. The other side of the V, which is uppermost because of the weight below it, is attached by a leader to the bait or lure and looks much like the spinnerbaits. Baitwalker rigs can be obtained at tackle shops or by writing Gapen's World of Fishing, Highway 10, Big Lake, Minnesota 55309. Ingenious readers can make them from melted lead and wire following the wirebending. The spoon-shaped portion isn't hollow; it is filled with lead and tapers toward the front end. Since this should need no more explanation, let's discuss the upper arm to which the bait or lure is attached.

To the end loop of this upper arm of wire we attach a monofilament leader between 1 and 2 feet or more long, with the hook or lure attached to this. First, let's bait the hook with a worm or two, or a baitfish hooked through the lips, and see how it works.

We can cast this out and let the weight act as a sinker, stillfishing it with the bait activated by the current. If nothing happens we can fish it in. By this means we can "walk" the weight over obstructions with minimum danger of getting caught up. The bait trails at a slightly higher level, so is kept out of danger. It always is close to or a foot or so off the bottom, well within the range of gamefish lying there.

An excellent lure for this purpose is a very small floating plug imitating a baitfish. Since this wants to rise, but is kept down by the weight, it rides free in the feeding zone of fish lying on or near the bottom. Lacking suitable current it should be cast out and fished in. This rig can be "walked" up or down inclines, ledges and over obstructions where ordinary tackle might become snagged. In addition to natural baits, plastic worms, small plugs and so forth, try streamer flies or bucktails, particularly in the currents of streams. Small floats can be used to raise lures higher, if need be, as the drawing shows. Salmon eggs or clusters are useful in rivers at spawning time.

5

SPINNING

In the 1940s, fishermen were startled to see a few rugged individualists using an underslung reel attached to a light rod with outsize guides which could accurately cast even the tiniest of weighted lures amazing distances. They derisively called the reel a "coffee grinder."

As time went on, however, the principle proved so practical that it has become a major angling method, perhaps more popular than any other. Its secret is in the open-faced reel from which the line uncoils from a stationary spool with negligible friction, thus permitting longer casts with much lighter lures. We can compare this with the principle of the much older baitcasting reel where, in casting, the reel spool must revolve. This causes considerable friction, which decreases the distance of the cast unless much heavier lures are used. Also, if the baitcasting reel is not properly "thumbed," or controlled, its spool may revolve so fast as to overrun the line and cause backlash.

Following the advent of the open-faced spinning reel, manufacturers developed another reel which is a compromise between spinning and baitcasting. This is properly called a spincasting reel and should not be confused with either of the other two. Spincasting calls for a closed-faced instead of an open-faced reel spool. It virtually eliminates the backlashes so common to beginners in baitcasting, but, owing to increased friction, it requires stronger lines and heavier lures to cast as far as the open-faced spinning reel. The next chapters will discuss spincasting and baitcasting, their advantages and disadvantages. Here, we are concerned only with spinning.

Since writers about spinning have stressed the ease of learning the method, anglers generally have assumed there's nothing to it.

They have fallen into unnecessary trouble by thinking they can buy any kind of tackle and start without instruction. They should realize that there are numerous tricks which, when learned, will make spinning more productive and enjoyable.

When choosing spinning tackle, beginners should avoid low-cost "bargains" in unknown brands. Bargains or discounts are all right if one can get them, but many bargains should be viewed with suspicion. Why the lower price? We all know who the firmly established tackle manufacturers are because our friends use and like the equipment, and we have seen it advertised consistently in outdoor publications. Select current (rather than discontinued) rods and reels made by the firmly established manufacturers because they wouldn't have remained in business so long by offering inferior merchandise. True, we find occasionally excellent imports which may not be well known but, generally, the vast volumes sold by the big and wellknown makers assure obtaining the best equipment for your money.

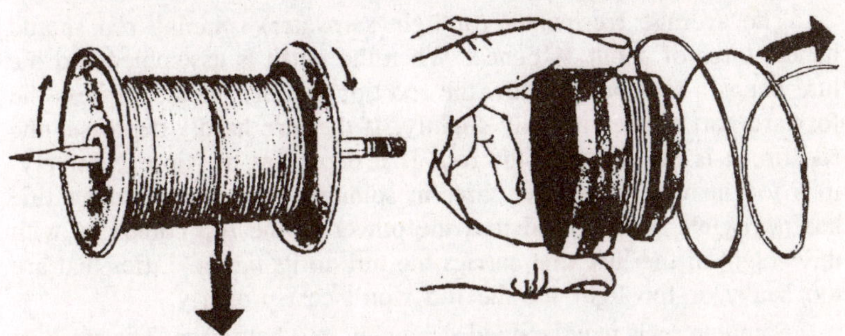

Principle of the spinning reel as opposed to the baitcasting reel. During cast, the weight of the lure pulls line from the baitcasting reel like thread from a spool *(left)*. The revolving spool creates friction, and may overrun the line and cause a backlash. Spinning reel has a stationary spool *(right)*, from which the line uncoils as it is pulled by the weight of the cast lure.

ASSEMBLING A SPINNING OUTFIT

Although one angler in many still may want a spinning rod made of split bamboo, the most sensible choice is one made of glass fiber or graphite. For average freshwater fishing the rod will be 6½ or 7 feet long, with a large butt guide and five or six smaller ring guides properly spaced and graduated in size to the tiptop. The grip may be all cork, or it may have a built-in locking reel seat. While the beginner may prefer the latter, some experienced fishermen using rods in the lighter weight ranges prefer the all-cork reel seat with sliding rings because they consider it more comfortable and it allows the reel to be properly positioned for the best balance.

Sometimes a reel doesn't fit the reel seat on a rod, thus causing it to wobble. Take along a piece of inner-tube rubber and cut part of it to the shape of the reel's foot. Putting this between reel foot and reel seat should prevent wobble.

In cold weather the metal on fixed reel seats may be uncomfortable to handle, or the rings on sliding reel seats may tend to work loose. These difficulties can be eliminated by binding the reel seat with plastic adhesive tape. It's not very pretty, but it can solve the problem!

In buying rods offered by established manufacturers one need not be concerned very much with rod action. Most manufacturers standardize the action that provides greatest casting efficiency. However, the action should be felt well down into the butt of the rod; the tip, when oscillated, should come to rest without excessive wobble.

The average freshwater (or light saltwater) spinning rod should handle lures of about ¼ ounce. When the outfit is assembled and the lure hangs a foot or so below the rod tip, the lure should depress the forward part of the rod only slightly. If the lure hardly depresses the rod tip, it is too light for the rod. If it depresses the tip excessively, it is too heavy. Remember that, in spinning, the monofilament line has negligible weight, and it is the power of the rod combined with the weight of the lure that carries the lure to its target. Lures that are too heavy or too light for the rod won't cast properly.

Spinning reels may be divided roughly into four sizes. The smallest reel normally is used with light tackle, which is lighter than that now being discussed. The next-larger size is for medium freshwater (or light saltwater) fishing. The next-to-largest reel would be used for heavy freshwater or medium saltwater fishing. The largest reel mainly is for heavy saltwater fishing, although it has freshwater uses, primarily for very large fish, especially in fast currents or deep in lakes.

The medium-sized reel should hold at least 200 yards of 6-pound-test monofilament line; a bit more or less, perhaps, when lighter or heavier lines are used. We used to think that 3- or 4-pound-test monofilament was about the right strength for medium freshwater tackle; that lines of 2- or 3-pound test, or perhaps less, were proper for light gear; that 6- or 8-pound test worked best with reels in the next-to-largest class, and that 10 or 12 pounds was as strong as one should go, even with heaviest equipment. However, modern monofilament is strong and limp for its diameter so many fishermen now find 6- or even 8-pound test suitable for medium tackle, even with lures as light as 1/4 ounce.

You should use as light a line as possible, considering the weight of lures to be used and whether or not you'll be fishing in open water or in water with snags, weeds or other obstructions where an entangled fish could break a light line. Remember, the lighter the line (the smaller its diameter), the farther you can cast with it. For extra-long casts, use the finest lines consistent with safety and with the tackle.

Thus, except for ultralight tackle (which will be discussed later) we can work out a table of tackle combinations. We should, however, agree that these are generalities; that there can be intermediate sizes; that lure weights are influenced by their size and streamlining, that rod lengths are influenced by their power, and that line strengths depend upon the type of monofilament used and its degree of limpness.

Reel	***Line Test***	***Lure Weight***	***Rod Length***
Light	4-8 lbs.	1/8 oz.	6½ ft.
Medium	6-10 lbs.	1/4-3/8 oz.	7 ft.
Heavy	10-15 lbs.	1/2-1½ oz.	7½ ft. (or more)
Extra-heavy	12-20 lbs.	2-4 oz.	8 ft. (or more)

The main point is that spinning tackle should not be assembled in a haphazard manner. If each element is matched with the others, the tackle should function efficiently. The final test is in using the tackle itself. Then we can determine the ideal combination of line and lure weight that functions best with the rod and reel.

We note that there is a line strength range which should be used with each size of tackle. When buying the reel it is very helpful to order at least one extra reel spool for it, and to put on the extra spool(s) lines of various strengths in the appropriate range. Thus, with the medium-sized outfit, we can have spools containing 4- and 8-pound-test line, or even 4, 6 and 8. We would use the lighter line for longest casts and the stronger line for fishing near obstructions or for casting lures which may be somewhat heavier than normal.

MORE ABOUT REELS

Unlike the early days of spinning, modern reels now are so efficient that one can make his selection from those offered by established manufacturers almost on the basis of how much a reel's appearance appeals to him. Modern reels usually have a rate of lure retrieve of between 3½ and 5 revolutions of the pickup for each turn of the reel handle. While the lower ratio usually is very satisfactory, a higher one provides better lure control, especially in upstream fishing. If one casts out slack line, the fast retrieve picks it up quicker. One may need to retrieve very fast to get strikes from certain species of fish. And, even if the reel has a very fast retrieve, one doesn't have to use it. He can crank in the line as slowly as he wishes.

Antireverse Lever

This device prevents the reel from backwinding when a fish is taking out line. It is conveniently located and easy to engage. But check the reel you propose to buy to be sure the lever is convenient for you.

The antireverse lever should be disengaged when casting for greater ease in handling the tackle. It is engaged at the option of the angler: sometimes as soon as a cast is made; sometimes only when one has a big fish on which starts to make a run. The antireverse is especially valuable in trolling, and always should be engaged to prevent the reel's backwinding on a strike. If one enjoys stillfishing, the bait can be cast out and slack line taken in. Then the antireverse is engaged and the rod can be set in a Y-shaped stick, or otherwise. The device also is useful to keep the tackle tight when not in use. Hook the lure into a rod guide, reel up slack, and engage the antireverse.

The Brake, or Drag

This mechanism is very important. A brake without wide latitude of adjustment, or one that is jumpy, is often the cause of lost fish. This, in part, can be tested at the dealer by screwing down the brake knob until the spool is almost locked. Then back it off until the spool revolves almost freely. One should be able to make at least two complete turns of the brake knob between these two positions, and at least three turns are even better.

Wide latitude of brake adjustment is necessary in setting the drag of the reel spool easily and properly.

The second element of brake examination is to determine how smooth it is. A good way to do this is by assembling the tackle, setting the brake to moderate tension, hooking the lure to a stump or something similar, and then walking backward with the tackle and allowing the reel to pay out line. By doing this it is easy to find out whether the brake operates smoothly or

whether it is jumpy. If it is jumpy, this may be due to lack of lubrication, but a smoothly operating brake is of major importance.

This testing method also is important in learning how to set the brake to proper tension, because beginners invariably set it too tight. The proper adjustment should put a moderate bend in the rod when line is being payed out. (We'll learn of auxiliary braking methods later.) Secondly, doing this before fishing pulls out the coils which are common when monofilament line has been stored on the reel for a length of time. Stretching the line to remove coils is an important aid in easy casting; also in testing the line for weakness.

Old hands at the game can determine proper brake adjustment merely by pulling some line from the reel. Beginners are advised to do this more exactly and then to pull out some line to become familiar with the amount of tension the brake should have.

If the strength of the line used is commensurate with the power of the rod, a fairly exact method of brake adjustment can be made by hooking the assembled tackle to something solid, as before. Back away, releasing line from the reel until you stand three rod lengths from where the lure is hooked. Tighten the brake screw nearly to a locked position. Hold the rod at an angle of about 55 degrees above the horizontal (a bit more for an unusually limber rod and a bit less for a stiff one) and regain line (with the antireverse lock of the reel engaged) until the rod tip is arced sufficiently to form a continuation of the line. Holding the tackle in this position, release the brake screw gradually until the reel pays out line when you walk backward, but not when you are standing still. This should be the proper brake setting when the line is of proper strength for the rod, or even stronger. If the line is abnormally light, set the brake correspondingly lighter than this method calls for.

If one can obtain a spring scale, a more exact method is recommended. Anchor the scale and tie the lure's end to it. At the same distance as above, and with the rod held similarly, slowly release the brake until the scale reads one-third of the line's published strength. For example, in using a line of 6-bound test, the brake should hold at 2 pounds when you hold the rod as described above, but should start releasing line when you walk backward.

Why not set the brake stronger than this? Because when a spunky sport fish hits a lure he increases the strain on the tackle for an instant or two before the reel gives out line. This temporary strain often is double that at which the brake is set, or two-thirds of the line's strength. By setting the brake in the above way we eliminate the danger of a fish snapping the line on a strike. We also have the insurance of about a third of the line's strength in reserve, which is good for two reasons: there may be a weak spot where the line has been rubbed or scratched; also, the knot which fastens the lure to the line may be weaker than the line itself.

Another way to set the brake to proper tension is to hold some line while the reel (without rod) is dangling from it. Loosen or tighten the brake until the reel drops *very* slowly. If it drops *very slowly* and steadily under its own weight it is an indication of a smooth brake set correctly for usual purposes.

Once the brake is set properly, do not change it under any circumstances. If you need temporary added braking pressure, there are two additional ways of applying it, as discussed later in this chapter.

Line Pickup Devices

Another thing to decide is the kind of pickup desired. This is the device on the revolving cup which collects the line and winds it back on the stationary reel spool. The most common is the bail type, which you open before casting and which flips back into position when the reel handle is turned. This type is recommended for beginners and it often is preferred by experts.

The second type of line-pickup mechanism is the trigger or automatic type. Neither term is quite accurate because the line is not actually picked up by the trigger, nor is it picked up automatically. What we're dealing with is a refinement of the bail-type pickup. The system consists of a trigger mechanism which is affixed to the rotating cup of the reel. The trigger activates a cam mechanism which in turn opens the bail.

In use, the angler reaches forward with the forefinger of his rod hand to pick up the line, continues coming back with the finger, which hits the trigger and opens the bail in one smooth, single-hand operation. The system is quite fast and easy for a beginner to master. Advanced anglers also like this system because they are able to make more casts faster and easier over a day's fishing. Tournament fishermen find this fact can sometimes give them an edge in competition.

Early designs of the trigger frequently caused some finger irritation over a full day of hard use, but new, easier operation and better trigger configurations have eliminated the problem. If desired, the trigger need not be used, and the bail may be opened and closed using the non-rod hand in the usual way at any time.

Fill the Spool Properly

A rule often ignored is that the line spool should be filled to the lip only. If it is only partially filled, it penalizes us on casting distance. If it is overfilled, too many coils may be cast off prematurely, causing snarls.

Probably the most frequent mistake of beginners is in allowing line to spool too loosely on the reel. This often is caused by bucktailing a lure; that is, by not respooling the line under equal tension. If loose coils are combined with tight coils on the reel, the tight coils may pull off several loose coils, causing a snarl. Proponents of spinning often say one can't get backlashes with the equipment, but one can get them this way, and they are an infernal nuisance. Avoid them by retrieving line under equal tension. This doesn't mean that we can't do a bit of bucktailing if we want to but, if we do, we should notice whether the line has been respooled loosely. If so, make the next cast a short and easy one, regaining line under even tension. If loose coils still are on the reel, a longer, easy cast and the proper retrieve should pack the line tightly and evenly on the spool. We stress the point that the line always must be packed tightly on the spool.

Insufficiently filled reel spool *(left)* prevents line from uncoiling smoothly and cuts casting distance. Overfilled spool *(center)* can cause snarls. Spool should be filled so line is just below the lip *(right)*.

In buying a new spinning outfit, reels often can be purchased with line already on them. If lines aren't on the spools, dealers usually will put them on with line spooling equipment for doing the job quickly. If the angler has to put his own line on a spool, an easy way is to assemble rod and reel and to remove the reel spool. Run some line off the manufacturer's spool and thread it down the rod guides, then fasten it to the reel spool. Open the bail *(pickup)* and seat the spool on the reel, setting the brake rather tightly. Have someone hold the manufacturer's spool(s) on a pencil or dowel so the line can be reeled off without twist. By cranking the reel, the line can be spooled on it under moderate tension. There are many obvious ways of spooling the line under tension. One way is to let it pass from the manufacturer's spool between the pages of a telephone book, on which a weight may be put, if necessary. The idea is to spool it on the reel tightly and evenly, without line twist.

Since monofilament line often comes on 100-yard spools, we may need two connected spools to fill the reel. These can be taped together until the first spool is used. Of course we could save money by using only one line spool and by filling the base of the reel spool with string, but this is not recommended. Spinning line is inexpensive and reels are made with ample

line capacity because, in many types of fishing, over 100 yards occasionally will be needed. If the forward part of the line should become frayed, or some of it should be lost, it can be repaired by removing the frayed part and by properly filling the spool with an added length of monofilament tied to it with a Blood Knot. Also, the line can be reversed by reeling it on an extra reel spool, as described above.

People often ask how much an undamaged monofilament line decreases in strength after a year or more of storage. If kept from sunlight a fairly strong line should remain strong enough over several years of use, but probably some brands of line deteriorate somewhat more rapidly than others. Lines should be tested at the start of each season to be sure they are strong enough, because they do lose strength in time. Particular attention should be paid to fine lines in the lower strength tests. Losing 2 pounds of strength in a 10-pound-test line may not be important, but losing it in a 4-pound-test line could be disastrous!

Reel Lubrication and Care

When buying a reel, if printed instructions don't come with it, the buyer should ask the dealer how it should be lubricated. Tubes of lubrication jelly are cheap, and it's well to carry one. When the reel spool is removed the spindle on which it sets should be greased lightly. Other lubrication points vary from one reel to another, but most of the places which need occasional oiling are marked on the reel, or are obvious.

Spinning reels are sturdy mechanisms which need very little care. The pickup device can be bent by careless use or transportation, which of course should be avoided. It should be opened carefully when in the proper position and should be flipped closed by turning the reel handle without trying to slam it closed. If damaged, it may not spool the line evenly, piling it up on the front or back of the spool. We need not add that reels should be kept clean and dry, particularly after fishing in the rain. If used in salt water, the spool should be removed as soon as possible and the reel and spool rinsed in warm water, then dried carefully. Salt sets up a chemical reaction with the metals of the reel which can bind the reel spool to the spindle and ruin the reel in a matter of hours.

Lures for Spinning

Any lure or bait of the proper weight can be used with spinning gear. These include weighted spinners, wobbling spoons, all sorts of plugs, weighted plastic worms, live or dead minnows and natural baits of many kinds. Some, which are too light to be cast properly, can have additional weight applied by using split-shot or strip-lead on the line at a suitable distance above the bait. While lures and baits are discussed in other chapters, there are a few tips especially applicable to their uses in spinning.

Weighted spinners work best in ponds and lakes and in streams where currents are from slow to moderate. In fast water they may be inclined to spin, and thus twist the line. This is more true of some types than of others. The revolving spinner blades tend to keep the lure near the surface, so they are less desirable for deep fishing than wobblers and other baits which sink more readily. Every spinner has its own ideal speed of retrieve, when the blades spin most efficiently. Try them nearby to decide the speed which works best. Spinners require a steady retrieve at their ideal speeds. Bucktailing them is less effective.

On the other hand, wobbling spoons can be bucktailed and can be allowed to sink for deep fishing. One often gets strikes while the lure is fluttering to the bottom. When on or near the bottom, it can be jigged. Wobbling spoons are less inclined to twist lines in fast currents, but they may do so if the current is very fast.

Some wobblers are large and thin; others are smaller and thicker. The small, thick ones can be cast the farthest. Bright lures usually work best in high or discolored water; darker ones in thinner and clearer water. You can let the lures corrode, if they will, and polish them to the desired degree of brilliance with crocus cloth or something similar; even toothpaste!

When fishing over a shallow, weedy bottom, lures that sink readily of course will become snagged. Here, spinners work better than wobblers, and it may be preferable to use floating-diving plugs that come to the surface at rest but which can be pulled under on the retrieve, or to use surface plugs such as wigglers or poppers.

Plastic Floats

Lures and baits (principally the latter) also can be used with floats of various kinds to keep the bait off the bottom and also to provide casting weight. Four typical rigs are shown in the accompanying illustration. If obtainable, colorless plastic bubble floats are very useful.

Water usually is put into the plastic bubble float to provide the desired amount of casting weight. If not filled full, the bubble will float. If filled full, being about the same density as water, it will drift slightly below the surface, the depth being influenced by current strength and the amount of pull on the line.

In using the bubble with bait, such as a worm or a grub, the trick is to add just enough water or mineral oil (with an eye dropper) so that it will barely float. Thus, when a fish takes the bait, it feels little or no resistance, and is less prone to mouth the bait and to refuse it. This is particularly true in quiet-water fishing, such as in lakes and ponds. Artificial unweighted flies and nymphs can be used on a short leader above or below the bubble, but the author has found it difficult to feel the stike and to hook the fish, especially when using a long line. This can be helped by watching the float. Any unusual motion is a signal for an immediate strike.

SHOCK TIPPETS

Many experienced fishermen enjoy using lines as light as possible because of thelonger casts obtainable and because light trackle is sportier and more fun anyway. But it is disconcerting to put on a heavy lure; to make a beautiful cast; and to watch the lure sail out, completely separated from the end of the line.

If one is using lures that are too heavy for the line, a shock tippet, sometimes called a "bumper line," may be necessary, although its use is more common in saltwater fishing than in fresh. This is essentially, is a level leader of much stronger monofilament than the line. It is long enough to allow a few turns around the reel spool when the tackle is in casting position. Thus one casts with the stronger but short tippet but still enjoys the advantages of the finer line for distance casting.

In addition to insuring that lures won't be snapped off, the shock tippet has other advantages. In landing fish the heavy tip section is helpful in preventing them from breaking off at the last minute; particularly when fishing from boats where the deck is a few feet above the water line. Many species of fish have sharp teeth which would bite through the fine line but can't sever the stronger one. In the case of bluefish and others with sharp teeth, even the shock tippet may not be strong enough. So a foot or two of even heavier monofilament is added to the tippet -even 50-pound test. An alternative to this heavier short tippet of course is to use a short single-strand wire leader or a plastic coated "Sevenstrand" twisted wire leader. These may be preferable in some cases, although some fishermen avoid them due to their tendency to kink or because they think the fish can see them more easily. Both types have advantages in certain cases, so the choice is the angler's.

When monofilaments which do not vary greatly in diameters are tied together, the Blood Knot is used most frequently. But, if the monofilaments do vary widely in diameter, the Blood Knot isn't strong enough, so we would use the Surgeon's Knot or the Double Strand Blood Knot. The latter (preferred by the author) is the familiar Blood Knot except that the finer strand is doubled before the knot is tied.

GETTING HUNG UP

All good anglers become hooked up occasionally. When it happens, violently jerking the rod rarely does any good. If the lure won't come free after a twitch or two or a moderate pull, avoid getting hooked up more securely. Go in the opposite direction, if possible, and often the lure will pull free. If one is hooked up securely and can't get to the lure, reel in as much as possible and then keep the spool locked by finger or hand pressure. Walk backward slowly with the rod pointed down the line. When the line is rigid, pull back some more, and something will have to give! Usually the lure will pull out with no more damage than a bent hook. If the line snaps, the lure was too solidly snagged, and nothing more could have been done about it anyway.

When a reel spool is filled with the manufacturer's line, usually part of a spool of it is left over. Take it along when you go fishing. It is valuable for piecing a broken line; for making leaders and droppers, and for other purposes.

TWISTED LINE

Although being watchful should avoid it, many fishermen have trouble with twisted lines, especially when trolling with spinning tackle. Using swivels may help, but probably not enough. Let out a moderate amount of line and check it for twisting after a few minutes. In trolling, the trick is to use lures that don't twist. If it is necessary to use a lure that twists, alternate a pair of them; one that revolves to the left and one that twists to the right. If one twists the line, put on the other to remove the twist.

Another method of lessening or preventing line twist in trolling is by using a plastic or metal keel put on the line just ahead of a swivel. Metal keels are for deeper trolling and usually are cut from sheet lead, often in the form of a heart. The heart is scored along the center line and crimped around the trolling line. Although such keels can be purchased, they are rather easy to make.

Angling writers sometimes recommend towing a twisted line behind a boat (with the lure removed), or letting it trail downstream. This may help a little, but not very much. If a line is not too badly twisted perhaps the twists can be stroked out, even though it takes patience. Probably the best answer is to discard the twisted part of the line and to tie in a fresh length. But burn up the twisted part, or take it home. Don't throw it overboard for somebody else to have trouble with.

ULTRALIGHT SPINNING

The French call it *lancer leger* (light casting) or *ultra leger* (extremely light casting). Of course the term is relative because all the tackle we have discussed previously could be called ultralight if the fish are big enough and the conditions sufficiently challenging. Here, let's refer to ultralight as tackle which uses monofilament lines of 2-pound test or less -and they can be obtained very much less!

The reel is a tiny one, with a spool about the size of a half-dollar, or not much bigger. Rods are tiny wands 5 feet long, or so, and rather whippy, to cast lures about 1/8 ounce or less. Needless to say, the tackle is precisionbuilt, and only can be obtained from a few of the quality manufacturers.

This form of spinning is exciting and challenging; somewhat similar to using a trout-size fly rod for Atlantic salmon. One has to pay extra attention to the smallest details of brake setting, knot strength, and so forth. Except for its diminutive size, the tackle is similar to stronger gear, but the angler is challenged to play his fish on the upstream side, and to keep it away from trouble. The name of the game, if there is one, is to land a fish weighing ten times the line's breaking strength. This means a fish of 10 pounds or more on 1-

pound-test line, or one of 50 pounds or more on 5-pound-test line. Either could be classed as ultralight in anybody's language!

CASTING WITH SPINNING TACKLE

Preparation

Before actually casting a lure with a spinning rod and reel, there are a few preparatory actions that you must perform. When you have practiced sufficient time with a spinning outfit, these moves become automatic.

The spinning rod is normally held with two fingers ahead of the reel's leg and two behind it. But if it's easier to place the forefinger tip firmly against the front face of the reel spool, for control, one or even three fingers can be placed ahead of the leg.

With the rod held in a comfortable position, reel in the lure until it hangs about a foot from the rod tip. This distance between lure and rod tip is needed for accurate casting. With fingers of the left hand, turn the cup of the reel to bring the bail nearest to the rod. Pick up the line with the forefinger of the rod hand, holding the line on the pad of the finger's first joint, not in the cleft, as shown in photo at right.

The next step is to grasp the bail with the thumb and fingers of the left hand, and flip it across the face of the spool, locking it in position. This opens the spool to allow the line to uncoil freely during the cast.

(If the reel has a hook type of pickup, turn it so the roller is nearest the forefinger and hook the finger under the line. Then turn the pickup counterclockwise to the lower part of the reel so it is out of the way and comletely disengaged from the line. Open it by pulling down at its base, if there is no opening lever on the reel.)

SIDE CAST

The side cast is the easiest one unless nearby bushes or other obstructions make some other type of cast necessary. The advantage of the side cast is that the lure travels in a low trajectory and the line is not caught as much by the wind as in the overhead cast. The photos show casting instructor Tom Stouffer performing this and other casts.

Lower the rod to the right below a horizontal position. Snap it back slightly and *immediately* sweep it upward and forward until it points over the target at an angle of about 45 degrees above horizontal.

At the instant the rod points toward the target, point the forefinger toward it also. This releases the line from the forefinger and allows the lure to soar toward the target.

The forefinger is the key to accurate control of the lure, so let's forget to engage the bail or automatic pickup for the present. Remember how important the forefinger is! It is pointing toward the target as the lure sails through the air. To slow down the lure, merely move the forefinger toward the front face of the reel spool. This will cause the uncoiling line to slap against it, thus slowing down the lure. The lure can be stopped instantly by touching the reel spool with the forefinger. The forefinger controls the line, slowing down the travel of the lure at will or stopping it instantly if it comes too close to any obstruction. As the lure reaches the target, bring the forefinger slowly into contact with the reel spool. If you do it right, the lure can be stopped within a few inches of where you want it to land.

An important advantage of forefinger control is in being able to stop the lure accurately. Slowing down the lure as it nears its target and stopping it as it reaches the target takes all unnecessary slack from the line. Thus, if a fish strikes as the lure lands, you're in control of your tackle and can hook him instantly. You can also keep your lure on the surface if you wish, instead of fumbling with slack line while it sinks. In comparison to this method of forefinger control, stopping the lure by closing the bail or pickup arm with the reel handle is much less desirable.

After the lure strikes the water, you can put the line under control of the bail or pickup arm by turning the reel handle counterclockwise. The first partial turn closes the pickup, which thus engages the line by taking it away from the forefinger. Pausing before turning the reel handle allows the lure to sink as much as is desirable.

Now begin to retrieve the line. The speed of turning the reel handle brings it in as fast or as slowly as you wish. By working the rod tip you can give any type of motion to the lure that you desire.

I am often shocked to see fishermen using a spinning rod as if they were beating a balky mule. My advice is to take it easy, because nothing is gained by excess effort.

When reeling in a fish, you should pump it in, if it's a big one. Raise the rod nearly to the vertical and reel in under steady tension as the rod is lowered nearly to the horizontal. Keep repeating this action, as necessary. This is the most comfortable and the easiest way to bring in a big fish. It uses the rod's action to the utmost and prevents twisting the line by cranking against the drag when little or no line is being recovered.

OVERHEAD CAST

This cast is made in the same manner except that the rod is held pointing vertically. It is snapped back and immediately forward in the same manner as the side cast, and the line is released when the rod is stopped and pointing over the target. The overhead cast has the advantage of greater accuracy because your are casting in a single plane. The variation in direction which often plagues the beginner when side caswting, because he does not release theline at the right instant, is largely eliminated. The overhead cast has the disadvantage of a high trajectory, which puts a good deal of belly in the line.

BACKHAND CAST

A cast from the left can be made by crossing the forearm across the body below the chin and pointing the rod downward, casting forward. This cast is ghandy when there are bushes or other obstructions to the right and above.

FLIP CAST

This cast will deliver a lure accurately, but only a short distance. It is useful for fishing nearby spots in small streams or on ponds and lakes where the lure must land between obstructions and where a long cast is not needed.

The cast must be made slowly. At the start, the lure should be about a foot from the rod tip. Point the rod toward the target and slightly downward. From this position, snap the rod upward through a very small arc and then immediately downward again, to put a bend in it. Using this bend, snap the rod upward again, releasing the lure as this last upward sweep directs it toward the target. This cast requires a bit of practice in timing, but it is a handy one to use. With it, you can poke a rod through a hole in the bushes and send a lure 50 feet or more from the most difficult of casting positions.

ARROW CAST

This is more or less a trick cast. It is useful for shooting a lure into an opening in the bushes. Be sure to hold the lure carefully lest. a hook become embedded in your finger.

Let out about as much line as the length of the rod and hold the line with the forefinger of the rod hand, the bail open and ready to cast. Hold the lure by the bend of the hook between the thumb and forefinger of the reel hand, and pull the lure back to form a bend in the rod. Aim the butt of the rod directly at the target. Quickly release the lure and, an instant later, release the line from the forefinger. If you do it right the lure will shoot toward the target, although not a great distance.

6

SPINNING FOR TROUT

Many fishermen think that spinning or spincasting always is the best way to take trout. Others, more competent with the fly rod, while acknowledging that these two methods usually are very effective, also know that many situations come up where they can beat the weighted-lure addicts hands down. Since greatest success in fishing hinges on using the most appropriate method for the specific situation.

Since these were rather general in scope, it may be helpful here to round things out by being more specific about trout, especially the brookies, brownies and rainbows which most of us seek so often. These can be grouped to-gether if the slight differences in their characteristics, recently discussed, are remembered.

Anglers who fish near populated areas, as most of us do, primarily are confronted with "put and take" conditions where the trout are stocked and usually are small. Light spinning or spincasting tackle is preferable to heavier gear for two reasons; it is more fun to handle smaller fish on light equipment, and more of them can be hooked with it. But how light is light?

Light tackle is a relative term but it could be an outfit handling monofilament testing not over 4 pounds and using lures or baits weighing no more than ¼ ounce. When one becomes hung up and has to break loose he will realize that this really is pretty strong stuff. He may prefer to go toward ultralight gear with lines testing 2 or 3 pounds and lures of about ¼ ounce. This really isn't "ultralight," but it is sensibly so. Real ultralight tackle runs quite a bit lighter, or finer. The whole idea behind going lighter is that it is more fun, more challenging, and perhaps more productive than merely pulling in fish with equipment that doesn't give them much of a chance. On the other hand there are big, fast and deep waters holding big fish where 4-pound line may be too light, or where heavier lures are needed for longer casts. The strength of the tackle must be decided by existing situa-tions.

Every rod performs best under a specific average lure weight which pulls the tip down only slightly. Use the weight best suited for the rod. If the lure doesn't depress the tip noticeably, the lure is too light to be cast properly. If it depresses the tip excessively, it is too heavy and may result in broken tackle.

To start with, only about a dozen artificial lures are needed for trout. These could be a shallow-running spinner and a deeper-running one in both flashy and duller finishes, plus the same of wobbling spoons, for a total of eight. About four midget plugs of various types can provide a set which can be deadly on trout when other lures are ineffective. They often take big brown trout when they will strike at nothing else! This initial dozen doesn't provide for spares. The favorite lure usually is the first one that is lost.

TROUT IN DEEP POOLS

Having suitable tackle and having learned how to use it in, let's apply this to some typical fishing situations. We might start with a fairly deep pool because, regardless of conditions, most fishermen seem to think that's where the big ones lie. This isn't always so, however.

Too many people waste too much time hopefully casting and retrieving lures 2 or 3 feet down when they should know there are no fish there. The cross-section of a deep pool in the accompanying drawing shows why, and it bears stressing.

In such a pool the trout feed on or near the surface only when there is surface food there. This usually means emerging nymphs or a hatch of in-sects. Even if the trout are feeding on top, they may not take spinning lures because they are selectively feeding. The only chance is along shallower edges of the pool, and this isn't a very good one. At such times the top is fly fishermen's water, and the spinning fisherman usually does better to leave it alone.

Beneath the top water of a deep pool is a barren area which extends down nearly to the bottom. This is the area too many fishermen waste too much time with. Why should trout be in it? The food supply is better else-where; there are no places where they can rest without combating the current, and there is no protection near which they can hide.

So, in a deep pool, let's face the fact that trout most often are close to the bottom in protective places such as are indicated in the drawing. If surface water is very warm this is all the more the case, because the bottom water is much slower and much cooler. Trout lying on or near the bottom may rise up a foot or two to take a lure when the water is in the 50s, but they hardly will move for anything when it is too cold. Thus the answer to deep pools usually is to fish very close to the bottom. This may cost a lure occasion-ally, but it is the way to take trout when they are not actively feeding.

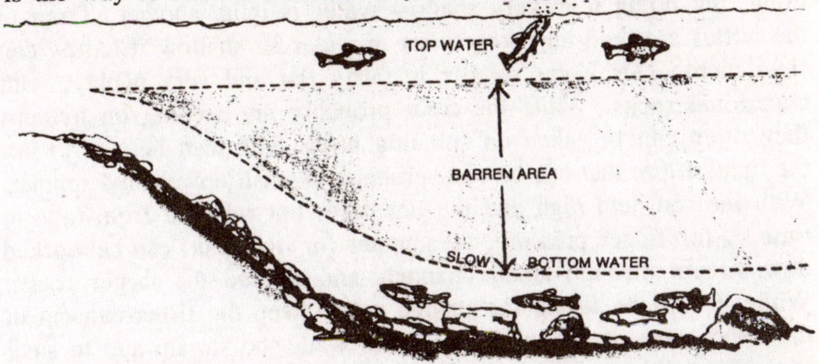

Cross section of a typical trout pool showing good and poor water. Trout feed in surface water only when a hatch is on; otherwise they're on the bottom and lure or bait has to be fished deep. Middle water is usually barren.

We can cast lures into a deep pool; wait for them to sink, and then fish them in under the illusion that they are scraping bottom. Actually, the cur-rent may take control of them and sweep them downstream far up from the bottom. It may be best to cast almost directly upstream and, by regaining line as the lure comes nearer, wait until we feel it touch bottom. Then, in fishing it in, an occasional pause should make it sink down again. This bottom-bumping technique is especially important in fishing for winter steelhead, but it works for other species of trout as well. Experienced anglers can sense or feel the lure bumping bottom, and they learn to fish it without getting snagged too often. It is a method which requires experi-mentation for complete success.

When actively feeding, which may be around noon when the water is cold, or in the very early morning or late evening when it is warmer, trout may leave their sanctuaries deep in pools and travel into the riffles and shal-lows to feed. Many trout, even big ones, do not lie deep in pools. They often are in more accessible places where the top water joins with the bottom water without the relatively barren middle stratum. This is water varying from shallow to only a few feet deep, with rocks and other obstructions which break the flow and provide places where fish can hide. When water temperatures are in the 50s, this water can be ideal for spinning.

THIN WATER

Spinning for trout in deep or moderately deep streams is one thing, but doing it in very shallow water is quite another. Trout in the riffles are feed-ing, sometimes in water so shallow it hardly can float them. This water usually is fairly flat and very pebbly, with occasional rocks. While the trout probably are feeding on nymphs they often can be taken on spinning lures. The idea here is to use the lightest lure that can be cast; probably a small unweighted spinner. With the rod held high and the line free, but released from time to time by forefinger pressure, the spinner (or light bait) can be worked downstream through small channels and around the larger rocks. While it must be fished far enough off to keep the fisherman out of the fish's cone of vision, it is easy to wade the stream and to steer the spinner into spots that could harbor trout. Under such conditions even the lightest spinners will get hung up occasionally but it is no problem to flip them off or to wade downstream to free them. The smallest hooks which will hold the fish should be used.

STREAM MOUTHS

Little needs to be added to spinning for trout in ponds and lakes in addition to what was provided in Chapter 15. There it was noted that fishing the shore toward which the wind is blowing should be best because winds blow surface foods such as spent insects in that direction. If surface water is too cold for trout they also will seek the shallows of that shore because winds blow the warmer top layer of lake or pond water in that direction.

In any event, when the weather is warm enough to hatch insects, which can be very early in the season at times, something happens that is of particular interest to trout fishermen.

Since water-borne foods for trout and for baitfish are washed down streams by their currents, stream mouths usually provide good fishing. When the wind, or even a slight breeze, blows toward the stream mouth, floating foods which have dropped into the water are wafted in that direc-tion. The opposing forces of stream current and wind therefore concentrate surface food to confine it in a relatively small arc which usually is where the stream's current spends itself at the stream's drop-off. The stronger the wind, the nearer the stream mouth this concentration will be, but it will be seen as a scummy area which, on closer inspection, contains an abundance of fish food such as insects.

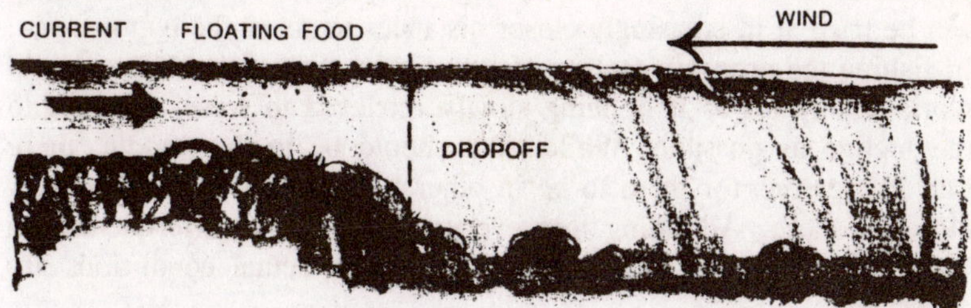

Opposing forces of wind and current can sometimes concentrate food near stream mouths. Food—usually in the form of insects—appears as a scummy area just before the dorp-off.

At these times stream mouths provide a double bonanza because trout collect there for the surface food and also for the greater abundance of baitfish which are there for the same reason.

If trout are surface feeding this may be noticed by slight dimpling (made even by big trout) when they come up to sip in insects. One should not ig-nore such action under the impression that, because he can't see trout splashing and boiling on the surface, the dimpling is made only by baitfish. Regardless of surface commotion, or the lack of it, it can be assumed that trout are there.

If there is any evidence of surface activity, spinning lures (or flies) fished shallow should get results. Otherwise, the trout may be lying deeper, resting between feeding sprees in the slower water over the drop-off. This may be under the floating food concentration, but a fairly strong breeze may indi-cate that it is somewhat farther out.

If the area is fished from shore the lure could be cast more or less up-stream and allowed to sink to work near bottom over the drop-off. A fast current might require a compact wobbler rather than a bigger, thinner one, or a weighted spinner. If this isn't productive, an alternative is to make a high overhead cast with a compact wobbler as far beyond the food concen-tration as possible. The falling lure on the bellied line then will plummet down and can be fished in after it has sunk deep enough. Correctly done, this will let the lure drop to, or near, bottom, from where it should travel upward along the, incline of the drop-off as it is being retrieved. Since trout may be concentrated in one area or another below the drop-off, numerous casts could be made.

If the area is fished from a boat, casting could start at a considerable distance from the food concentration because the trout may be schooled farther out than might be presumed. Fanwise casting can be tried at in-creasingly closer distances to cover the bottom area. In fishing the drop-off itself, a slowly fished compact wobbler should work downward as it is being slowly retrieved to travel as close to the incline as possible. While trout should be in the middle, there may be reasons for them to be in other hold-ing positions even very close to shore. Admitting that every situation is a different one, a knowledge of probabilities should help when actual condi-tions are observed.

SPINNING WITH FLIES

Illustrated ways to use sunken baits with plastic floats. A question often asked is whether or not these little floats are practical for use with flies for trout. We could give the question a quick brushoff by saying that artificial flies should be used only with fly-fishing tackle, but this may not suit spinning and spincasting fishermen who want to try them. I did considerable experimenting on the subject when American spinning was in its infancy. While artificial flies of course are most suitable for fly fishing, they also are practical in a limited way for spinning when used in connec-tion with the plastic float. Let's not say it is an ideal way to take fish but, under suitable conditions it can be a lot of fun!

In the top drawing at left, a float is tied to the end of the line. The float contains enough water to provide adequate casting weight. About a foot up on the line from the float we have tied in a dropper a foot or so long rigged with an artificial nymph, a sinking fly, or a bit of bait. A foot farther up on the line is a fairly short dropper to which is attached a dry fly. The droppers should be of the finest monofilament which is practical to use.

The fisherman casts this rig upstream and across and takes in enough line so the dry fly is above water. By raising and lowering the rod tip he causes the dry fly to touch the water lightly and frequently as the float drifts downstream. Actually he is "dapping" the dry fly; a method as old as fly fishing itself. The fly dances on and just over the water, and trout very often rise up to take it. The fisherman usually must hook the fish, but that isn't difficult in this case.

Meanwhile the nymph or wet fly is drifting down in the current several inches below the surface, and it is being given very slight motion due to the dapping of the dry fly. This is exactly the motion a nymph should have as it evidently is struggling to reach the surface. If a fish takes the nymph, and the take can't be seen, it is signaled by unusual motion of the float. The fisherman should strike hard upon seeing any unusual motion of the float

because it is difficult to hook a fish under these conditions. If a worm or a grasshopper is on the end dropper, the fish probably will take it in and will hook himself, but striking would do no harm anyway. On completion of the drift the float will swing with the current, but it can be left out there as long as it covers ground. Meanwhile the rig can be steered around rocks and other holding positions such as have been described.

I would be inclined to let the idea go at that, but there are two less efficient ways of rigging the float with flies (or bait). One is much like the former except that the two flies or nymphs are on longer droppers for under-water drift. A small split shot can be pinched on about halfway up each dropper if necessary. The trouble is that fish are very hard to hook with this rig when flies are used. Of course streamers and bucktails are unsuitable. Small baits provide a more sensible arrangement.

Lastly, for deeper fishing, a leader 2 or 3 feet long is attached to the float, with the float attached to the line on the other side as shown in the drawing at left. A nymph or bait is rigged to the end of the leader, per-haps with one or more small split shot clipped on midway up. This is a good, rig for drifting down pools and runs but, as before, one has to be quick, clever and lucky to hook a fish on the nymph when any unusual motion of the float is seen. The rig can be cast long distances, sometimes under con-ditions where fly fishing isn't practical. Due to the difficulty in striking, the use of natural baits such as grubs, hellgrammites, worms, etc. would be better. Then, a trout takes in the bait and usually hooks himself.

Another question is about weighted flies, such as streamers and buck-tails. Since a casting weight is needed, why not wrap the shank with plenty of wire before dressing the body? The idea has been tried in all its varia-tions, but with insignificant success. The heavy body deadens the action of the fly to such an extent that only the most ignorant trout will bother with it. It can be tried for other kinds of fish, but even the poorest sort of plug would be better.

While reviewing this section on spinning with flies for the second edition of this book I was inclined to discard it, but am leaving it in because some anglers may enjoy fussing with it. The use of floats, of which there are many kinds, seems to have fallen into partial disfavor since its surge of interest when spinning was new. This indicates they aren't among the best ways of employing spinning equipment. Readers who want to use flies would do better with a fly rod. As a prominent pioneer at the inception of the spinning method in America, I was condemned by some purists for advocating using flies with this tackle. While I favor the fly rod to the extent that in recent years I rarely have used a spinning rod in fresh water, it pleases me to see how much pleasure spinning has brought to American fishermen, many of whom, I am glad to note, have graduated to fly fishing. Spinning still has its place, but the fly rod always will mark the peak of the sport, and non-fly fishing readers are urged to review the parts of this book which explain its basics.

TACTIC FOR FISHING A TYPICAL STREAM

Having discussed spinning tackle and how to use it, let's now follow up the last two chapters about the trouts and how to find them in streams and brooks with some advice on tactics. On reaching a promising spot on a moderate-sized river, what should a sensible angler do first, and then how should he fish the stretch?

Long ago, an expert who took me stream fishing in what was then part of the Maine wilderness, made a point I never forgot. He taught me to sit quietly for as long as it takes to "read the water" I was about to fish so I could determine where the fish should be and decide on the best way to fish the stretch. Such minutes are far from wasted. They pay off in more fun and more action.

The stream spills into this pool from gravel riffles at the left. It is fairly shallow on the near side, with much deeper water across. We could wade easily to the rock at *X-2,* but let's plan to fish it first from *X-1* because we notice a darker, deeply scoured hole just below it which may contain trout; probably small ones.

This rock is a good place to fish from and may offer concealment if the water isn't too deep for us to get behind it on the upstream side. If the pool's surface isn't too disturbed, this is the best place to fish from because most of the pool can be covered from here.

While sitting to estimate the situation, we notice a deep run at upper left. This could hold trout, especially rainbows, because the water there is deep and fish food is being washed down. If a near-surface spinner doesn't get results, let's plan to change to a small and heavier wobbling spoon to work close to bottom. This place is good for several casts, the lure allowed to arc downstream. We have noticed two surface boils near the middle of the pool, indicating subsurface rocks. We can assume there are other smaller ones in between, and good-sized trout may be lying in their protection.

We also notice while planning our fishing that some bushes shadow the deep run along the opposite bank. Trout should be lying under their protec-tion, so we may wish to risk a few casts there.

In all these it usually is well to let the lure swing until it hangs down-stream. That point often is a strike zone because fish may be following the

lure but won't take it until they detect a change in its motion. Let it hang downstream for a bit, and give it some added action there before retrieving it. If the water is deep, and if a wobbler is being used, it may be good to raise and lower the rod tip once or twice during the lure's swing to keep it running deep. This is also a good tactic when using bait, which is most effective near bottom.

After fishing from *X-1* and *X-2,* we could move to the midway spot marked Y, to cover the surface boils from this position. If the stream is wider than the drawing indicates we may need to wade out a bit. From here we also can cover the near side of the midstream rock and also the back of the rocky point marked Z.

From position Z, a long cast might reach the downstream edge of the big rock on the bank across the river; an ideal spot for a trout or two. From here several casts should be made to the holding water back of the mid-stream rock, also a promising location.

Across the river a tree has toppled from the bank; part of it is sub-merged. Trout like to lie in the shade of such obstructions and feed on ter-restrial insects that drop from the branches. At the risk of snagging our lure, we could give it a try or two.

So far our strategy for fishing this area has included casting up and across, across stream, and quartering down and across. If we surmise that our lure hasn't gone deep enough, we could try the two surface boils, and perhaps other spots we may have noticed, by casting upstream. If we cast above one of the boils, the lure should wash downstream into the pocket, but this depends partially on the speed of the current. If a spinner is used, it must be recovered faster than the speed of the water. We might do better with a wobbling spoon.

Besides spinners and wobblers, other suitable baits or lures described in Chapters 4 and 5 can be used. Worms are always reliable. A split-shot or two should be crimped to the line a foot or so above the hook. The weight aids in casting, and ensures that the worm will sink and drift properly. Worms work best when they bump bottom, but they often are effective when fished slowly on a hook attached to a small Colorado spinner. What trout take this for is a mystery to me. I know of no worm that ever won a swim-ming competition!

Now that we have planned our strategy for fishing this pool, let's go out and fish it. Only part of the water in any trout stream or river is productive, and time spent reading the water will indicate potential hot spots as well as eliminating useless areas. That's a big part of what separates the experts from all the others!

AIDS TO CONSERVATION

Most fishermen are coming to realize that trout are in short supply, even wild ones, even rainbows in the Alaskan wilderness, as noted in the preced-ing chapter. Thus, "catch and release" is a practice rapidly gaining popular-ity. There is something very satisfying about carefully releasing a trout so it may be hooked again-perhaps several times. I know of streams once nearly barren which are now simply boiling with big, beautiful trout because of the rule of "catch and release." The fun of the fishing is such that no one seems to mind returning home fishless.

But we can't safely release trout when they are hooked on the trebles of most spinning lures. That's why anglers who deserve the name remove treble hooks in favor of singles. The best sportsmen among anglers who use spinning lures even go a step farther. They pinch down the barbs of the single hooks. Few fish are lost by this means if one keeps a tight line. In "put and take" areas a trout or two for the table once in a while may be in order, but let's not overdo it. Give a second thought to a gift of trout for Aunt Emma. Let her go to the fish market!

Finally, let's be sure the trout we release stay alive and healthy. Bring them to hand as quickly as possible; that is, avoid tiring them excessively. Keep them in the water if possible and keep hands away from their gills. Remove the hook carefully with needlenosed pliers. Hold them upright in the water (by the tail if they are large) and move them back and forth until they regain enough strength to swim away.

7

FLOAT-TUBE FISHING

Not far from the town where the author was raised there was a small pond in which the author often noticed rises and splashing, indicating feeding fish -probably small-mouth bass. These couldn't be reached fishing from shore because the little pond was lined with grasses and pads, making effective casting almost im-possible for a neophite fisherman with the tackle we had then.

Back in those days used tire tubes were readily available and often were favored by kids as an aid to swimming. You just encircled your body with an inflated tube, put your arms over its circumference, and paddled around in shallow water, using your hands as oars.

One day, on inspiration, the author decided to use such a tube to get out into the deeper water of the pond where the author could fish the rises. the author rigged the tube with a seat of clothes line and ventured out beyond the weed line into deeper water. The ropes had a habit of sliding together, binding and chafing me in intimate places. With a rod in one hand, locomotion proved difficult, espe-cially when a breeze was blowing. In spite of all this the author hooked a few fish, but finally gave up the idea as being more bother than it was worth.

Nevertheless, such experiments led others to perfect the modern float tube. Anglers now can sit comfortably in a low-slung seat, can carry all necessary equipment, including lunch, and can move about as silently as drifting shadows with both hands free for fishing. A kick or two quickly turns the tube into proper casting direction for flicking a fly or lure accurately to a rising fish. Compare this with what we often have to put up with when casting from a boat!

Today's float tube consists of an inner tube and a sturdy cloth covering, held in place with a zipper, which has pockets for gear and a low-slung seat. Some even are equipped with backrests, rod holders, line-stripping trays and other optional conveniences. Weighing only about 10 pounds, the float tube is light and compact. It costs much less than a boat and is easier and quicker to launch. It presents a lower profile, allowing anglers to sneak up on fish with less chance of detection. In fact, float-tubers often see big fish cruising nearby. For these reasons they rack up trophy scores or at least enjoy suc-cessful fishing when others in boats or fishing from shore hook little or nothing.

The difference among float tubes is in the cloth cover, which may be of canvas, nylon or other material selected for strength and the ability to resist tears. The cover helps to protect the tube from damage, but it does a lot more than that. Slung low in the hole in the middle is a comfortable seat of canvas or nylon mesh. The seat usually is adjustable for body shape and weight and is equipped with a quick-release buckle so it can drop away to make it easy to get out of the tube. If you plan to walk with the tube around you, shoulder straps are available. Some tube and cover combinations are lightweight, weighing 5 pounds or so; others are heavier for rougher use.

Tube covers are equipped with various zippered pockets for storing tackle, rain jackets, lunch, etc. Nylon zippers are superior to metal ones because they won't corrode. Some tube covers contain a large pocket in back for storing jackets or other bulky equipment. This pocket also can contain a small inflated inner tube (such as the Volkswagen type) for added safety which doubles as a backrest. Covers also have two or more D-rings to which anchors or fish stringers can be attached.

Float-tube covers come in many colors, including camouflage, green, tan, yellow or orange. As far as float tubes are concerned, fish do not seem to mind bright colors; in fact, we have observed that fish pay little or no attention either to the tube or the person in it. Bright tubes may be safer when boats are on the water, so other people can spot you more easily. This may have a particular advantage for night fishing. Float-tubing can be very productive at night but it also can be dangerous. Obstacles and shallow areas may be invisible. If you decide to accept the risk it is prudent to be very familiar with the area to be fished and to have friends nearby. Many float-tubers always wear life vests, but they are particularly important at night.

Proper clothing of course depends on water and air temperature.

Anglers usually wear waders. In warm weather you may be tempted to wear next to nothing, but remember that water that seems warm enough at the start can feel uncomfortably cold in a surprisingly short time. Many veterans at this sort of thing wear thermal pants under their waders, and add a thermal top on cool days. Keep in a float-cover pocket a lightweight waterproof wind-breaker for wind and rain. Such a jacket (with hood) can add surprising warmth. Many veterans also forego the usual fishing vest in favor of a shirt with breast pockets. This tucks inside the waders, out of the wet and out of the way. It is difficult to reach some of the pockets in a wading vest when fishing in a float tube.

Since temperatures vary so much, we won't go into more clothing sug-gestions here, but recommend asking regional experts for advice. They may even advise wearing a wet-suit, obtainable from stores where diving equip-ment is sold. Such a suit is warm and less bulky than waders. Usually, though, you'll wear waders. Avoid the lightweight ones which collapse in the water and cling to the legs, making fishing uncomfortable.

PROPULSION DEVICES

There are two types of propulsion devices available that attach to the feet and enable you to glide through the water by moving your legs.

Swim-fins, commonly used in diving, are worn only on stocking-foot waders, instead of wading shoes, or on bare feet. If you use them, wear heavy socks over the waders for greater comfort and to prevent chafing them. The one disadvantage of swim-fins is that you can only paddle back-ward.

If you wear boot-foot waders, you'll need a pair of paddle-pushers, also shown in the accompanying drawing. They have one advantage over swim-fins-they allow you to propel yourself forward. They can be worn with either type of waders. When you move your foot backward, the vanes stick out at about right angles to the path of travel, propelling you forward. When you move the foot forward, the vanes trail backward, out of the way. Paddle-pushers are so easy to use that you can forget you're wearing them. All you have to do is sit up straight in the float tube and tread water rhythmically to allow the fins to open and close properly, almost like walking.

Hal Janssen, the prominent angler and writer from Petaluma, California, has made an extensive study of float tubes, and offers this comment: "If you merely use the float tube to stalk fish, then paddle-pushers are probably more than adequate for your needs. But if you fish a long line, either by cast-ing or by stripping the line out as you move along, allowing the line to settle and then stopping to retrieve, swim-fins are the method of propulsion to use."

These added notes may be of value:

Avoid overinflating your tube. When a tube is left on the beach, exposed to the sun, the heat expands the air in it and it can rupture. Use only enough air to inflate the tube to its proper shape. It's easy to inflate the tube at a service station, but it usually is handier to carry a hand or foot pump so this can be done at waterside. Such pumps are small and inexpensive. Campers use them for inflating air mattresses.

Damage to tubes is infrequent because of their sturdy construction and covering, but it can happen. Check your wader repair kit to be sure it con-tains enough materials for both purposes.

Since float-tubers are close to the water, the danger of sunburn is even greater than when fishing from boats. It is prudent to use a head covering that shades the eyes and protects the back of the neck. Periodic applications of suntan lotion also help, particularly in southern climes. On bright days sunglasses are important for two reasons: They protect eyes from glare and help you to see better beneath the surface. There's another reason for wear-ing a protective hat and sunglasses when flycasting. Since you're so low in the water, the danger of hooking yourself is increased. If this happens (and even experts are not immune to it), it is better to get the barb in your cloth-ing than in your hide. Such accidents usually can be prevented by keeping the backcast high.

Usually you'll keep the float tube in motion, but there may be times when you want to fish a spot from a stationary position-a tough trick when a breeze is blowing! You then can use anchors, which can be carried in a pocket of the tube covering. Usually, you'll need two. They can be attached

with parachute cord (or something similar) of the right length to the D-rings on the tube. Drop one anchor, guide yourself to where the line becomes taut, and drop the other. Two anchors will prevent the tube from moving.

Care for float tubes as you would for boots and waders. After use, check the tube for damage. Be sure it is dry before storage. Keep it in a dark, dry place, preferably partially inflated. If it is deflated, try to avoid creases.

Finally, a personal note. The float-tuber gets into his waders, settles himself comfortably in his little craft, and paddles out to where he thinks the fishing should be good. He gradually feels a compelling and increasing need to go ashore to obey a call of nature. How inconvenient, when we think of what such an interlude entails! You can guard against this by resisting the intake of liquids for an hour or two before you launch. Nibbling dry crackers also helps. And you can leave that can of beer in the cooler until you're finished fishing.

8

Dry-Fly Fishing

Before the development of the dry fly, anglers found that they could take trout by fishing wet flies on the surface. The wet flies had to be false cast to dry them, after which they would float only momentarily. To improve their floatability various British anglers made several changes in wet flies, including winding them with collars of stiff hackles and adding tails that would make hooks ride properly. All this was reported in several classic books written by the great English angler, Frederic M. Halford, around the turn of the century.

An American angler named Theodore Gordon heard of this and made himself and his favorite trout streams, New York's Willowemoc and Beaverkill, everlastingly famous by writing to Mr. Halford about it. On February 22, 1890, Mr. Halford sent to Mr. Gordon about three-dozen English dry flies-and dry-fly fishing in America was born.

Since British insects aren't exactly like American ones, Mr. Gordon used the English flies as models for American patterns, thus giving birth to the renowned Quill Gordon and perhaps the Light Cahill, among others. Thus started, in this area of New York State, the development of many of the classic dry-fly patterns as we know them today.

Most anglers who are familiar with all fishing methods consider dry-fly fishing to be the peak of the angling art. Fishing the dry fly requires a fairly close imitation of the actual insect, delicate tackle, and precise presentation. Unfortunately, too many anglers shy away from dry-fly fishing under the impression that it is difficult and complicated. As in nymph fishing, one can make dry-fly fishing extremely complex, but it doesn't need to be so. One can have fun fishing dry flies for trout with only the basic knowledge set forth in this chapter.

WHAT DO DRY FLIES IMITATE?

When the nymphs described in the last chapter rise to the water's surface or crawl up on rocks or logs, their nymphal skins split, allowing them to leave the skins and to transform themselves into flying insects. When this is done on the water's surface, the transformation can be almost instantaneous, in sort of a "pop, and away they go" manner-a hastening developed by nature to keep some of them from being eaten by trout. When it is done on land the process can be much slower. The skin splits and the insect crawls away from the shuck, still not looking very much like a fly. While one watches, however, the pulsating body pumps fluid into the almost invisible wings, which rapidly spread and become larger. Very soon, with wings fully extended, dried and hardened, the newly born insect flies away. We have seen that these flies are in many different species and subspecies, including mayflies, caddis flies and stoneflies, to mention only three.

Mayflies, the most important of all species to anglers, are water-born flies with folded upright wings like little sailboats. After emerging, the fly is called *a dun* because the wings are dun-colored; a dull grayish brown. The duns may spend hours or days perched in trees or on bushes before a final transformation takes place. When this happens they molt, thus becoming *spinners,* more brilliantly colored and somewhat different in size and shape, with three long tails. The spinners, often in dense multitudes, dip swiftly downward and glide more slowly upward in a dancing flight while they mate. The females then skim the water to lay their eggs. Then, evidently exhausted, their wings become spread and uncontrollable, and they fall to die as *spentwings*.

When this happens, usually just before dark in May, June and July, trout seem to indulge in a frenzy of gorging themselves, swirling and darting about to eat all they can while the supply lasts. Strangely enough during such times of abundance, trout take reasonable imitations in dry flies avidly. Anglers try to imitate not only these three phases of the mayflies' life, but also the many sizes and colors of the various subspecies which hatch from time to time during the season, thus giving birth to a bewildering complexity of fly patterns. Those most faithful to the "match the hatch" school of thinking go to great pains to do so. Less meticulous anglers feel that only a few general representations are sufficient, and this often is true if the selection is a reasonable one and is presented properly.

Caddis flies, second in importance to anglers, look like small moths, with long feelers and tentlike folded wings. Their various types emerge between April and July, flying over the water in erratic zig-zag swarms. They are called miller and sedge flies. After hatching and mating, the females swim or crawl down the branches of bushes and the stems of grasses and plants to lay their eggs in the water, after which they die. Trout are in the shallows of the grasses waiting with noses near the surface to sip them in. Anglers use flies of the sedge or miller type with heavy hackles so they won't be caught on the grasses. The idea is to cast to the grasses or to the bank and to pull the fly off into the water. Caddis flies also zig-zag over fast water stretches of streams, almost touching the surface. The idea then is to use heavily hackled flies of similar size and color and to skitter them on the top.

Stoneflies, perhaps third in importance to anglers, appear even before ice leaves northern lakes. In shades principally of gray, brown and black, they average about an inch in length and have two pairs of wings which are folded flat against the body when not in flight. They crawl beneath the water's surface or dip the ends of their abdomens into it while flying in order to lay their eggs. Imitations include flies such as those of the same name and the Adams, Brown Sedge and Grannom. The gray, brown or black Wooly Worm is presumed to be taken for a stonefly.

When I was fishing in northern Saskatchewan one year in June, I noticed a large, ragged black cloud, probably a few miles away. Changing shape rapidly, the cloud became denser as it approached and descended. When it arrived, the car was pelted with a hail-like storm of flies which were presumed to be stoneflies. The fly-storm lasted for many minutes and, when it was over, the ground in a vast area was covered with dead and dying insects. Vast hatches occur in many regions, and the insects often become lost and unable to find water. When hatches occur they can be very widespread. On another occasion, for example, we had to stop many times during a drive of a few hundred miles to clean the windshield of mayflies.

No list of the best dozen dry flies would be widely acceptable because (again) no two anglers would agree on it, and it also would vary with the area being fished. We can, however, start with one of which many anglers might partially approve and the reader can go on from there.

Black Gnat	12-16	Leadwing Coachman	10-14
Blue Dun	12-16	Quill Gordon	12 only
Cahill (Light)	14-18	Ginger Quill	14 only
Grannom	12 only	Royal Coachman	10-14
Hendrickson	12-16	Spentwing Adams	12-18
Irresistible	6-12	White Wulff	10-14

About half of these flies are old classics, and seem to be perennials. In summer, when the naturals are around, good representations of the bumblebee, the small green caterpillar and the grasshopper are very

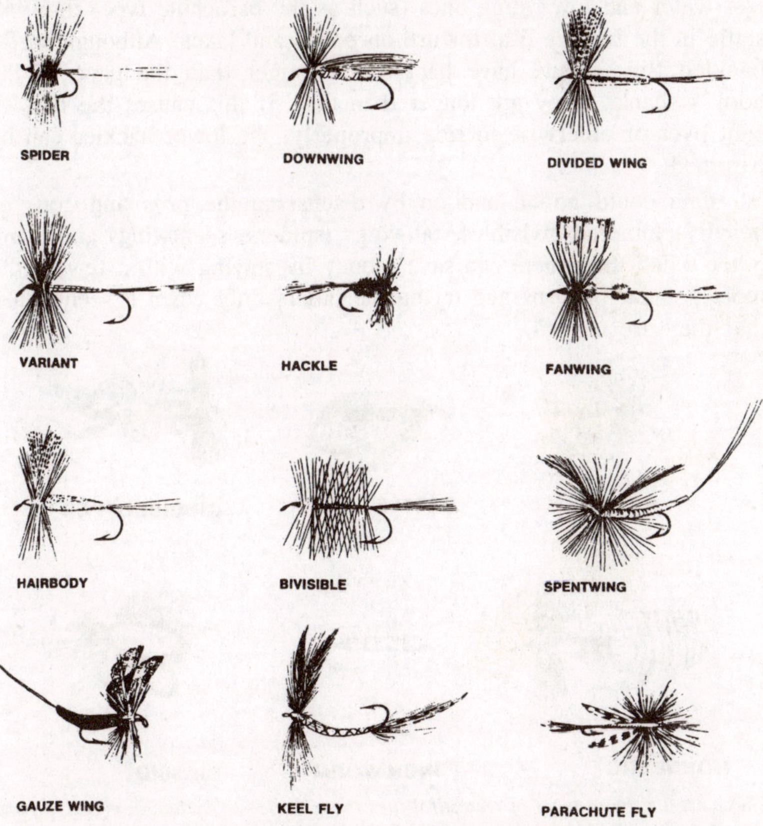

The basic types of dry flies are shown in this illustration. Each type is tied in many different patterns, or colours. For example, the divided-wing fly may be tied with blue, brown, or cream-coloured hackles and matching wings. These flies usually imitate actual insects. Other types, such as the spider, bivisible, variant and fanwing, attract trout more by their colour and shape than by their exact likeness to a specific insect.

useful. Purists may frown on these being included as "dry flies," but there are many times when nothing can beat them. Tiny flies fished in the surface film often do well on long, fine leaders. These include midges, ants, grubs and a terrestrial called a Jassid. This is about the size of a house fly and often is winged with two jungle-cock tips tied on flat.

Part of the never ceasing argument about dry flies is how they should be constructed. Some like them collared with soft hackles, others with stiff ones, and a few with no hackles at all. Here, the stiff hackle advocates would win. I generally prefer high-riding flies for fast water and low-riding ones (such as the parachute type) that will settle in the surface film for use on ponds and lakes. Although stiffly hackled flies should have hackles no longer than the gape of the hook's shank, many are longer than that. If this causes the flies to cant over or otherwise to ride improperly, the lower hackles can be trimmed.

One could go on and on by discussing the pros and cons of heavily palmered bivisibles, fanwings, spiders, spentwings and many other types. Beginners can save money by staying with a few of the most popular patterns and trying the others only when it seems sure that they are needed.

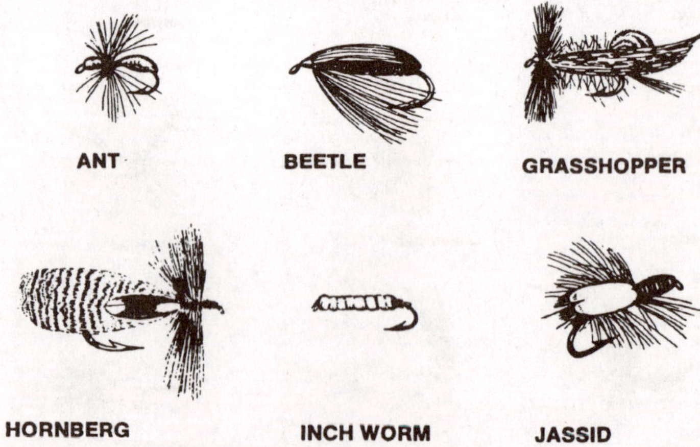

These are the basic types of terrestiral flies that are tied to imitate insects that jump into the water from land or overhanging trees. After the hatches of aquatic insects have tapered off. terrestrials are extremely effective.

DRY-FLY TACKLE

Since dry-fly fishing is more pleasant with light tackle, and since long casts rarely are needed, the most popular rods are in 7- to 8¹/

2-foot lengths. We have noted that makers of glass rods usually standardize on mediumaction ones, and these all-purpose sticks handle dry flies very nicely. The dry-fly specialist, however, would prefer a rod with faster tip action (often called "dry-fly action") because it is superior for making the many false casts necessary to dry the fly in the air. By the same token, users of bushy flies such as streamers and bucktails would prefer a slower-action rod (often called "wet-fly action") because it takes more time for bushy flies to travel through the air during casts.

By gripping a rod with both hands and holding it extended forward, the tip can be oscillated from side to side and the extent of the deflection can be observed. Since it seems to be agreed that fast action is between medium and fast, and that slow action is between medium and slow, medium-action rods cover all requirements quite well. A slow-action rod obviously is not ideal for dry-fly work, but old rods passed down in the family often are of this type. Some may have been good dry-fly rods originally, but they may have softened through time. Another argument in favor of light rods for dry-fly fishing is that the lighter lines which fit them help to present flies more delicately.

Since light reels are appropriate with light rods, the single-action types are preferred over automatics and multipliers.

Delicate presentation of dry flies requires a floating line with a double taper rather than a level line or one with a heavy taper. We have seen that two can be had for the price of one by cutting the double-tapered line into two equal parts and putting half of the line on a reel with Dacron backing.

Although double-tapered floaters usually lie on the surface film without sinking, they handle better when kept clean. Put a little line dressing on a cloth and rub the line briskly with it. Then pull the line through the closed fist to remove excess dressing, which otherwise would scrape off to gum the rod guides. We need to dress only the part that will lie on the water. A clean line rides higher and will pick up better, with less drag, than one which has become stained by dried surface film.

The line can be picked up for recasting very smoothly and with almost no surface disturbance. Lower the rod tip while removing slack. Before making the backcast raise the rod tip quickly and smoothly until only the leader remains on the water. Then make the backcast using an upward and backward flick of the arm.

A properly balanced leader, is most important in presenting the dry fly delicately. It should be as long as can be handled properly, but need not be more than 12 feet. Shorter ones, at least as long as the rod, often are considered adequate, but specialists agree that longer ones increase the number of strikes. The leader, or at least the forward 6 feet or so, must sink. If it sinks the fly this can be corrected by false casting. The tippet must be as fine as will provide a secure connection; the general rule being "the smaller the fly, the finer the tippet." Many anglers like very long tippets of 3 feet or more. These probably will drop in coils, which is why some don't like them. However, the drifting fly will uncoil them and result in a longer free float.

Experienced anglers have a trick for hooking trout which is particularly useful when using dry flies and nymphs. Instead of raising the rod tip, they flick it *downward.* This provides a faster strike and, if the fish isn't hooked, the fly only has been moved a few inches and can continue its drift. We often miss by striking too soon as well as too late. This is a knack learned only by experience. Some wait until they feel the fish in the belief that he will turn and take the fly down before ejecting it.

Handling hooked trout on the dry fly is a delicate operation which usually requires only moderate tension while the fish tires himself out. When too much tension is applied, either the fine leader tippet will snap or the small hook will pull out. Pull-outs are caused either by faulty hooks or because the angler has been too heavy-handed with his fish.

Perhaps the most important maxim in dry-fly fishing is that the floating fly must be given an absolutely free float. We often think that the fly drifting in the current is floating freely when it isn't. This can be checked by watching other things, like bubbles, floating nearby. If the fly isn't traveling at the same speed, the drag must be corrected because the trout will know that something is wrong with it.

UPSTREAM CASTING

Casting upstream is the most important dry-fly method. Since trout lie facing into the current, the angler approaching from behind is less likely to be seen. The fly, drifting down over the fish, usually can be given a longer free float before drag sets in. Leader and line are less obtrusive, and probabilities of disturbing the water are less.

Upstream casting doesn't mean *directly* upstream, for line and leader then would pass over the fish, and the angler would have a problem taking in slack in fast currents without pulling the fly.

Upstream casting means putting the fly partly upstream and partly across, on an angle that will provide the best float over the best water.

The position from which a cast is made is very important. Before casting, study the water to decide the best casting sequence. First, look for holding or feeding positions where trout should be: spots above and below big rocks, deep holes in sharp bends of the stream, deep runs, obstructions such as fallen trees, deep, shady places along brushy banks, bridge abutments, and so forth. Bridge abutments are important because they break the flow of water and provide shade. Trout invariably seek shady places in moderate flow.

Having determined the trouty spots, you must then decide how casts should be made to put the fly over them for the longest float that will be free from drag. This demands a study of water currents. When you have made these two determinations, you should know where to stand to make each cast, the casting sequence necessary to cover the water, and the spot to which each cast should be made. Old hands at the game can sum this up almost at a glance. Newer ones should take the time to do so. A few minutes invested in "reading the water" pays off in strikes. Inept fishermen ignore most of this. They wade in where trout should be, flail the water with random casts which send trout scurrying to safety, and then wonder why they don't catch anything.

Figuring out the best casting positions and the best targets is important, but two other things are equally so. The cast must be completed in the air so that the fly will drop as lightly as an insect lands -with no splash from leader or line. The cast must be delivered so the fly will drift at least within a foot or so of where the fish is supposed to be.

It is greatest fun when trout can be seen, so one or more can be selected as targets. When the fly goes over them, one of three things will happen. The fish will rise and take the fly; it will frighten him away; or he will ignore it. If he ignores it, you may have another chance, particularly if he showed interest by rising to inspect the offering, or even quivering slightly. There are several things that you can do: try a smaller fly, or a different fly, or another method. Probably most experts would recommend resting the fish for a few minutes while changing the fly and then using one of the first two alternatives. Others, perhaps not wanting to bother with this, think other methods work better. When the orthodox free float doesn't work, they give a little motion to the fly. Remember that live flies don't always drift motionless. They often swim fast enough to take off. If a little motion doesn't work, try skittering the fly, pulling it over the trout so it leaves a small wake. Spiders are preferred for this tactic.

When trout can't be seen, of course blind casting is necessary. This means floating the fly over all positions previously discussed, usually by wading slowly upstream so good positions can be reached. On windy days accurate casting may be difficult, but dapping may pay off. This means letting the wind handle the fly. Let enough line out so the fly blows in the wind, alternately being in the air and touching the water. This seems to simulate an insect skimming the water, and it takes but an instant for a trout to flash up and smash it.

Mending the line has been discussed, and it often is vital in obtaining a long, free float. When the current starts to put a belly in the line, this should be corrected before the belly gets too big by using the rod tip to flip the line into an upstream curve. Sometimes this may have to be done repeatedly, and it is a bit of a trick to know how much of a flip to give the line to remove the belly and yet not pull the fly. In casting to a good position, such as the moderate flow below a rock, a long float may not be necessary. You can drop the fly on target, let it float for only an instant, and perhaps repeat this several times, thus more or less simulating a hatch. This often pays off, and it often is the only way to reach a trouty target in fast water.

DOWNSTREAM CASTING

Many streams are too fast for comfortable or safe upstream wading. The dry fly can be fished downstream, but often broken water makes a free float impossible. However, a disturbed surface makes it difficult for fish to see you, so you don't need to be as concerned about a cautious approach. On this kind of water, any natural fly is buffeted about by wave and current action, so a tumbling or skittering artificial, or a fly that is being pulled, may appear natural if it isn't being towed too much. Obviously, good floaters are necessary. Heavily hackled Wulff patterns and flies with clipped deer-hair bodies are preferable. Padding the bodies of Wulff patterns with kapok before completing the dressings increases floatability. In fast water, it is easier for both fish and angler to see large patterns. Beyond a choice among light, medium and dark flies, the patterns are of minor importance. Trout must act instantly to take flies in fast water, and conditions are such that they don't see details clearly. By the same token, adding a little action to the fly usually helps to attract a trout's attention to it.

The slack-line cast helps to provide a good float in downstream casting, even if it is a short and quick one. Upon completing the forward part of the cast, aimed at the target, the rod is stopped quickly at 45 degrees and pulled back slightly. This causes the line to fall in S-shaped curves. While these are straightening out, the fly can get a free float to the position where a trout is presumed to be.

This slack-line cast can be extended with line that is off the reel by wiggling the rod tip from side to side, causing the extra line to pay out in curves. On completion of the float, action can be given to the fly as soon as it starts to drag. If the slack-line cast is made partially across stream it may pay to skitter the fly when the float has been completed. Thus, rough-water dry-fly fishing is done somewhat differently than upstream dry-fly fishing on more placid surfaces.

DRY-FLY FISHING IN LAKES

This rarely is the best method, and it only works when water temperatures are such that trout are feeding on top in lakes where insects are prevalent. These insects can be flies such as mayflies or caddis flies, or terrestrials such as ants, bees or grasshoppers.

The dry-fly leader must sink to eliminate too obvious a connection between fly and line. The fly can be given very slight motion to simulate the feeble action of a natural on the water, but this motion should be so little that it makes only the tiniest ripples.

Trout may be located cruising about or feeding in certain places on top, but this usually isn't the case. Other chapters in this book have attempted to describe where to find them. This may be in spring holes in summer. It usually is along the shore toward which the breeze is blowing, partly because drifting insects are blown there. It often is near stream mouths on the leeward shore where drifting food from the stream and from the lake collect together in a floating line or mass.

Dry-fly fishing for trout usually is done under warm-water conditions when terrestrial insects such as bees, ants and grasshoppers are prevalent. Although it is traditional to use imitations of water-born insects, the landborn ones often are more productive -so let's not be too traditional. How to tie a bee with a black and yellow clipped deer-hair body, and the little lure is dynamite at times. Most tackle catalogs offer representations of grasshoppers, and most of them are very effective. Caterpillars usually work well. Artificial ants have many advocates, but this author hasn't had much success with them. Along the shorelines of lakes and ponds in the summer, representations of terrestrials such as these can do much better than imitations of water-born insects.

9

FISHING IN STREAMS

A skill that distinguishes the expert angler from the novice is the ability to find fish in streams and brooks. We will start with the trouts because they are most prevalent in such places in most parts of the United States. Knowing the preferences of trout helps in catching many other species whose individual peculiarities will be dealt with as we go along.

EDGES, AND WHY FISH SEEK THEM

Trout usually lie near the edges of currents and in protected areas of relatively quiet water where there is moderate flow. While they *may* be in the very fast water or in the dead water of streams, this is so unusual that we can dismiss it as a probability. Although the various species of trout have slightly different water preferences they can be lumped together to start with. Rainbow trout like water a bit faster than brook trout, and brown trout generally prefer it a bit slower, but for all general purposes there isn't very much difference.

What are the edges of currents? A way to find out is to put on boots or waders and to stand knee-deep in fairly fast water, facing upstream. Trout usually lie facing that way. If you put a hand on either side of your boots, you'll notice that the water will be very fast. But if you put your hand in front of or behind a boot, the water will be much slower. There will be an area of quiet water just above and below the boot. Notice that between the fast water and the quiet water the current makes sort of a streak, which separates the two. This is the *edge* of the current. Since it usually requires too much exertion to hold in the fast water, fish habitually lie just inside the edge, where they can rest and at the same time can see food drifting down to them. If they dash out to gather the food, they will return to the area just inside the edge.

The larger the rock or other obstruction, the longer is the edge, and the more quiet water there is between the two edges trailing downstream from the rock. A small obstruction, such as a football-sized boulder in a small stream, may have edges that can accommodate only one small trout, or perhaps none. A very large obstruction could accommodate several. If you catch a fish in such a place, try it again later on, because other fish will take the same position after it has been vacated. Desirable edges can harbor large fish because bigger ones will drive smaller ones from them.

The drawing shows such a position beside a rock in a fast stream. Looking down upon the rock, we see that the current divides to flow around the rock. The dashed lines indicate the edges, and there are areas of quiet water in between them which provide resting places, or "holding positions" for fish.

The current not only divides to flow around the rock, but that it also is forced upward to pass over it, if the rock doesn't extend much above water. Thus, the holding positions above and below the rock more or less resemble a cone of quieter water. Of the two, the one below the rock usually is the best one. Fast streams normally sweep out depressions on the upstream and downstream sides of the rock; these usually being deeper on the upstream side because the stream's force is greater there. Fish like to lie in the quiet flow of these depressions.

Edges are found in other places in streams also. The current of a fast brook entering a river will form an edge with that of the river itself. The edge will trail downstream and outward from the mouth of the brook, and it usually can be seen as a line marking the differences in current flow. Fish will lie just inside the edge, but the rest of the water inside the edge may be too slow for sportfish, and it may take the form of a quiet eddy.

A ledge or a big rock jutting out into a stream makes a very similar edge. When a river divides to pass around an island, an edge usually is formed on the downstream side. These are prominent examples, and all should be good holding positions for fish.

The photograph shows a well-defined edge caused by the river's current rounding a bend. The edge starts where the current hits the right bank at the 2 o'clock position. Notice that it is marked by a difference in surface turbulence taking the form of a rough line extending diagonally downstream to the 8 o'clock position. Gamefish should be lying just inside (to the right of) this edge. If they are not rising to take surface lures, we must try for them near bottom-not in the "in between" water.

The best place to fish this edge seems to be to wade out on the riffle just below the root mass at upper left and to make extending casts to the start of the edge (2 o'clock) so the lure will swing down current to cover the entire edge area until it hangs downstream, or nearly so. Extending casts can cover the entire area outside and inside the edge.

If one is on the right bank he can start fishing above the edge (2 o'clock) and let the lure swing from outside the edge to inside it. The very start of the edge may be a hot spot in the hole close to the bank. By extending casts all of the visible edge can be covered. If surface or near-surface lures don't work, try the area again with bottom bumpers.

WHAT'S UNDER THE FAST CURRENT?

Fish seek areas where the current's flow is moderate, but these areas may be in water that appears faster than it really is. Rocks, both large and small on the stream bottom, act as brakes to the current, so the current near the bottom usually is much slower than that on the surface. There are many depressions in the gravel where the current is still slower. These provide good resting (or holding) positions for fish. As most of the available food is near the bottom, and deeper water provides more protection, fish usually lie very close to the bottoms of streams. They may rise up to take food on the surface, as when there is an insect hatch, but it is safe to presume that they are very near the bottom most of the time. They also may be near the banks because the braking effect of obstructions along the banks reduces the force of the current there.

In shallow streams, the surface and the bottom may be so close to each other that fish lying on the bottom can see food on the surface, and rise to it. In deeper streams it usually is one thing or the other. They rarely are on top, usually are on the bottom, and almost never are in the area in between. Yet fishermen who are using sinking lures or baits usually are fishing in the area in between, and they wonder why they don't catch fish! If there are indications that fish are rising for surface food, of course surface or nearsurface fishing is indicated. If that isn't so, avoid the middle layer, which usually is fishless, and make sure your flies or other lures actually scrape the bottom. While this may result in occasional hangups and a few lost lures, it is the best way to catch the big ones.

AVOID USELESS WATER

Most water in a stream can be considered fishless. In deep runs and pools the useless water is everywhere except close to the bottom, unless fish are rising to surface insects. An exception in deep pools often is the shallower tail of the pool where the water gathers speed before passing over the pool's lip. The usually smooth surface of the pool becomes disturbed by the faster current as it approaches the lip. This rough surface provides a feeling of protection for fish and, since this area is narrower than the pool itself, surface-born food tends to concentrate there. The surface water may appear to be too swift, but, as the drawing indicates, water below the surface is more moderate, and this area usually is shallow enough so fish can lie in the slower water near the bottom and still see what food is being washed down near the top.

Very flat shallow stretches, where everything can be seen on the bottom, usually are useless, particularly when the flow is slower than moderate. An exception under conditions of moderate flow is when fish may be feeding in the riffles. Dead water usually is devoid of sportfish. Polluted areas should be avoided. For example, even in rural or wild regions sawmills which dump sawdust into the stream can ruin it for fishing.

MOUNTAIN AND MEADOW STREAMS

Many mountain streams, particularly in western regions, have long, swift, shallow stretches filled with small boulders as well as some bigger ones. Since this is not good holding water it should be passed by in favor of the occasional deeper, quieter stretches. If it is fished, the usual methods include drifting and bouncing salmon eggs, worms or other bait, or wobbling spoons, along the bottom. The deeper pools, which contain bigger fish, are suitable for flies. Large ones are used, such as streamers and bucktails, or Wooly Worms, and they are allowed to sink to the bottom before being retrieved. These fast, rocky streams often contain small pools, or pockets, which may contain small numbers of large trout. In general, though, the fast water only harbors small fish. This chapter attempts to explain where to find fish, rather than how to catch them. The characters of western rivers all are so different that many varied fishing methods are used, and visitors therefore are advised to go out with local experts who can show them how it is done there.

Western meadow streams (if one type can be differentiated from the other) may be deep and placid, since they often have been dammed by beavers. Mountain streams often become meadow streams when they reach lower country, and of course the latter have more insect life than the former, thus making the types of flies which imitate insects very popular. Favorite holding water includes deep pools and areas near undercut banks and ledges. Many of these meadow streams meander sc much that a few miles of their length can be contained in about a square mile area. The outside of the sharp curves often contains a deep hole, perhaps with an undercut bank, and such places as well as the riffles, glides and runs, should be fished carefully. These are not places where careless anglers usually get results. A stealthy approach and a correct cast with a suitable fly is necessary.

On the following pages are diagrams of a typical stream, with suggestions on how to fish it.

HOW TO FISH A TYPICAL STREAM

Nearly all streams harboring sportfish (especially trout) contain stretches similar to those shown on the accompanying diagrams of a typical stream. Some should be hot spots while others are barren, or nearly so. The angler must be able to distinguish the hot spots from the fishless stretches. The following comments apply to situations designated by letters on the diagrams:

(A) In this deep run bordered by undergrowth, fish usually feed on bottom, but the bottom probably is muddy and brushy. Since paths and litter indicate overfishing, we should move downstream after a few quick casts along the bridge abutments.

(B) Except possibly in remote areas, spots near where cars can be parked are fished too often to be productive. Fish may be near or under the dam, but currents make them hard to reach unless they are out in the pool feeding on or near the surface.

(C) The deep run by the ledge and the edges below the three big rocks should be productive if lures are fished deep. The eddy on the left bank and the shallow water below it probably harbor only trash fish.

(D) The best bet is to start at the riffle and to fish upstream on the righthand side. If trout are lying near the tail of the pool, this method may take some without disturbing others farther up.

(E) By wading down the riffle, an angler can cover the edges below the rocks. He can also cover them by fishing upstream and across from spots farther down. If the water is deep, the fish usually will be lying near the bottom.

(F) This is shallow, flat water providing no protection. If the bottom is of gravel, fish could be feeding in the deeper stretch along the right bank.

(G) This deeply undercut bank provides moderate flow, shelter and food, so it should be a hot spot. When fish aren't showing, work lures deep and as far into the pocket as possible.

(H) Uprooted trees with branches and trunks in the water also offer ideal situations in moderate or swift flow. Trout often are under the submerged trunks and usually take lures only when they are fished within inches of the wood.

(I) This forested high bank provides afternoon shade. If water is reasonably deep, with a rocky or gravel bottom, fish should be in the shade near the bank. In shallow areas trout seek shady positions to the exclusion of sunny ones.

(J) The ledge across the stream should be productive. Fish should be lying just inside the edge, but they may be deep. The quiet water farther inside the edge is poor for sportfish.

(K) Since the entering stream brings food, its mouth may be a haven for trout all season long, but especially when the main stream is warm. Then, trout should run up the cooler tributary, and again, regardless of water temperature, during their spawning periods. In the river, look for them along the edge marked by the dashed line. The quiet eddy inside the edge looks unsuitable for trout, but it may harbor smallmouth bass, chain pickerel, suckers and chubs.

(L) Here the stream widens into shallow, flat water which seems to contain no "trouty" spots. If the bottom is of gravel, trout may be feeding in the current lanes.

(M) The way to fish this riffle seems to be by wading down it, first being sure we can wade out at the lower end. Pay attention to the right bank where it is overhung with bushes, and cast as close under the bushes as possible. If current and depth on the western side are sufficient, casts can be alternated from one side to the other, covering all the water while wading down.

(N) The large rock should be fished thoroughly, perhaps by

bumping bottom. The edge below the rock may be the best position, but the entire area should be productive.

(O) Just around the stream's bend, on the western shore, lush vegetation and trickling water indicate a spring hole.

This should be a hot spot in warm weather. Start casts well out before working them in closer. Since cold spring water flows downstream, the short distance from the hole to the rock should be explored.

(P) Outward from the spring hole is a deep pool with many rocks, so the pool's bottom probably is rocky also. If there is no surface activity, fish lures near the bottom, using weedless ones, or weedless Keel Flies, if necessary.

(Q) Here, the deep pool's outlet is constricted by rocks. The rapids may be too fast for trout, or for effective fishing. By sinking the lure deep, the edges of the rocks here and at R could be productive.

BRUSHY BROOKS

In the Northeast we have many brooks flowing through woodlands and valleys which are so overgrown with bushes that few people bother to fish them. This is a pity because many are bonanzas for brook trout. Being a New Englander, I learned some of the secrets when very young. The first secret is not that they are there, but that they usually hold fish.

There's no sense in poking through the bushes, trying to make a cast or two, and then poking back out to see if it's any better somewhere else. The trick is to put on hip boots, get into the brook, and stay in it. Because of the bushes, a long rod is at a disadvantage. Use a short one, no longer than 6 feet. Use bait, if you wish, but tiny streamers or bucktails, or nymphs, or even dry flies in the summer, may be even better. Wear dark colored clothing that blends with the landscape. Fish slowly, quietly and deliberately.

In fishing small brooks, remember that a brook is a stream in miniature. On a big stream a rock of football size would be ignored. On a small brook a good-sized trout may be lying back of it.

Even careful wading will dislodge a stone or two, perhaps sending a small cloud of mud downstream. It may be a good idea to make this happen occasionally because trout may interpret it as a sign that part of a bank has caved in and go on the feed.

If the day is bright, fish the shadowy places. Trout often lie in the shade over a patch of dead leaves lying on the bottom. Stop often enough to read the water just ahead. Look for an undercut bank that could hold trout; a dislodged tree whose disturbed roots leave a watery hole under it; a big rock against the shore with a tiny pool behind; a little riffle making a larger pool; or any other place a sizable trout could hide.

If so, how does one get a fly or lure into the position without getting it tangled in the branches? With the short rod, a normal cast often can be made, or perhaps a miniature roll cast. If not, hold the rod out and let line through the guides, so the current can carry the lure downstream. If it sinks before getting to its target it can be pulled to the surface by raising the rod tip, thus letting the current take it down some more. When near enough to the target, a small twitch or two should start some action.

Another trick is to cut a green leaf with an inch or two of the branch connected to it. Nick the barb into the edge of the leaf and let this float the lure or bait downstream, trying to steer it to wherever it should go. If the place is a good lie for trout the current probably will take it there anyway. A twitch of the rod breaks the lure from the leaf, and another twitch or two (if it's a wet fly) should tempt a strike.

In using flies on such brooks there may not be enough room to operate small streamers and bucktails in a lifelike manner. Drifted nymphs work well, but summer favorites are natural imitations such as grasshoppers and caterpillars. My secret weapon at such times is a small, fat bee made of clipped bands of black and yellow deerhair, with brown deerhair wings.

MEADOW BROOKS

Another type of brook, too often ignored because it doesn't look like much, is the kind that flows through meadowlands. It may be well that such brooks are so often passed by in favor of larger streams, because they usually are nurseries and breeding grounds for trout.

Some meadow brooks appear to be so small that they would seem to be useless for fishing, but this often is far from the case. They can look like deep ditches, bordered perhaps by a few bushes and lusher vegetation left uncut by mowing machines, and one often can step or jump from bank to bank with ease.

Unseen below the narrow exterior may be deeply undercut banks where the stream perhaps is several times wider, and much deeper, than it would appear to be. This furnishes cool shade and protection, plus an abundance of small baitfish and quantities of caterpillars, grasshoppers, flies and other living things which fall from the foliage into the waters below. Fish live here year-round, and others come up from larger streams during spawning migrations to perpetuate their kind in the ideal and constant temperature of the gravel, unseen and unbothered except by the few adept anglers who understand this hidden secret.

Where the heavy turf of the topsoil becomes thin the banks may cave in, leaving slanted slabs of turf whose underwater nooks and crannies also are hiding places. Here and there may be a small riffle or waterfall emptying into a tiny pool, or perhaps even one quite large. But one specializes mostly on the undercut bank areas to find the biggest fish.

These places demand a quiet approach because an angler, thoughtlessly clumping along the bank, can send his echoing footfalls down to telegraph danger, thus forcing all fish to take cover. Such a man would leave the place in disgust, calling it fishless. These places demand a stealthy approach because, if the angler is seen, the fishing is done for. Experts, trying a new brook for the first time, make it a practice to stroll along it and map the good places, deciding how each cast should be made. They come back on another day and fish the brook efficiently because they know where and how to do it.

Such fishermen crawl when close to brookside, rather than walking directly to it, and they take their time. To the casual observer this may look rather silly, but it gets results. Their rods are long, rather than the short miniatures preferred for brushy streams, because some of the fishing is done by "dapping"; dropping the fly or other tiny bait time and time again onto or into the brook's surface with little or no line out. Short casts also are in order, and the idea often is to flick the fly to grass overhanging the brook, and then to flick if off and into the brook as an insect would drop in. Needless to say, considerable accuracy is required. More than likely, however, a large trout, waiting below for such a tidbit, will grab it instantly.

Another way, instead of dapping or casting, is to lower the fly or bait into the brook, with the rod held high. By lowering the rod, several feet of drift can be obtained. After doing this two or three times the angler sneaks or crawls along a few feet more and repeats the process.

Meadow brooks, between one meadow and another, may turn into brushy brooks, and so can be fished a bit differently, as has been described. In general, however, a brook is either one kind or the other, and the techniques usually used on one are somewhat different from those used on the other.

Meadow brook fishing is at its best in the summer, when fish run up the cold-water brooks to escape the warmth of larger streams which are more exposed to the sun. Also, in summer, the bordering fields breed quantities of insects favored by fish for food. In flying or hopping from one place to another, these insects often land in the brook.

While some fishermen prefer to use small worms, perhaps impaled on hooks in the weedless manner, many find it more successful to use artificial flies that imitate these insects. These include imitation grasshoppers, ants, bees and caterpillars of one kind or another. The last two most often have spun deerhair bodies, clipped smooth and in natural shapes, sizes and colors.

In fishing brooks, particularly, one often has to be so close to his quarry that he wonders if fish can see him, whether or not he can see the fish. This is a question I never have seen answered satisfactorily in books.

WHEN CAN'T FISH SEE FISHERMEN?

Those who wade sandy flats to spear flounders, or who enjoy bowfishing, know that, when they aim the spear or the arrow directly at a fish, they don't hit it. They must estimate where the fish actually is in relation to where it appears to be. This is due to the principle of the refraction of light, explained by the fact that light rays bend at an angle when they pass from the air into water or from water into the air. Because of this, fish also can see fishermen even when fishermen think their silhouettes are so low they can't be seen. So when can and when can't fish see fishermen?

There are two points to consider, the first being a fish's "cone of vision." Light rays entering the water from the vertical are not refracted by the water, but all rays entering at an angle from the vertical are refracted, or bent. The greater the angle, the more they are bent until they reach about 48 degrees from the vertical. Then, light no longer is reflected back into the atmosphere, but is bent downward from the water's surface very much like the ricochet of a bullet when it is deflected by a solid object (and even by water).

Although the study of optics as it applies to refraction is complex, let's try to simplify it by referring to the accompanying diagram. Since light rays which are reflected back toward the atmosphere from below the water's surface are deflected downward at angles greater than 48 degrees from the vertical, the vision of fish is conical, the point of the cone being at the fish's eyes and extending upward and outward at a 48-degree angle to the surface. Thus the diameter of the cone at the surface is about 97 degrees, the circle of vision growing larger the deeper the fish is. The ratio is a little more than twice the diameter of the circle in proportion to the depth of the fish. Therefore, a fish at a depth of 3 feet, for example, can see through a circle on the surface which is about 6[1]/2 feet in diameter. Outside this circle, light rays are bent downward, so what a fish sees there is similar to a steely mirror reflecting the bottom, or the depths. This cone of vision is called the fish's "window."

The vision of fish is somewhat different from that of humans in that their eyes are on the sides of their heads, allowing them to see to the right, to the left, and above, at the same time. Their eyes protrude and can turn in their sockets, all this affording a very wide range of vision. Thus, unless the angler crouches down, out of sight, fish usually can see him before he sees them. An exception is when the surface is disturbed by wave or current action, but even then some vaguely seen surface motions may spook fish. Trout are particularly shy and usually won't strike if danger seems to threaten. Their blind spots are behind and below, which suggests an advantage in fishing upstream.

The second point is that when light rays enter water (in which the velocity is different) their direction is changed, or refracted. A pole driven into a lake bottom at an angle demonstrates this clearly. This refraction, or bending, forms an angle between two straight lines, the apex of which is the water's surface. The extent of refraction increases as the angle of entry or emission increases. In the diagram on page 356, we more or less correctly

This diagram shows the limitations of a fish's vision due to the refraction of light waves on the water. The angler at A sees the fish at F, but due to light refraction the fish appears to be farther away and nearer the surface than it actually is. The fish cannot see the angler, but can see the motion of his rod. Again, due to light refraction, what the fish sees of the angler appears to be at B rather than at A. assume the angle of refraction to be about 32 degrees. The fish at F, looking upward through its window, can see far outside this cone of vision because of refraction, which widens its vision by an added 32 degrees or so, allowing it to see the rod of the angler at A, but not the angler himself, although he would be partly visible if he were standing. Because of refraction, the angler at A appears to the fish to be at B, but this is unimportant to us. The important thing is whether or not the angler can be seen.

Thus, if the angler is crouching below an angle of about 10 degrees from the point on the water's surface under where the fish is, he is hidden from the fish, or appears only as an indistinguishable object because he is outside the fish's window. An object 21/2 feet tall at a distance of 15 feet, 5 feet at 30 feet, or 10 feet at 60 feet wouldn't be seen at water level. The simple fact is obvious: if you don't want fish to see you, you must lower your silhouette more and more as you draw closer to the water. While this is accepted as a fact, we have noted that it becomes less important when water is discolored or when its surface is broken by waves or current.

If an angler must move into a fish's vision in order to cast to it, he often can do so by approaching slowly and cautiously while keeping his silhouette as low as possible and wearing dull-colored clothes, at the same time making use of the concealment of rocks and other obstacles whenever possible. Since fish usually are concerned with feeding, they may not be spooked by a cautious approach.

When an angler does spook fish he should remember that their memories are short. After sitting quietly by streamside for several minutes, the fish may reappear and start feeding again. Then, a cautious cast without excess motion may attract them. Older fish usually are more cautious and remain spooked longer than younger ones. Of course, the angler's shadow should be kept off the water. Polarized glasses cut through surface glare and make fish and other underwater objects more visible.

WHAT DOES A FISHERMAN SEE?

Ordinarily, perhaps this isn't very important, but it may be worth mentioning. Due to the refraction of light rays entering the water, fish seen other than directly from overhead are not exactly where they appear to be and, the deeper they are, the farther they are from where they appear to be. The fisherman kneeling on the bank in the diagram on page 357 sees a fish at position A, but the fish actually is a bit nearer and deeper. This may not

make much of any difference in how one casts to the fish, but it must be considered during activities such as spearing or bow-fishing.

By drawing a straight line from the angler (A) to the fish (F), we note that he can see the fish, but the fish can't see him. A standing angler, however, almost always is spotted by fish before he sees them. At best, as far as most gamefish are concerned, the most an uncautious angler can see is the flash of fish as they streak for cover. Spooked fish rarely take lures.

10

ICE FISHING

Equipment for catching panfish and sportfish through the ice has blossomed since daddy was a boy. He froze in long johns, while our quilted, insulated clothing and thermal boots keep us snugly warm. He towed his gear in a wooden box on junior's Flexible Flyer sled while we streak out to the sets on snowmobiles, or perhaps in heated recreational vehicles. Spudding holes in thick ice was a time-consuming chore; now it's easy with power drills. Dad and his pals built bonfires on shore and crept back there periodically to thaw out. We have gasoline or propane heaters and stoves where hot beverages steam and bubble in tents, shacks, vehicles and windbreaks right where the action is, and on which complete hot meals can be prepared when needed. Dad's common sense in knowing where to spud holes has been superseded by electronic fish locators which can read through the ice to tell exactly how far down bottom is and what kinds of fish are cruising down there.

In spite of modern improvements such as these, knowledge still is required to hit the jackpot. Too many innocent ice fishermen merely drill a few holes somewhere, lower baited hooks, and try to keep warm while waiting for the action. No one knows all the answers, but there are rules and tricks which help to improve the score.

BASIC EQUIPMENT

Tip-ups can be homemade, but commercial ones are so cheap it isn't worth the effort. These have a reel spool and line held underwater (to prevent freezing) by a pair of cross braces set on the ice over the hole. The upright stick to which the reel is attached has a red or orange flag on a flexible trip wire which is hooked into a trigger so it can be released to pop up when a fish pulls line from the reel. "Tip up!" is the usual cry when this happens, whereupon the fisherman cautiously tests the pull on the line, attempts to hook the fish, and to pull it in. One fisherman can operate several tip-ups in as many holes, the number often being restricted by state or regional regulations.

Holes are cut by a spud (or ice chisel) or by an ice auger, which resembles a carpenter's brace, and which can be powered by a gasoline motor. Holes usually are about 6 or 8 inches in diameter and should have smooth edges to prevent chafing lines. The holes should flare outward toward the underside of the ice layer to diminish chances of fish being knocked off while being drawn through them. Since mush collects in the holes, a skimmer is needed to ladle it out.

Minnows are the favorite bait for most fish. Keep them alive in a minnow pail in which the water is oxygenated by oxygen pellets or other means, and kept from becoming cold by insulation or a source of warmth. Proper sizes of minnows vary with sizes of fish being sought: 1 inch for smelt; 2 inches for crappies and perch; 3-inch shiners for walleyes; 4- to 6-inch suckers for lake trout and pickerel; 12-inchers for pike. Since live bait is cold and slippery, it is more comfortable and does less damage to the fish to handle them with cloth work gloves. Have a few pairs so dry ones are always available. Use lively minnows for still sets. The dead ones can be fished with jigs.

Minnows usually are fished about a foot off bottom. This depth is found by attaching a sinker, such as a dipsey weighing 2 ounces or more, to the hook and lowering it until it touches bottom. Pull it up a foot or so and mark the line at water level. A good marker is a very small button strung on the line through two holes so it can be slid up and down for various depths. A piece of a pipe cleaner wound around the line also will do.

Falls on slippery ice can be avoided by wearing ice creepers strapped to soles of boots or shoes. These are rectangles of steel plate, slotted for straps, with corners bent down to form prongs. Sunglasses reduce glare and add to vision. A compass for use in snow squalls on large lakes and a coil of rope in case of accident are good insurance.

When a big fish is hooked on light line it can be lost while being pulled through the hole. A small gaff can be made from a 6/0 hook, or larger, and a wire coat hanger. Use wire-cutting pliers to remove the curved part.

Straighten the rest of the wire and put an end through the hook's eye. Using strong pliers, to hold the hook, bend the wire around the shank for a secure connection, and clip excess wire closely. Bend a loop in the other end for a handle, and sharpen the hook's point, if necessary.

Every fisherman will want to add to these basics several other necessities and conveniences such as a Thermos of hot beverage, a lunch, and many other things of personal choice. If lightly equipped, a knapsack or packbasket should hold most of them.

Those who go out with only a few tip-ups are less than half prepared for good fishing because a jigging stick or rod, with some bait and lures, can more than double the catch and the fun. Jigging sticks are of many kinds. A typical one is a dowel, such as a broom handle, about 2 feet long with a screw eve under the tip for the line to go through. A cleat fastened under the dowel forward of the grip is used to wind on or release line. The cleat width is measured to indicate how much line is out. For example, if the cleat is 6 inches wide, each loop released from it would be a foot long. Jigging sticks can be made from rod tips, with handles and reels attached.

Many fishermen prefer complete spinning or spincasting outfits, and some like stronger baitcasting gear for the big ones. The strength of the tackle suits the size of the fish. For example, it's more fun to catch panfish on light monofilament of between 2- and 4-pound test, partly because such light lines give better action to small ice jigs and small baits like dead shiners. On the other hand, we may need a strong rod equipped with a reel holding 15- or 20-pound-test braided line for fish like lake trout down deep. Usually this terminates with about 6 feet of leader of suitable strength, with a split shot or two about a foot above the hook, if needed to get the bait down.

ICE FISHING LURES

Lures for panfish start with tiny jigs weighing less than 1/16 ounce with hooks between sizes 10 and 14. One side often is nickel plated; the other is painted red, yellow, orange or green, often in fluorescent colors. Some jigs, usually called ice flies, are dressed with bits of marabou, like a wing, or are wound with small hackle. Other types, more like spinners, have tiny silver or gold blades and perhaps one or more small colored beads. These need not be baited, but they often are, with small grubs or worms.

Ice flies are easy to make. Pinch a small split shot back of the eye of a number 10 to 14 hook and dip it in paint. Tie on back of it a small, soft wing of marabou or hackle, or wind the shank with two or three turns of a small hackle.

Lures for medium-sized fish are similar but larger, including dartlike jigs with hair or feather tails. These vie in popularity with many shapes and sizes of wobbler blades of metal or pearl, either with fixed or loose hooks. For example, select an oval or fish-shaped wobbler weighing about ¼ ounce. Remove the hook and tie the line on the end where it was. Put a split ring in the other end, with a single hook on it and perhaps also an attractor such as a fishtailed piece of red plastic. Spinning lures with revolving blades also are used as jigs. Sometimes these are baited with any of the things to be mentioned, including a dead minnow impaled crosswise. A different vertical type is minnow-shaped, with the eye balanced in the middle of the back. Rigid single hooks point upward on each end and a loose treble one hangs below the eye. This type swims in a circle when jigged up and down.

Lures for big fish, such as lake trout, are similar, but even larger. These include jigs of the diamond type and other elongated ones like the popular Swedish Pimple.

BAITS

Baits for panfish comprise everything small enough to be edible-a range covering such a wide variety that a few examples should suffice. These include corn borers, grubs, meal worms, goldenrod gall worms and especially the larvae of the syrphus fly, called the rat-tailed maggot, or "mousie." Mousies are found in mud, or in decaying wood or vegetable matter and are so-called because they have "tails" which actually are breathing tubes. Two of these half-inch-long (not including "tail") tiny creatures usually are put on a single number 10 or 12 hook, the upper one being squashed in order to attract fish by scent. Preserved single salmon eggs are popular, as well as canned kernels of corn, presumably because they resemble fish eggs. Add roaches and crickets to the list, plus anything else that wiggles and many things that don't, including miniature marshmallows.

Garden worms are not very effective, probably because they are not in evidence in winter. Except for some of the smaller panfishes, the best baits are minnows and larger baitfish for everything else on up the scale, starting with crappies and perch. The proper way to hook a live minnow with either a single or a treble hook is just under and slightly to the rear of the dorsal fin, with the hook's barb angled toward the tail. This is because fish usually take minnows head first but, since larger fish may take the whole thing at a gulp, this point may not be very important. The important thing is to hook the bait above the spine because lower hooking will kill it. Dead minnows are hooked through both eyes or through the lips in order to provide maximum and more lifelike action when they are jigged.

Hook sizes 6 and 8 are excellent for crappies and perch, 6 for walleyes, 4 for pickerel, and size 2 or larger for pike, lake trout and other big ones. Short shock leaders of monofilament, or wire leaders, are necessary for sharp-toothed fish such as pickerel and pike.

WHERE TO CUT ICE HOLES

Whenever possible, smart ice fishermen plan the locations of their ice holes during open-water months when they can probe the bottom, or see it.

Many lakes have been mapped by fish and game departments or other sources, with depth contours included. Lacking these, sketch maps can be made to indicate submerged reefs, steep *drop-offs, underwater brush* piles, shallow *or deep* weed beds and other *preferred feeding or hiding* places for the species to be sought in winter. Since these hot spots will be hidden under the ice, the places should be identified on the map, using shore landmarks to line them up when necessary. Sonic fish-locating devices are ideal for this because they not only reveal exact depths but also the nature of the bottom, whether it is sandy, rocky or muddy and where there are weed beds, brush piles, and so on.

Although we know that every species of fish prefers its own favorite water temperature, it has to put up with the narrow one between freezing at 32 degrees and the maximum density of water at the lake bottom of 39.2 degrees unless it can find relatively warmer inlets or underwater springs. Since this range is only 7.2 degrees, water temperatures are not of major concern in winter. Fish seek food, protection and oxygen; three basic requirements as necessary in winter as in summer, even though the need for food is somewhat less when very cold water makes fish relatively dormant. It can be assumed that some species of fish, such as lake trout and smelt, will seek the slightly warmer water of the depths, but they won't be in the deepest water if protective and oxygen-generating plant growth doesn't exist there due to insufficient penetration of sunlight. Shallow-water species, such as panfish, pike and pickerel, seem unconcerned by water temperatures, and inhabit weedy areas of relatively shallow water at all seasons. Ice holes cut in coves or over weedy reefs should locate them.

Lacking more specific information to go on, and knowing that some species are roamers and that most of them cruise at definite depths, select a spot where travel must be concentrated. In the accompanying drawing, such a spot would be between the island and the point of land. Cut the first hole over about 10 feet of water on a line between the point of land and the island. Jig the hole for a few minutes. If nothing happens, put a still set there and cut another hole about 20 feet farther on, continuing this procedure as shown in Series 1. An alternative is to line up the holes toward the deepest water, as indicated by Series 2. In this case, which is an actual one, the best fishing was found at the 40-foot depth, so additional holes were cut there, as indicated by Series 3.

Some fish, such as perch and smelt, roam about in schools which remain in the same general location. When a hole has been productive and then becomes fishless, a few more can be cut in the same area, one or more paying off as the school returns below it. Since a hooked fish tries to rejoin its school, the direction in which it wants to swim indicates the one where another hole should be dug. Like oil wells, some holes produce little or nothing, and shouldn't be bothered with. On the other hand, a productive hole or two can provide more fish than even the greediest fisherman desires. Even the experts cut useless holes, but they do it less often than novices do. Experts usually go out for a specific species and, knowing its habits, can locate their holes with considerable accuracy.

HOW TO FISH ICE HOLES

Since lures or baits should be fished within about a foot of the bottom, or just over the weeds, put a sinker on the hook of a rod or a jigging stick and lower it until bottom is felt. Then raise it about a foot and mark the line at water level or wind it on the line holder so fishing at that depth can be maintained. A lure or bait or both is put on and lowered, using a split shot or two about 6 inches above it if necessary. The lure is jigged very slowly for panfish, more actively for other species. Just how the lure or bait should be jigged depends on water depth, the species being fished for, and the type of lure or bait used, so this must be found by experimentation. Tiny lures are jigged so slowly that they barely show motion. Larger ones, particularly when fished deeply, are given much more action. When lures weighing 1/4 ounce or more are fished deeply, a proven method is to raise the rod or jig stick sharply to jerk the lure 3 feet or so upward, then allow it to flutter down. This is alternated by a slow twitch or two, imitating the action of a wounded minnow. Strikes often occur while the lure is settling.

Split shot or a sinker or two usually are needed to quickly take light baits or lures deep, but no more should be used than is needed to do this. A fish mouthing a bait will drop it if he feels too much resistance from leads. Diamond-type jigs, with hooks removed, often are tied into the leader or between leader and line to act as attractors as well as sinkers when baits are used. Sinkers can be scraped to brighten them.

After jigging a hole for a few minutes, this method is given up if no action results. The hole then is baited with a still set, probably using a minnow fished close to the bottom. A split shot or two usually will be needed to keep it down. After baiting the hole the fisherman goes on to the next one and jigs it, proceeding to new holes in the above manner.

Several tricks are helpful. One is to cut two holes fairly close together and to jig one while a still set remains in the other. The jigging often attracts fish to the bait in the nearby hole.

Another trick is to attract fish by chumming, or priming, a hole. This is especially useful when several holes are near each other. Chum includes salted shiners (one or two being dropped into the hole from time to time), crumbled egg shell, chopped fish bits, oatmeal, canned corn, split peas, cracker crumbs, rice, white beans, and fish scales or guts. Since most kinds of fish need to be scaled, this might as well be done over the holes so the scales can filter down to attract more of them. The sooner fish are cleaned, the better they will taste. If the weather isn't uncomfortably cold, why not do it on the lake and use the innards as priming? A taste of blood in the water often draws big fish to the bait.

Still sets should be given frequent attention. It helps to go from hole to hole and to raise each bait a few feet, then let it flutter down. This often attracts strikes. Baits should be inspected periodically because a dead or nearly lifeless minnow won't interest fish.

Bobbers are helpful with still sets when rods and reels are used instead of tip-ups. Rod stands, which can be made or purchased, hold the rod angled upward with its tip over the hole. Bobbers are useless in very severe weather but, when holes freeze over very slowly or not at all, there are very small ones made of sponge rubber which can be squeezed to break off ice film. These are available in several sizes and, like sinkers, a variety should be carried so one can be used which is barely buoyant enough to support the bait. The little fish must be hooked when the slightest motion is noticed. There's a knack to it which, once learned, can quickly fill the skillet.

Fishermen who forget their tackle have no problem if they can borrow a few yards of line, a couple of split shot, and a hook. Cut and trim a limber green branch about 4 feet long and split the end so a loop of line can be pinched lightly into it. Tie the line to the pole with about 10 feet of slack extending to the loop. Support the pole by embedding its end in snow or mush ice, as shown in the drawing. Bait the hook, preferably with a small live shiner, adding one or two split shot on the line to keep the bait down. A fish which takes the bait will pull the loop from the pole's tip and the slack line should give it time to swallow the bait.

THEY DON'T HIT ALL THE TIME!

Successful ice fishermen have a combination of persistence, knowledge and patience! Regardless of whether or not one believes in Solunar Tables and similar theories, there are major and minor periods when fish go on the feed. Bluegill fishermen know that when bluegills hit they do it all at once, the feeding period lasting for a certain time. Fishermen after lake trout and similar species know that they cruise deeply in winter, but that they also come up on the shoals and around the ledges about twice a day to feed. One period usually is early in the morning. Hours around noon probably will be dull. A second feeding period should occur in the afternoon, about 2 o'clock or later. No one can predict when these periods will occur, but be ready for them.

In mid-winter, as well as in summer, underwater springs can provide bonanzas most of the time. Subterranean water emptying into a lake can be in the 40's or 50's, while that in the rest of it is in the 30-degree range. This, to fish, is like our coming out of the cold into a warm kitchen for dinner, because baitfish also collect there, and the oxygenated water can be lush with grasses. Use a depth thermometer to locate such places in the summer, and mark them down for winter fishing.

Ice fishing methods for the various species are so varied from place to place that it is impossible for a chapter in a book to cover them all. Experts have their own ways of doing things, and don't need any help. The basic suggestions just covered may, however, be of value to novices and to casual fishermen who want to enjoy an exciting and exhilarating day outdoors when the ice is safely thick.

Select a lee shore where the sun shines down. Take along a few congenial friends and something hot to eat and drink. Cut a few holes, set up the still sets, and try some jigging. It's sure to be an enjoyable day, and one thing is certain: the fish won't spoil.

11

FRESHWATER FISHING

Who can blame beginners for being bewildered when viewing the myriad lures offered by tackle dealers? Millions, bought on impulse rather than for reason, clutter tackleboxes from coast to coast. Of course there are some beauties in every bunch, and a few hot innovations make their debuts every year, but many that still are the best were old standbys when this author was a boy. We don't need as many as we might think, and no one can provide a list of essentials that will hold up everywhere. However, there are plenty of rules to go by, so let's look at the principal types and see why they are effective. This can save money and increase fishing success.

WOBBLING SPOONS

The story is that the wobbling spoon, or "wobbler," was invented by a man who dropped a teaspoon overboard and saw a fish rise to take it. The man sawed off the handle of another spoon and bored holes in both ends of the bowl. He put a split ring into the rear hole, with a treble hook on it, and added a snap swivel to the forward one. He then had a wobbling spoon, and all modern wobblers descend from this idea.

If any readers should try this, do the same thing to the handle. Such a lure may be too big for small fish, but it's a killer for larger ones such as striped bass and pike. Experiment with bending the handle into various degrees of curve until the desired action is obtained. Effectiveness usually can be improved by tying small bunches of dyed bucktail around the shank of the treble hook, or by using a large bucktail fly tied on a ringed-eye hook (which is one having an eye bent neither up nor down).

If we ignore color for the moment, wobblers fall into four basic types: heavy and light (which means thick or thin in proportion to size) and slim or wide, or somewhere in between. Of course they come in various sizes, with proportionately larger and smaller hooks, selected according to the sizes of fish to be caught. If in doubt, the smaller sizes usually are better.

Heavy, or thick, wobblers have two purposes. Since they are compact they offer less wind resistance and so can be cast greater distances. Being heavy, they sink faster, and usually are used for fishing deep. Quite obviously a fast-sinking wobbler will give trouble when used in shallow water or over weeds.

Light wobblers are not intended for long casts. On windy days their lack of weight in proportion to size inclines them to "kite," or to be blown off course during the cast. They work best in relatively shallow water where the heavy ones might snag bottom. They are excellent when worked downstream in shallow, fast currents, and they usually are preferred for trolling.

Wobblers work best with a slow, fluttering motion in imitation of baitfish whose sides flash occasionally as they dart about in search of food. They are much less effective when fished too fast and then are inclined to spin, which twists the line. In slow, deep water an ideal use is to cast them out and to let them sink on a tight line. As they flutter toward the bottom, fish may flash in to take them. If not, let them sink to, or near, bottom and then jig them up a few feet, repeating the process during the retrieve so they cover different depths. In faster, shallower water, try to keep them drifting and flashing with the speed of the current. They are much less effective when the line, swept into a bag by a stream's flow, causes them to be pulled so fast that they will "whip."

In these many sizes and thicknesses wobblers vary in shape from the familiar teardrop design (like the famous Dardevle) to oval patterns (like the Johnson Spoon) and to irregular patterns cut into the outlines of baitfish. Readers who are inclined to be technical might say (correctly) that size and general shape should resemble those of the baitfish where they are being used. The author would add that the amount of flash of the lure, and its action, are even more important.

The pressure of water on the concave surface of a wobbler is what throws it from side to side. A deeply dished one will fish more erratically than a shallow cupped one. Many wobblers (like the Dardevle) are dished more deeply toward their wider rear and have an additional slightly bent planing surface at their front to increase erratic action.

In general, fish take lures because of their lifelike form, flash and action. In selecting wobblers, size is more important than form. Flash and action are of major importance and color (except for flash) may be overrated. Since manufacturers want to sell as many lures as possible, it is natural for them to provide the widest range of color combinations.

Flash is what attracts (or repels) gamefish, since it is the simulation of the shimmering sides of baitfish. It is easy for fishermen to use lures with too much or too little flash, but this shouldn't be a problem because we know what we are trying to represent. The relative brightness of the day influences the amount of flash of a lure in the water, so it must be considered. If one uses a bright, new lure on a sunny day the lure's flash (like a mirror reflecting the sun's rays) may be so great that fish will be repelled by it. Conversely, fish may not see a dull lure from very far on a dark day or in discolored water. Give them reasonable flash to suit conditions; brighter on a dull day, and duller on a bright one.

The water's depth also enters into it. When fishing a wobbler deep, especially when the water is somewhat discolored, select a brighter lure. Many experienced anglers allow their wobblers and spinner blades to tarnish. They may use them that way on bright days, and they may use a polishing cloth (like crocus cloth) to burnish them to the desired flash when need be. Lacking a polishing material, a bit of mud will do. Metal lures usually are lacquered to preserve their brilliance. This mainly is to keep them looking new in the stores. The lacquer will wear off in time, or it can be removed with preparations like nail polish remover.

As for when to use what finish-nickel, brass, or copper-nickel is probably the best for dull days and brass for brighter conditions. Painted-on colors are of doubtful value. The oft-imitated Dardevle sells best in red, with white stripe. Some prefer colors like green or black, but many will say it makes little or no difference. The underside of the lure is bright metal, and its degree of brightness seems to be what counts most.

A tip on luminous or fluorescent finishes may be helpful. It has been proved (to the author's satisfaction, at least) that a bit of fluorescent material on a fly (such as its tag) attracts strikes. If wobbling spoons don't come that way, a stripe of fluorescence can be painted on. The intensity of fluorescence decreases as sunlight decreases. It is easy to put on too much. Little or none is needed on bright days, but more can be used on dull days or when fishing deep.

An advantage of wobblers of the right size and weight is that they can be fished erratically. They can be jigged deep as described above. They can be fished upstream in deep pools to dredge bottom. They can be fished downstream and, by letting out a bit of line from time to time, can be worked around rocks and other obstructions which provide resting places for fish. They can be cast cross-stream and allowed to swing with the current, bringing them in or letting them out a bit during this process to reach good runs and holding water. Learning how to select and use wobbling spoons properly is an important key to catching fish! If in doubt as to size, use the smaller one.

Using chunky metal spoons for deep jigging has taken on added importance with the development of flasher-type depth sounders and the advent of chart recorders. These electronic devices are expensive, but they pay off handsomely in finding suitable structure in lakes where fish congregate. In fact, jigging spoons bounced off deep dropoffs have accounted for more than their share of winners in professional bass fishing classics. Mr. Champ (or Kastmaster) and Hopkins lures, illustrated here, are examples. The Bomber Slab Spoon and Luhr Jensen's Krocodile are others. They are narrow, flat or rounded, and thick; usually cut from stainless steel or molded from lead. They sink rapidly, silvered finish glinting and fluttering in whatever light there is, even in the semidarkness of 20- to 40-foot depths, and produce strikes with amazing frequency. The trick is to know where and how deep the fish are and to work the lures near them from a position directly above.

There are other tricks of this trade in addition to those mentioned above. Tie in a small swivel about a foot and a half above the spoon to prevent line twist. Use a strong, stiff rod to provide a lot of snap to the relatively heavy lure. Let it drop to bottom, or to whatever depth seems appropriate, and jerk it upward, with every jerk a little higher, for 10 or 15 feet, then drop it again and repeat the process.

Why do so many different species of fish, including striped bass and Pacific salmon, strike such lures so enthusiastically when they bear only slight resemblance to baitfish? A silvery jigging spoon fished actively seems to resemble a flashing baitfish with the fits, and its very activity prevents fish from determining its shape or appearance.

We know that gamefish will pass up ordinary bait to strike at something which seems to be in trouble. That is a law of nature. No further evidence seems necessary because the value of jigging spoons has been proved time and again. If I am so lucky as to be near the edge of a school of surfacing stripers, I would use the same lure to cast into them and would retrieve it as fast as possible. Such action usually guarantees a heavy strike on nearly every cast.

SPINNERS FOR CASTING

Spinners for casting have a revolving blade attached by a clevis to a springy wire on which are strung a few colored beads or a weight, or both, and a hook. Since this rig is an attractor in itself it usually is not baited. It can be baited (normally with a worm), or a bucktail or streamer fly can be attached to it. Flies used on spinners shouldn't have a wide wing because the purpose of a spinner and fly combination is to imitate a baitfish. They should, however, have pronounced hackles to give added pulsating action.

Spinners differ from wobblers in several ways. They should be fished at constant speed, perhaps with an occasional short pause, rather than very erratically, as a wobbler should be. This speed must be enough to make the blade revolve actively. They can be cast cross-stream and allowed to swing with the current. They can be cast downstream and allowed to hang in the current where, by letting out and taking in line, they can be worked around holding positions for fish. Unlike wobblers, spinners essentially are topwater lures although there are weighted ones which can be fished deep. They are excellent for near-surface fishing in lakes and streams. Another advantage is that they can be cast in to the shoreline where fish often lie and where a wobbler would be more likely to become snagged in the shallow water. In fishing upstream and across stream, hold the rod high to keep as much line as possible out of the water to prevent the lure from whipping, an important point in fishing both wobblers and spinners. A pause in retrieve after casting allows the lure to sink.

In fishing a pool, it usually is better to cover its tail and lower sides before fishing the inlet because hooked fish can be coaxed downstream for playing and netting probably without disturbing those farther up. In fishing deep water a few split shot can be pinched on a foot or more above the lure where they won't interfere with its action. A few small split shot are better than one or two bigger ones.

Spinners are equipped with a wide variety of blades, usually attached to a clevis which is strung on the wire shaft to make them spin easily. Some of the blade designs are the Figure-4, Willow Leaf, Kidney, Indiana, Colorado, June Bug and Fluted. Some spinners are fitted with small propellers rather than blades. All of these fall into four basic types. The standard type, such as the Kidney and Indiana, is attached to the shaft by a clevis which helps it to revolve freely. The Colorado often is hitched to a ring which is attached to a swivel and perhaps also to a second trailing swivel holding a dressed or undressed hook so that the swivel (s) allow rotating action. In spinners of the June Bug type the shaft passes through the eye and also through a brace extending from underneath to hold the blade at a fixed distance from the wire. The fourth type is the propeller, which obviously is double-bladed to rotate around its center on a shaft. Every spinner addict has his own preference in blades, but there are definite rules for guidance.

Since fish are sensitive to underwater vibrations, the amount of vibration emitted by a spinner can be an attractor for them, as it has been proved to be with other lures such as "sonic" plugs. As proof of this, we know that fish can locate, and will strike at, spinners in complete darkness. (This could be a tip for fishing in very discolored water.)

The shapes and thicknesses of spinner blades have effects on their actions which are of interest. Roundish patterns (like the Colorado) rotate nearly at a right angle from the shaft; more elongated ones more or less at a 45-degree angle from it, and long, narrow ones (like the Willow Leaf) very close to the shaft. The heavier or thicker the blade is, the slower it will spin. Some heavy blades of poor design may drag and have to be twitched into motion. A good spinner starts revolving the instant one begins to fish it. Roundish blades provide more water resistance and so are better in slow currents and in lakes and ponds. Long narrow ones provide minimum resistance and can be used in fast water without usually causing the lure to spin. Thus there are times when slow spinners should be used, and others for faster ones. Sometimes spinners are productive when wobblers aren't, and vice versa. Certain ones may be favorites one season and nearly forgotten the next.

When either spinners or wobblers don't seem as productive as they should be, action often can be increased by adding one of various kinds of embellishments to the hook. Try a bright worm or nightcrawler, or part of a plastic worm with the cut-off end strung on to cover the shank but to expose the bend and barb. Preserved pork strips work well, either cut into a "V" or any sort of single strip not too long or too big for the hook. Even strips of red balloon rubber cut into a fluttering skirt, or strips cut from a discarded white kid glove, will add action to the lure.

SPINNERBAITS

As an inquisitive angler interested in trying anything that seems to make sense, the author admits to an error about spinnerbaits. They didn't seem to make sense! Why would a self-respecting bass, or any other fish for that matter, bother to hit such a contraption when there are more lifelike goodies to masticate, such as artificial flies, plastic worms or most any kind of plug?

So the author was completely dumb about spinnerbaits until he had had a few hours to kill on a dock amid the Thousand Islands, which separate New York State from Canada on the wide St. Lawrence River. There were middling-sized bass off the dock, and they showed a mild interest in the various flies, poppers and plugs. Some spinnerbaits in my tacklebox (this because someone gave them to me) caught my attention. Wouldn't this be a good time to prove they don't work? The author decided to find out. The result was a revelation! Bass banged the funny lures with supreme gusto, but the thing that surprised me most was that they always took the hook and not the spinner.

Spinnerbaits come in a myriad of types of which a few suit most purposes and belong near the top of your list. As the illustrations show, they resemble an open safety pin with a spinner or two on the upper arm and some sort of a weighted lure on the other. Being no expert, I'll not try to recommend what should be on the other, as it can be most anything. I'll opt for a jig with a hula skirt, but only because I've had best luck with it -and that's because I use it the most. Everyone has opinions, which is why there are so many types. You'll have yours, after you have tried a few.

Note that there are three ways to connect lure and line: the inside loop (like a safety pin), the outside loop and the open loop, all with no swivels. In use check often to be sure the line isn't fouled in them. This rarely happens with the open loop. Attachment by the Improved Clinch Knot is usual.

The upper arm ends with one or more spinning blades, such as the popular Colorado, in silver or gold plain or hammered finishes or (more rarely) in colors. Some spinnerbaits have two upper arms, perhaps on the theory that, if one is good, two are better. Maybe so, because their purpose is attraction by flash and sonic vibration.

The lower arm consists of a jig whose weight (usually between ¼ and ½ ounce) keeps the lure on an even keel with the upper arm always on top. The hook often is dressed with a colorful shredded skirt of hair, rubber or plastic, sometimes with a plastic worm, plastic twister tail or a baitfish imitation. Choices of color are controversial; some users have definite opinions while others feel that color makes little difference. Yellow and chartreuse are popular.

Note that jig heads are provided in several shapes. Round heads are used in open water but may hang up on brush. Pointed heads are more apt to ride over obstacles.

The spinning of the spinner blade (or blades) signals by rhythmic throbbing that the lure is working properly. For this reason high-grade lures have blades connected to them by American-made ball-bearing swivels. It pays to check this feature. You can use the countdown method to let the lure settle before steadily fishing it in, or you can "buzz" it in by reeling fast enough to keep the lure near enough to the surface to make a slight wake. If fish don't come up for the lure, try the deeper method. It is more fun to see them strike near the surface!

With a small investment in spare parts, you can make your own spinnerbaits and enjoy the added satisfaction of hooking fish on homemade lures. The basics of lure making are explained at the end of this chapter.

SPINNERS FOR TROLLING

So far, we have discussed spinners, which are intended to be cast although they also can be trolled. Another type, usually unweighted, is fastened to the line's end and a streamer fly or bucktail, or a baited hook, is attached to it for trolling only. These rigs come in all sizes, with one or more revolving blades. Glass beads often are strung on the wire on which they are rigged. The flashing of these spinner combinations attracts fish to the lure, very often with great success. Fish never seem to strike at the blades, even though the lure may be very close to them.

Some spinner rigs for trolling have as many as six large revolving blades; these usually being trolled very deep in summer for cold-water species such as lake trout. An advantage of all such rigs is that they work well near surface or at any depth. Keel leads help them to run deeper while lead-cored or wire lines can take them even farther down. Two tips may contribute to success. Troll the rigs slowly; probably more slowly than may seem correct. If in doubt between selecting a larger or a smaller rig, use the smaller one. Fish can see these slowly flashing blades even in very murky water, and they usually will be attracted to whatever lure is offered.

Do-it-yourselfers need little equipment or ability to make their own spinners or to refashion and rebuild those they find. Ways of doing this are described later in this chapter.

PLUG FOR EVERY PURPOSE

Many fishermen, especially bass addicts, treasure voluminous tackleboxes fitted with several cantilever trays and a well bulging with hundreds of plugs. These are fun to own and to paw over, and some of them are useful to fish with. The trouble is that most of these collections are so mixed up that, when one type of lure is needed, the whole business has to be inventoried to find it, and in the search an even more suitable goodie may be missed. The suggestion is to put all plugs of a type together for easy finding, which would include the four types we will discuss. If we want a plug of the floating-diving type, for example, then we merely look in the appropriate compartment.

How many plugs or similar lures does a fisherman need? If we agree there are four principal types, perhaps three varieties of each type should suffice, which makes a dozen. Probably we'll want each one in a light and a dark color, which makes two dozen. Perhaps we'll want each in two sizes, which adds up to forty-eight. That should be enough. But we'll also include some spinners, jigs, wobblers and so forth, and we'll acquire extra plugs in various sizes, types and colors. Hence a fisherman who owns a tacklebox containing about a hundred lures doesn't think he has too many.

Obviously anyone just starting in can use some advice about the sequence of his acquisition schedule before he buys an excess of this and perhaps none of that. If we divide plugs into four types they will include surface plugs (or floaters), floating-diving plugs, sinking plugs and bottombumpers. Let's discuss some of those in each class to decide what may be needed:

Surface Plugs

Since these stay on the surface they are the favorites of many because the smashing strike can be seen when a fish boils up and hits. Many have concave heads which pop a scoop of water when a quick twitch is given to them. These poppers, of which Arbogast's Hula Popper is a good example, have heads of various sizes, with shallow or deep cups, to provide a variety of popping noises and splashes. Some, like the one mentioned, are equipped with plastic skirts for added motion. Others, like Arbogast's Jitterbug, have wide cheeks providing both a spluttering and wiggling effect. Some of these "splutterers" are minnow-shaped, with fore or aft propellers, or both. Other minnow-shaped plugs have small lips which provide seductive wriggling motions. Many simulate frogs and other fish foods, as well as minnows. These are merely examples, and most are used for bass. In addition to wood and hard plastic models, many of the newer ones like Burke's Pop Top and Flex-Plug are offered in colorful soft plastic. The idea here, which is a good one, is that fish will hold them longer, thus increasing chances of being hooked. All these are in such a profusion of varieties that the fisherman feels like a kid in a candy store; hard put to decide among them if he limits himself to an initial three. Of course he will want more than that, but three should do for a start.

Floating-Diving Plugs

This is the type which floats at rest but which dives when being retrieved and which rises to the surface when the retrieve is stopped. Good examples are Creek Chub's Pikie Minnow, Bomber's Speed Shad and the expensive balsa-wood Rapala, which has been widely imitated by excellent lowerpriced plugs such as the Rebel. All these accurately imitate baitfish in their long, slim shapes and lifelike silvery surface markings. Some feature the ability to give off underwater vibrations which tempt strikes by sound as well as sight. These may be called "sonic" plugs. Sonic plugs usually have no lips, and the attachment eye is on top of the front of the body. This gives them an intensive wiggle that throws off sound vibrations. These lures should be fished fast.

Floating-divers come in two basic types. One is the shallow diver, usually identified by a small lip. The other is a deep diver, equipped with a long scooped lip. The latter, on retrieve, can go down as far as 10 feet.

Thus, if we are to be limited to an initial three, a shallow diver, a deep diver and one of the sonic type would provide the widest variety.

Floating-divers come in two basic types. One is the shallow diver, usually identified by a small lip. The other is a deep diver, equipped with a long scooped lip. The latter, on retrieve, can go down as far as 12 feet or more. These deep divers, usually known as *crankbaits,* have highly buoyant bodies and long, broad plastic lips. They float at rest but dive sharply on retrieve when the big lip digs into the water. Long casts with lightest sensible lines and fast retrieves get crankbaits down deep where big bass rarely see lures and thus are more vulnerable to them. Proper use of crankbaits has won many fishing competitions. More will be said about them in Chapter 23.

Sinking Plugs

These sink after being cast and reach depths in excess of a few feet, but they are not intended to strike bottom. Most have large diving lips, as illustrated. If a floater made to dive deep, such as the previously mentioned crankbaits, gets hung up on an obstruction a slack line often induces it to back off and rise to the top; an advantage not found in sinking plugs.

Some of the sinking plugs are of the "countdown" type, made to sink at known depths per second. For example, if our testing of water temperatures indicates that the ideal depth is 20 feet, we would cast the lure out and count the seconds as it sinks. If it sinks at one foot per second, we would count 20 seconds before starting the retrieve.

Plugs are provided in a bewildering multiplicity of colors. Pale ones, often in exact baitlike coloration, do well in clear water and on bright days. Red-headed models with white bodies usually are successful when visibility is only moderate. Black or dark colors do well on dark days or late in the evening. Admitting that color preferences from year to year are as fickle as fashions, the redhead with white body is most popular as this is being written. The other three most popular colors or combinations are natural scale, yellow and black. A rule is to use darker colors the deeper plugs are being fished, with black or purple on the bottom.

Plugs with white bellies don't have as much going for them as one might think. Divers in World War II observed that the bottoms of ships most difficult to see from below were the white ones. Probably fish looking up at plugs can see the darker bellied ones more clearly.

Finally, let's give a word of praise to the small plugs used on light tackle. They are extremely productive when fish aren't deep, and they are fun to use. Improvements in light spinning tackle and the greater strength of fine lines have made the use of the midget plugs very practical. We can take at least as many fish with small plugs as with bigger ones, and any fish with a small plug in his mouth is more fun to handle than one hooked with a bigger one.

These midget lures in the 1/8 to 3/8 ounce class often are exact miniatures of popular larger ones which have stood the test of time.

They land with a tiny "plop" instead of the heavier splash of their bigger brothers. They include darters, poppers, splashers, divers and wobblers for surface to medium-depth fishing for walleyes, trout, crappies, pickerel and bass. Remember that some fish, such as crappies and smallmouth bass, have mouths which are too small for some of the bigger lures. Also, miniature lures can be worked around pads and obstructions more easily than the bigger ones can.

If plugs are to be changed frequently, a small snap swivel may be helpful, but most small lures work better without one. Snap swivels rarely are used on surface lures.

Miniature plugs are not intended for use only on ponds and lakes. They can be deadly when worked properly on rivers and, when big fish are present, even in small streams. They often will take sophisticated big brown trout when nothing else will.

In recent years lures (especially plugs) that make a noise have become increasingly popular because they add the attraction of sound to that of sight. Of course *propeller baits* have been around for a long time; usually plugs with metal propellers fore and/or aft which make a splashing noise on retrieve. *Buzz baits,* such as spinnerbaits, do the same by means of revolving single blades, of which there may be more than one on a lure. Sonic items such as shot are added to the bodies of plugs to provide ticking, clicking or rattling sounds known to attract fish. I presume that there are so many aquatic sounds in nature that fish don't distinguish between minor ones even though they don't seem natural to us. I've never heard a baitfish or a frog, for example, tick, click or rattle; nevertheless some gamefish, especially bass, seem to go for such vibrations.

How to Vary Plug Action

Basically, there is only one best way to fish any plug, the way that makes it swim best or that gives it the action common to whatever it is supposed to represent. A plug imitating a baitfish usually should be made to swim like one. A plug imitating a frog would be allowed to sit on the water until the ripples of the cast have ceased, and then it would be made to act like a frog floating on the surface and occasionally making slight swimming motions.

But, when a plug doesn't attract takers by its natural movement, we have two choices. Either change to something else, or add unusual action to whatever is on the line. Sometimes fish, especially bass, react to a slow or to a medium retrieve, but, at other times, they respond only to a fast one.

When the proper action of a baitfish-imitating plug doesn't work, try fishing it more slowly and lazily. Then work it fast-perhaps so fast that it skips and splashes over the surface. There are times when bass will ignore slow plugs but will chase and attack something going unusually fast. A plug made for floating and diving doesn't always have to be fished that way. Let it lie on the surface for a minute and then give it a quick pop or two by jerking the rod tip.

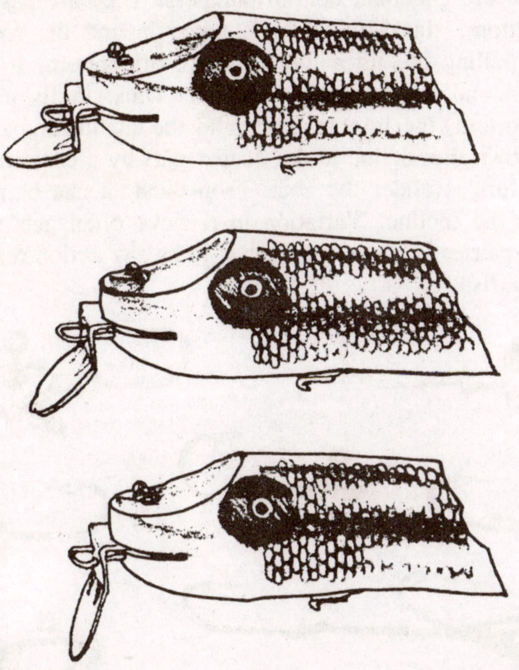

Plug action can be vaired by adjusting the diving lip. Floating-diving plug (top) will dive deeply on retrieve with lip in normal position. To make it dive less deeply, bend down lip as shown in center drawing. To use it as a surface splasher and popper, bend the lip down completely, as in the bottom drawing.

Some plugs have little propellers fore and aft and commonly are used as splashers to create a commotion when fished along the surface. At rest, the tail probably will sink, with only the tip showing. This looks something like a frog, so it can be left lying motionless for a while and then made to bob and cause slight ripples. A bass may be eyeing it from down below, and may come up to take it then, or when it finally splashes away.

These instances illustrate that, when orthodox methods don't work, it may be productive to try something else. Bass usually react to lures because of hunger. Like other fish, they also react for different reasons. They may attack a plug because its unusual action makes them angry; because they are curious to find out what it is, or because the way it is being fished induces them to grab it in the spirit of play.

JIGS

These are excellent bottom bumpers. They are cast and allowed to hit bottom, slack line then is taken in and the rod is twitched upward, pulling the jig a foot or more off bottom. It is allowed to sink again and the action is repeated. Thus, in fishing a ledge or sloping bottom, the jig can be cast to the shallower part and can be hopped down the incline to the deeper part by a continuation of jerks until the lure is under the boat. From land, it can be cast deep and fished up the incline. Variation in retrieve often gets results: quick twitches, pauses, slower and longer jerks-any action that simulates a frantic baitfish.

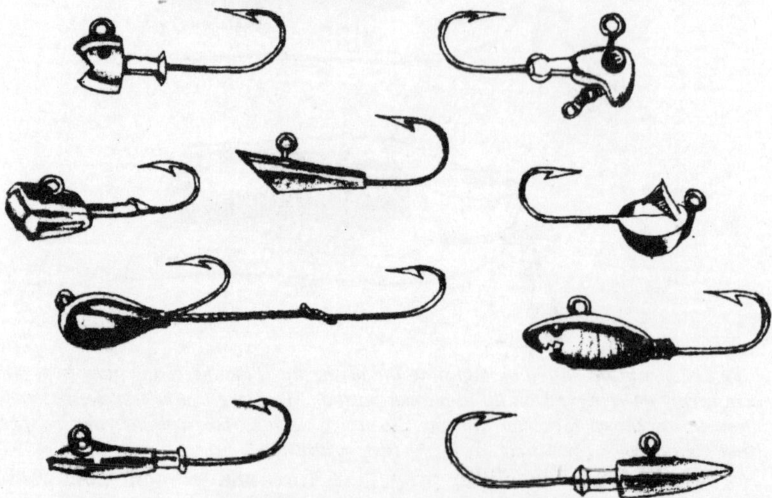

An assortment of typical lead-head jigs.

Jigs don't look very baitlike, but when they are energetically fished the illusion fools gamefish consistently. Two jigs fished together often do better than one, probably because they simulate a small school of fish. Tie in a dropper 18 inches or so from the line's end, preferably by making a Blood Knot extension about 8 inches long. Tie the heavier jig to the line's end and a smaller one to the dropper.

Jigs are available in a wide range of weights or sizes and with various shapes of metal heads. These are so made that the hook's bend curves upward with the jig resting on its head to help prevent snagging. Oval-shaped jigs with vertically flat heads are used mainly in currents because the flat heads help the current to drift them. Some jigs are keeled more or less like the prow of a boat so they can be skidded along the bottom. Others for general use have roundish or bullet-shaped heads.

Most jigs are dressed in one way or another, usually with a skirt of hair, nylon fibers or feathers. Those dressed with marabou feathers are very popular because of their greater fluffing action when fished. Undressed jigs can be baited, such as with a live worm, or can be fitted with a whole or part of a plastic worm. Pork strips or something similar often are added to the hook. It is economical to buy undressed jigs and to dress them as one pleases. It also is easy to paint or repaint them by dipping their heads in any color of enamel or plastic paint.

Sharpness of hook is very important in jig fishing. The triangulation method of hook sharpening is recommended.

Jigs are obtainable in many color variations, but color seems to be less important than the size and type of the jig and the way it is fished. Colors such as white or yellow, or light colors in combination, can simulate those of baitfish. Darker colors, such as red or black, are used for visibility and contrast, such as contrast with the color of the bottom.

PLASTIC WORMS

Every experienced plastic-worm fisherman is a specialist who knows there are rules of the game, many of which can be broken. In other words, one goes by the rules until he develops his own system. Most experienced worm fishermen will agree that the method is more deadly, especially for taking big bass, than using any other kind of bait or any artificial lure.

Big worms 8½ or 9 inches long are preferred when going for the lunkers on the theory that fewer of the smaller fish will bother with them or that fewer smaller ones will be hooked. Hooks used for big bass usually are the Sproat pattern in sizes from 3/0 to 510. When fishing for crappies and similar smaller fish, sizes go as low as number 2, with the worms proportionately smaller.

Each expert also has his favorite way of hooking his worm. Some of the favorite methods are shown in the accompanying illustrations. One of the more ingenious of these is the Mean Machine jig head and the patented weedless worm hook-up. This jig head has an action-lip which provides a diving effect coupled with a sinuous swimming action giving the lifelike motion of a snake, worm, leech or eel-all favorite bass foods. Note that it is available in four sizes in kits with plastic worms to suit each size.

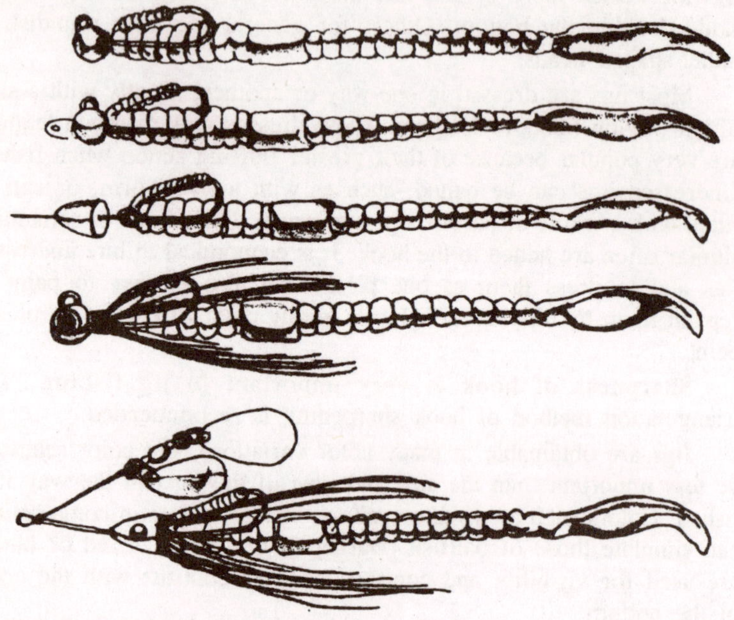

Burke Lure's Mr. Macho Weedless Worm can be rigged in various ways. From top, three Texas-type rigs, with a jig and a spinnerbait.

I particularly like the Mr. Macho weedless attachment on these Burke plastic worms because the small-diameter "arm" releases the hook point easier than if it were embedded in the worm itself, and therefore does less damage to it, thus allowing longer use. Stringing the worm on the hook is equally simple. Press the hook point into the head of the worm with the shank over the arm. Continue threading through in the usual manner, bringing the hook out near the end of the arm. Press the hook point and barb into the arm and you have a weedless hook which frees itself easily when a fish strikes. Part of the arm can be cut off if necessary.

An alternative to the jig attachment is to use a worm-holder hook with a conical sinker strung on before it. This is called a Texas rig. A Texas rig is a method of weighting a plastic worm and making it weedless. I like to crawl this rig over lily pads, let it sink into open spaces, and continue the crawling and swimming retrieve until a fish hits. It is fun to see them burst through the pads to take the lure.

Any tackle using monofilament line testing between 10 and 20 pounds can be used for fishing plastic worms if the reel is able to pay out line freely. Fishing without drag is necessary if the bass is allowed to run with the worm while swallowing it, which usually is the case.

First, let's get acquainted with how the worm acts by casting it nearby into shallow water where it can be watched. The lead will take it to the bottom but since the worm is a floater, its tail will rise and wave naturally in the current. Now pull the worm very slightly. The lead will kick up a little mud and the worm will actively appear to "walk" and to be grubbing along the bottom. That is how it should be fished-very, very slowly.

Three types of hooks for rigging plastic worms. At left is special worm-holder hook; center, a bait-holder hook especially bent for weedless plastic-worm fishing; right, an ordinary long-shank hook.

Now for a cast where the bass or crappies are supposed to be. If a fish hits the worm while it is sinking, the strike should be very hard and the hook should be set immediately. When the worm is on the bottom, fish it as previously described. In bringing it in slowly, with many pauses, a gentle nudging pressure or a few sharp taps may be felt. You must be able to distinguish the difference between when a fish mouths the worm and when it is pulled against some sort of obstruction. If in doubt, assume it is a fish. Point the rod toward the bait and let line run freely off the reel. If it's a fish it will make a run while mouthing and swallowing the worm. Let it go, with no drag from rod or reel; if it feels drag, it probably will drop the bait. Give it plenty of time to swim away; even as long as a minute or so. Then hold the line until tension is felt. When it is, put the line under control of the reel -and strike, hard.

Some fishermcn strike immediately upon feeling a fish mouthing the bait. Others feel that the fish may take the bait by the tail and that it must be given time to swallow the whole thing before it can be hooked. The only way to learn is by experience. If underwater cover is thick, you should strike sooner. If the fish is heading for a brush pile or weeds or a stump bed, you should strike anyway. However, when a fish has the lure, it probably will head for open water.

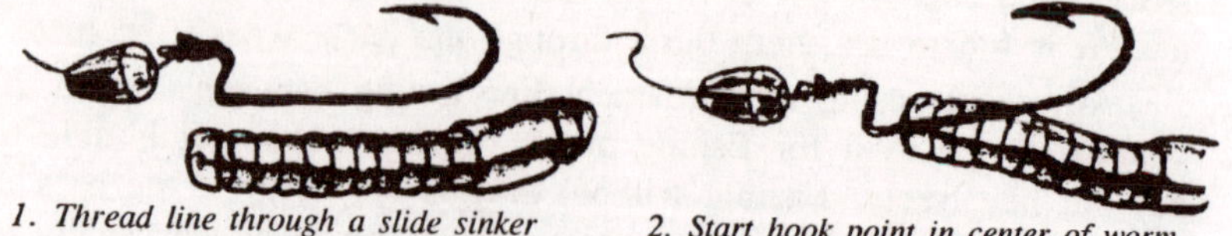

1. Thread line through a slide sinker and tie to hook, preferably by snelling method. Worm should be a floater.

2. Start hook point in center of worm head at a 30-degree angle. Push hook point through worm about ½ inch from head.

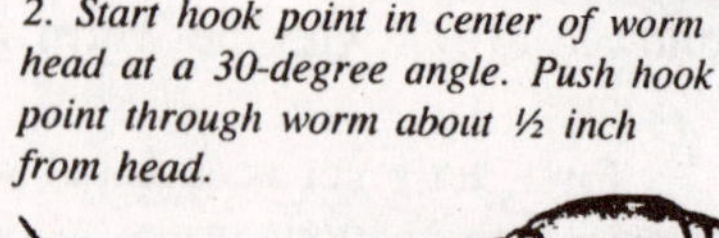

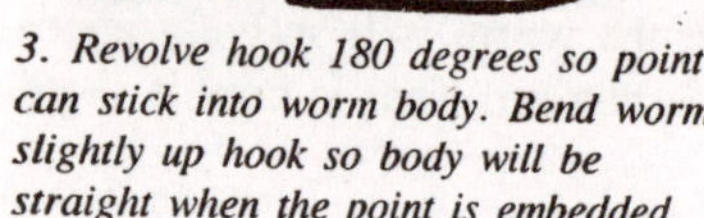

3. Revolve hook 180 degrees so point can stick into worm body. Bend worm slightly up hook so body will be straight when the point is embedded.

4. Push hook point almost through worm (1/16 inch from surface). This rig is weedless and will ride head up when fished.

Many fishermen think a double-hook rig is superior to a single. If a double is used, the lead hook is applied as described and the tail hook, snelled to it by 3 or 4 inches of strong monofilament, is hooked in near the middle of the worm. People who prefer single hooks think the tandem arrangement makes the worm act unnaturally, so fewer fish take it. People who like the tandem hook-up think it hooks more fish even though many will mouth it and drop it.

Many fishermen also use embellishments such as beads, skirts and spinners. Experts evidently agree that, as far as bottom fishing is concerned, they do more harm than good. Many plastic worms are sold with double hooks and other embellishments already affixed to them. If these are used they seem more adapted to near-surface fishing. Never discount the killing potential of a fancy plastic worm as it slithers through open spots amid lily pads or weeds!

Plastic worms are offered in almost every color and color scheme imaginable, and many are scented with preparations whose flavors seem to appeal to fish. While some experienced fishermen think that color makes little or no difference, most seem to prefer red or black. Purple and blue worms also are popular.

Finally, we should stress the importance of fishing plastic worms very slowly and naturally, but to keep them working. They won't catch fish all by themselves!

OTHER NATURAL LIMITATIONS

Beginners in fishing often are tempted to purchase lifelike imitations of natural fish foods such as tiny plastic baitfish, frogs,

1. *With knife or cutting pliers, 'cut ¼ inch off worm's head.*

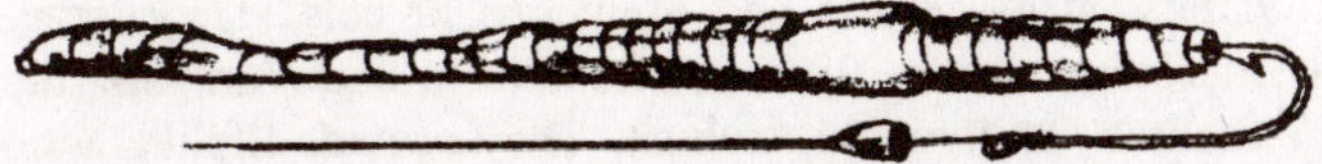

2. *Pss leader through sliding sinker and tie to hook. Push point of hook about ½ inch into center of cut-off head.*

3. *Bring the point of the hook out of worm ½ inch behind the head.*

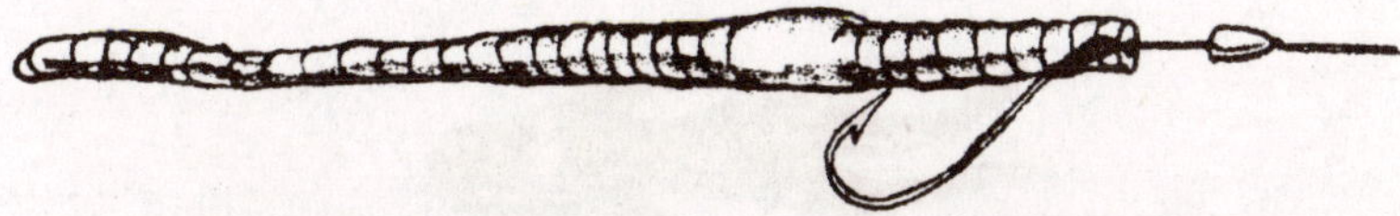

4. *Pull the hook back through the worm until hook's eye disappears into the head. Revolve worm 180 degrees on the hook until point of hook is up.*

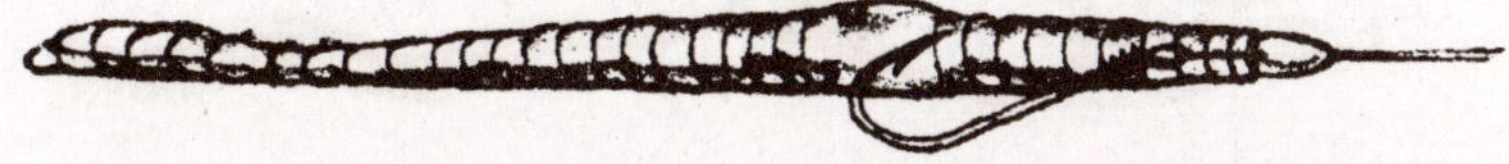

5. *Insert hook point into worm and, holding hook firmly, push forward on teh worm until the point is embeddded.*

grasshoppers, hellgrammites, crickets and so on. Since they are so realistic, how can fish refuse them? All these can be good lures at one time or another, but they may not provc as consistently efficient as you might think. If you want to present such lures to fish, why not use the real things, as discussed in the next chapter? Of course all of these fish foods also are reproduced artificially from hair and other substances for fly-rod use. Perhaps these are more productive because the substances they are made from provide a more pulsating and lifelike effect than their counterparts in plastic.

PORK RIND LURES

Why do some fishermen fish pork baits almost to the exclusion of everything else, and why do they catch so many big fish with them so consistently? These baits, or lures, come pickled in jars of brine, in a wide variety of sizes, colours and shapes including pork frogs, wigglers, chunks, strips, skirts and imitations of eels. They have a firm, soft, muscular texture fish seem to like to chew on, and they often come scented, or flavored, to boot. They provide lifelike action in tough cover when fished knowledgeably with weedless hooks, and some of them can be added to other lures such as spinners, wobblers and plugs, often with better results. The skin strips or chunks are so tough that one can be used all day or until the hook is lost. They also are cheap.

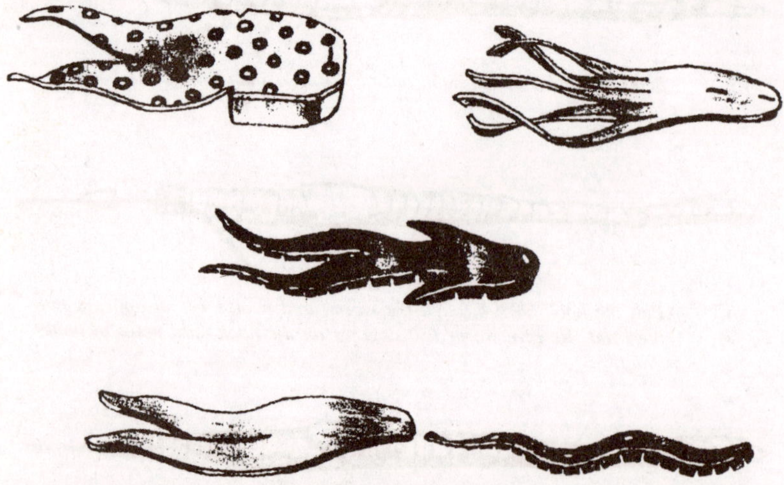

Pork rind lures provide lifelike action, come in variety of shapes, sizes.

Since these lures so often are fished in heavy cover such as pads, grasses and brush, we need a strong line to free the hangups which occur even with weedless hooks. The line should be 12- or 15-pound test, or perhaps even stronger, and can be used with any type of fairly strong casting tackle or even with the proverbial cane pole. In surface fishing the rod should be long enough to keep as much line off the water as possible. Big lures for big fish should have big hooks, as large as 3/0 or more.

Surface chunks which simulate frogs, pollywogs and lizards can be cast into avenues between clumps of grass, onto lily pads, or into open places between brush. Lily pad fishing can be especially exciting. Cast onto the pads; let the lure sit there for a few seconds, and then pull it off into the water. Leave it there for several seconds, and then give it a small twitch to start a few ripples. If nothing strikes, fish it in, over and between the pads in the same manner. One of two things sooner or later will happen. A fish will come up and smash it amid a geyser of water, or there'll be a wake from the side as one plows through the pad stems to reach the lure.

Eel-like strips as long as 9 inches can be fished like plastic worms. Since the rind is tough, perforations are stamped into the strips so weedless hooks can be inserted in the hole providing the desired length; the excess being trimmed off. If these can be cast without added weight for surface fishing, so much the better. Since an eel swims very slowly, make the lure slither along on or just under the surface amid the pads or weeds with a very slow crawling motion.

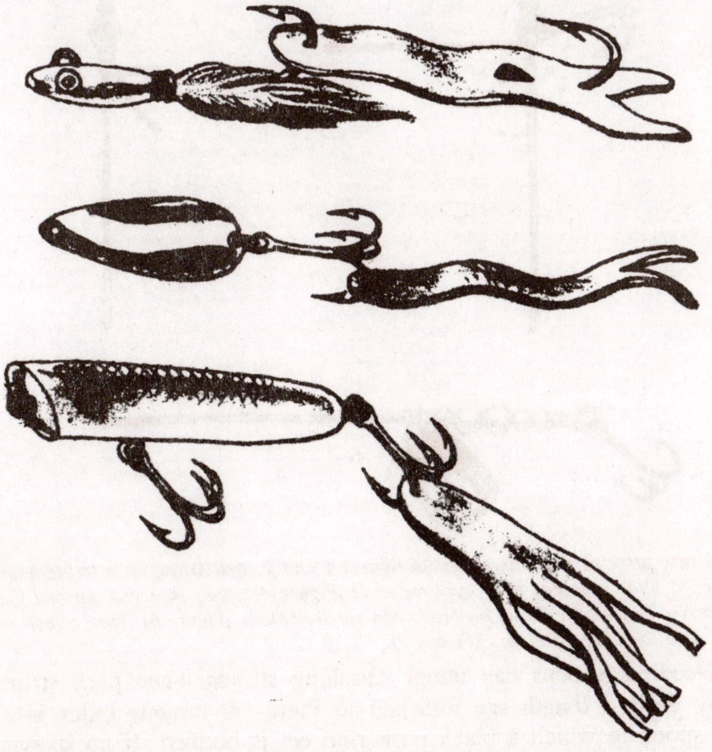

Pork rinds rigged on lures often bring strikes. Strong lines are needed to free hooks in heavy cover.

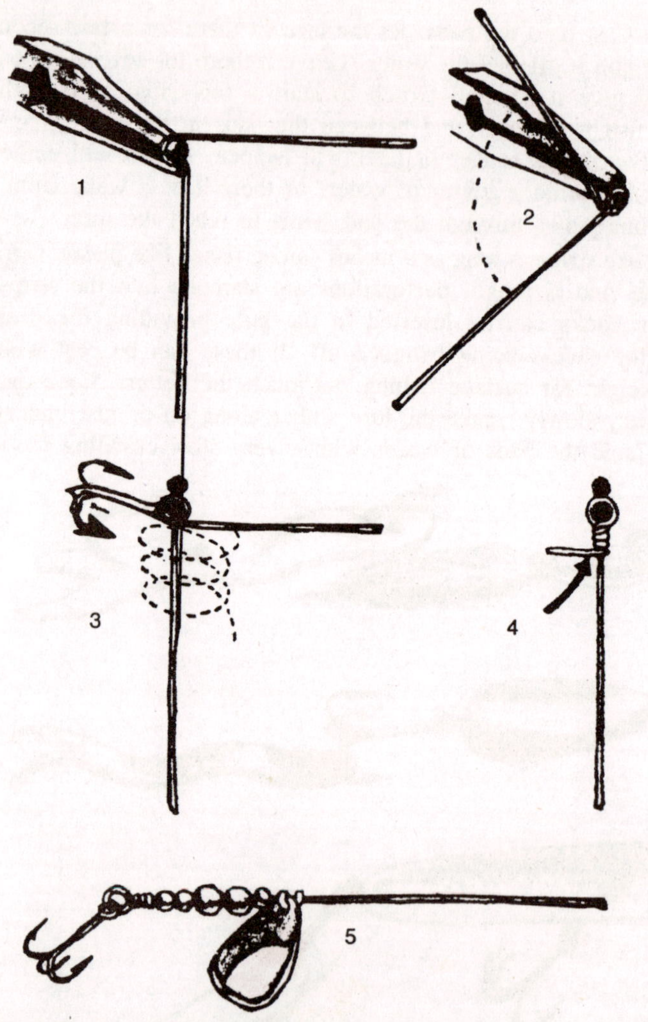

Bend a new piece of wire, as shown in figures 1 and 2, and string on a treble hook (Figure 3). Slide weighted body over the ends (Figure 4), bend shor end out and cut off at arrow (Figure 5). String on beads adn clevised blade (Figure 6). Turn a loop in end of wire, and cut off excess (Figure 7).

Weedless spoons can tempt smashing strikes when pork strips of any desired length are attached to them. A famous killer is a black spoon to which a black pork rind eel is hooked. If no spoons in the tacklebox are black, one can be sprayed with a pressure can of black automobile lacquer. Six inches is a good length for the eel.

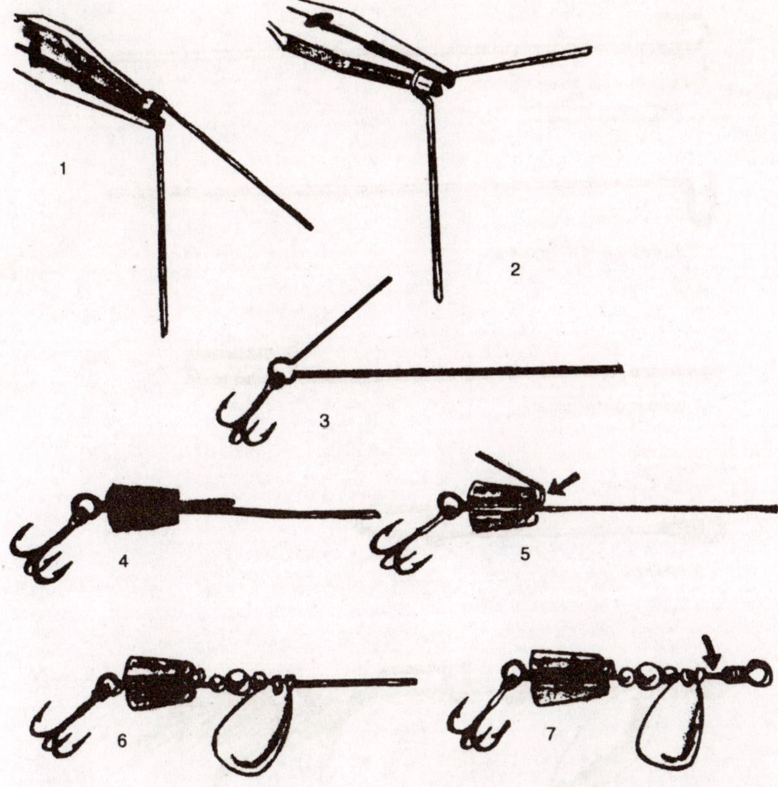

When forming the loop with pliers, begin as in figure 1, but then shift wire in pliers and bend until the end makes a 90-degree angle with the shank (figure 2). Slide on a treble hook (figure 3) and bend end around shank three times, maintaining the right angle in order to amke close turns. Cut off end closely (figure 4) and string on beads and clevised spinner (figure 5). Then twist another loop in end of wire.

Sometimes bright spoons with white pork strips do extremely well, or a pork chunk could be used instead of the strip.

Smaller pork strips add to the efficiency of spinners of various kinds. If the hook is a treble one, try three very small strips, one on each hook, for an enticing fluttering action.

Pill-shaped pieces of rind are perforated in their centers to fit over a hook's barb. They stay on the barb indefinitely, but slide up the shank when a fish is hooked. They make lures more weedless and, on small wire hooks, or jigs or spoons, are favorites for fishing through the ice.

Pork rind skirts or strips on the tail hook often increase the

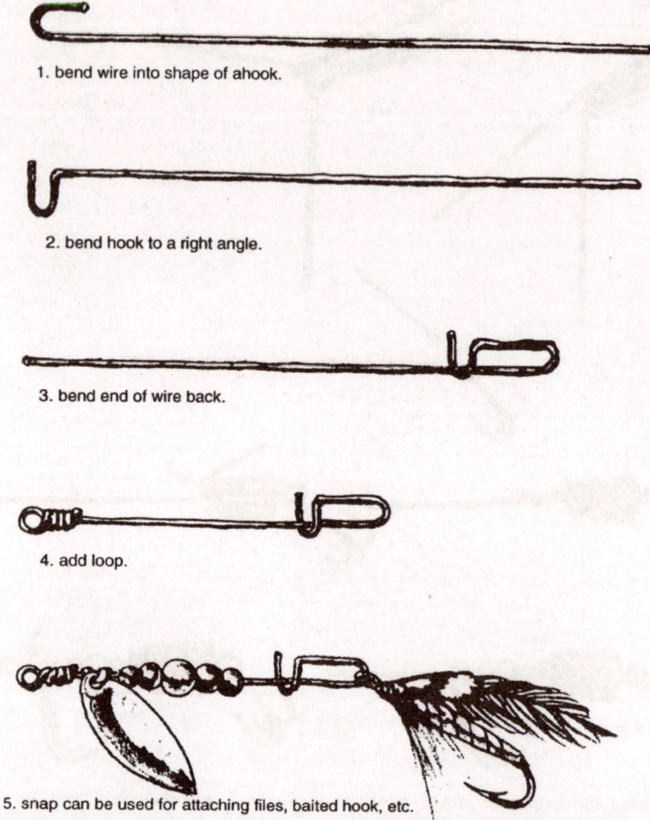

efficiency of plugs. If they are too large or improperly connected they can interfere with the action of the lure.

Since these pork rinds are preserved in a brine solution, lures used with them should be rinsed afterward. If the brine is spilled in a tacklebox, everything should be washed to prevent rusting.

Addicts of this type of fishing often carry various jars of several shapes, colors and sizes of pork rinds in a special tacklebox. After experimenting with them on weedless hooks and on various kinds of lures they settle on lucky combinations they find most effective. They learn by experience whether they should fish on top or deeper down, and they learn the action that gets best results with each lure and method. Non-addicts often overlook pork rind because these lures aren't attractive and may be somewhat messy. Try them until you understand their many uses. Pork rind lures are famous for hooking the big ones!

HOW TO MAKE AND MEND SPINNERS

Fishing lures are easy to make or to repair. Plugs can be put together from a variety of materials, from ballpoint pens to broom handles. Jigs can be molded from low temperature metal alloys, dressed with hair or feathers, and enameled with bright colors. You can cut spinner blades from metal cans or pie plates and hammer them into shape. Teaspoons and other bits of metal are transformed easily into wobbling spoons. Discarded necklaces provide glass beads. Tools for these little jobs usually are found around the house, so there isn't much to buy.

Those who enjoy such things can make lures exactly as they want them, turning the useless into the useful while saving money and enjoying the pride of accomplishment. Fly fishermen in vast numbers dress their own artificial flies and think it more important to catch fish on what they make than whatever can be bought in stores. So why not do the same with artificial lures?

You will need the following items:

Small round-nosed pliers

Small wire-cutting nippers

Coil or two of 9 or 10 stainless-steel leader wire Dozen clevises of various sizes Dozen split rings of various sizes Dozen assorted treble and single ringed-eye hooks

Cut apart or disassemble the unwanted lures. Discard the useless parts; polish and lacquer old blades; and separate everything worth saving in a compartmented box.

Following are illustrated directions for making three types of spinning lures from parts of old ones and the materials listed above. If you want to economize on lures, which often snag on underwater logs and are lost after a half-dozen casts, here is a way to do it and, at the same time, fill those winter evenings with a worthwhile activity.

When forming the loop with pliers, begin as in Fig. 1, but then shift wire in pliers and bend until the end makes a 90-degree angle with the shank. Slide on a treble hook and bend end around shank three times, maintaining the right angle in order to make close turns. Cut off end closely and string on beads and clevised spinner. Then twist another loop in end of wire.

12

ATLANTIC SALMON

Many anglers never have enjoyed the thrills of fishing for Atlantic salmon on the presumption that it is too expensive or too difficult. It can be very expensive, even to the tune of a few thousand dollars per person per week. On the other hand, licensed anglers can park their campers beside a remote salmon pool and enjoy miles of free (or small fee) fishing very economically.

While one is supposed to get what he pays for, based on the fame and productivity of the river and the quality of the accommodations, this isn't always so. For example, a friend of mine spent over two thousand dollars for a week's booking at a camp on a famous river and returned home, skunked. That same week two other friends camped at low cost on another river and, for a small daily rod fee (charged by the government for river upkeep), took their limit nearly every day. Both rivers are in the Province of Quebec.

That's the chance one has to take. The river may be "right," or it may be high and dirty. Runs of fish may or may not be in. Hooking one big salmon or two, or even a few grilse, seems to many to be ample recompense for occasional fruitless trips. The greedy and the unphilosophical would do better fishing for something else.

FROM RIVER TO OCEAN, AND BACK

Unlike the Pacific salmons, which die after spawning, the Salmo salar of the Atlantic (whose name means "the leaper") returns to the ocean (if he is lucky), perhaps again to come back to his river, bigger and lustier than before. His rivers include a few in Maine, many in New Brunswick, Quebec, Nova Scotia, Newfoundland, Labrador, Iceland, the British Isles, and others in the Scandinavian countries extending on the continent as far south as Spain. The world record is a 79-pound, 2-ounce fish taken in 1928 on the Tana (Teno) River, which separates Norway from Finland. The Canadian record is 55 pounds and the current record breaker in Maine is 26 pounds, 8 ounces. All fishing in North America is with the fly rod and artificial fly.

Atlantic salmon travel as far upriver as they can get, which often is to small, clear, gravel-bottomed tributary brooks many miles upstream. Their runs in spring, summer and fall vary from river to river, depending partly on water conditions. Anglers, knowing when these runs should occur, can visit various rivers at the appropriate time, thus extending the season and improving their luck. Salmon can find the river in which they were born, and usually return to it after spending between one and four years at sea. Fish returning after the first year are known as "grilse." They normally weigh between 2 and 8 pounds, although in some localities any over 5 pounds technically (or legally) are called salmon.

The fish spawn in late fall or early winter by laying eggs in a series of depressions they scour in the gravel. These beds are called "redds," the gravel scoured from one depression being used to cover the previous one. The warming water of early spring causes the eggs to hatch. When the baby fish have absorbed their yolk sacs, they wriggle between stones of the gravel to the surface and school amid whatever protection they can find. The fry which survive raids of birds and other fish exist on microorganisms of the river until they exceed finger length. They then develop vertical bars along their sides, are called "parr," and resemble small brook trout except for their deeply forked tails.

The parr, subsisting on nymphs, insects and smaller fish, live a year or more in the river, gradually drifting to sea. Reaching the estuary, their parr marks become concealed by a silvery sheen, at which stage they are called "smolts." The schools of smolts develop in the estuary until instinct sends them to sea. Their travels then take many of them to areas southwest of Greenland where, in recent years, far too many of the developing salmon have been caught in nets and on long lines, principally by fishermen from Denmark and its former territory of Greenland. When this book reaches publication, it is hoped that the practice will have stopped. If not, this and other hazards to the survival of the species may doom this noble sportfish to extinction.

When the survivors of these depredations return to their rivers, they must run the gauntlet of coastal and estuary nets and of pollution and poaching. The annually decreasing numbers that return diminish the sport of angling and of the tourist revenue it brings (or could bring) to the regions which for generations have been meccas for the sport.

Because of estuary netting, which, due to the size of the mesh, takes too many salmon while letting most of the grilse get through, some river systems, such as the once-famous Miramichi watershed in New Brunswick, are called "grilse rivers," meaning they contain far more grilse than salmon. Of course this unfortunate situation quickly could be cured by eliminating or drastically restricting estuary and river netting, an urgently needed remedy that would be a financial bonanza to all but a few commercial fishermen.

Whether or not salmon feed after returning to their rivers is an oft debated question. It isn't a matter of whether or not they do, but of how much.

While the salmon's instinct at this stage of its development is toward spawning rather than eating, we do know they consume various tidbits from time to time. Luckily, these include artificial flies. The smashing strike of a salmon of 20 pounds or so when it slams a tiny fly and erupts cartwheeling and gyrating high into the air will raise the hackles of even the most jaded angler and should sell him on the sport for life.

ATLANTIC SALMON TACKLE

If your strongest trout rod has moderate stiffness and power, it should be adequate for all but very big salmon in very fast rivers. Many a whopper has been taken on lighter tackle. The choice of rod power depends on the skill of the angler, how sporty he wants his fishing to be, and on the probable size of the fish and the force of the water. It is the custom in many other countries to use much longer and stronger rods. We don't think we need them, and we consider lighter gear to be more fun.

For North American Atlantic salmon fishing, I usually take about six rods. These are a Scientific Anglers in size 7, a Cortland in size 8, a Scientific Anglers, an Orvis and a Fenwick in size 9, and a Fenwick in size 10. The size 7 is a grilse rod, but salmon usually are no problem for it. If a few salmon, along with the grilse, are expected, the size 8 might be used. When rivers contain only (or mostly) salmon, the size 9 rods are selected. The size 10 is for big fish and fast water. Of these, two usually are rigged for the day's fishing-one with a floating line for dry-fly work; the other with a floating line with sinking tip for wet-fly fishing. Sinking lines are avoided because they are more difficult to pick up, but they may be necessary when rivers are in flood or when fish are deep in pools. Well equipped anglers have two extra spools for each reel, so a choice of the three types of line can be used when needed. Leader butts are semipermanently tied to lines with the Leader Whip Knot, to which tapered leaders are attached with the usual Blood Knot. Since salmon are not very leader-shy, tippets can be strong. Ten pounds is about average, but the tippet strength depends mainly on the size of the fly. Use the strongest tippet that will fish the fly correctly. The leader usually is about as long as the rod. It often is much longer under low water conditions.

Reels for salmon fishing should hold the line and at least 150 yards of about 20-pound-test backing. Line should not be packed on to full capacity because fast reeling may spool it unevenly and jam the reel when the salmon is brought in close. A smooth and easily adjustable drag is essential; never set it so tight that it could cause a pull-out or a broken leader when a salmon jerks the tackle suddenly. Multiplying fly reels are a bit bulkier and heavier than single-action reels, but they retrieve line much faster when fish are far out or making a run toward the angler.

ATLANTIC SALMON FLIES

In 1895 an English author named George M. Kelson wrote a book called *The Salmon Fly* which showed many of the then-popular patterns in color. Kelson had the peculiar notion that salmon rose most readily to butterflies, and he featured flies resembling them in his book. The error took hold and led to the popularity of hundreds of patterns of dreary sameness in style but of an excessive variety in color. Many of these old classics still are used occasionally. They include the Jock Scott, Durham Ranger, Black Dose, Thunder and Lightning, Green Highlander, Black, Silver, and Blue Doctors, and scores of others. If readers can locate any of these beautiful and complicated classics they should be preserved for museum pieces. Modem patterns are much better for fishing!

Most modem salmon flies are dressed with both featherwings and hairwings, the latter being more popular because they provide more action. I once obtained the records for several years on many rivers of the flies which had taken the most fish. In the vast Quebec area four patterns stood out so strongly that all the others were also-rans. These, in order of popularity, are the Rusty Rat, Green Highlander,

Silver Rat and Black Dose. While the second and fourth are old classic patterns, the modem adaptations are dressed as hairwings.

While styles in flies vary from year to year, the present pets in New Brunswick include the "butt patterns." These are so simple that any amateur can tie them. The basic dressing calls for a fairly fat black silk body ribbed with fine silver tinsel, a sparse and short black throat (sometimes brown) and a wing of native Canadian red squirrel, for which black bear hair or some other dark fur often is substituted. These flies have a butt (but no tail) of yarn or silk of one color or another; usually red, yellow, orange or green, and often of fluorescent material. (The fluorescence, if sparse, is very important.) A fly of this type with a red butt is called a Red Butt, and so on. Other popular patterns are a splaywing called the Butterfly, and a hairwing named the Cosseboom. These few patterns, in various sizes, should be sufficient for all North American Atlantic salmon fishing, as far as wet flies are concerned. A few bucktails and nymphs are handy on occasion, but are unnecessary for casual anglers. Two types of almost unsinkable dry flies stand out. One comprises the various Wulff patterns, of which Black, White and Royal provide good variety, the Royal Wulff often having a fluorescent mid-body of one color or another. The other type is the Bomber. A Bomber resembles a Wooly Worm, but it has a closely clipped cigar-shaped deer-hair body and a fairly short bunch of the same body hair extending forward at an angle of about 45 degrees to slant it upward over the eye of the hook. The Black Bomber is all black except for a white palmered hackle wound into the body from rear to front. The White Bomber is all white except for a red palmered hackle. These two quite different types are of primary importance.

Hook size 6 or 8 is a good average. Larger flies do well on big rivers, or on smaller ones during high water. Smaller sizes (which may be smaller flies dressed on larger light-wire hooks) may do better under very clear or low water conditions. While black salmon hooks are standard, many anglers think bronzed offset (kirbed) ones provide better penetration. Opinions differ on selecting color combinations of flies, but drab to dark ones should do better when days are bright and rivers are clear. Drab to bright ones could be selected for dull days or for high or discolored water.

A featherwing fly is opaque and its action is rather stiff. This may be satisfactory in rapid streams, but not as much so in slower water. A correctly dressed hairwing pulsates under all conditions. It has translucence and mobility. The trend seems to be to the hairwing as the best modem type for salmon.

During the 1970s anglers fishing in Iceland found that salmon often would take long bucktails when they wouldn't touch anything else, and that they sometimes preferred bucktails to more orthodox types. There is a theory that ...is is so because, in the ocean, they feed on immature elvers (eels) at the stage when they are about the size of garden worms. I say "theory" because I have no proof that salmon do consume elvers, although it seems to make sense. When salmon return to their rivers their stomachs are empty, so such proof would have to come from commercial ocean fishermen, who haven't supplied it, as far as I know.

Anyway, don't diminish the importance of the long bucktail, at least as a change-of-pace fly. These usually are tied on double hooks in sizes 4, 6 and 8. The wing is about three times as long as the hook, dressed flat and rather sparse. To these I would add an allblack and a red and yellow bucktail, both with silver bodies. In Iceland, where I have fished for many years, we find an all-black tube fly very valuable. The red and yellow pattern is most useful in discolored water.

The reasons for selecting specific salmon flies for one purpose or another are involved ones on which much has been written. Readers wishing to pursue them further are referred to books specializing on the subject.

TIPS ON LOCATING SALMON

Knowledgeable guides are valuable to expert salmon anglers and almost indispensable to non-expert ones. Even old hands at the game, proficient at "reading the water" of salmon rivers, are helped by guides so familiar with specific pools that they know exactly where salmon lie in them under all water conditions. A good guide often becomes a valued friend who contributes immeasurably to the pleasure and success of a trip while sharing his knowledge of salmon fishing, handling the boat and helping to lug equipment.

We can't presume that all guides are that kind because some are nonprofessionals, guiding on a parttime basis, whose knowledge of salmon fishing leave' much to be desired. It pays to know the basics of locating salmon, rather than depending on someone else. Salmon fishing differs from trout fishing in many ways. For example, salmon make no attempt at concealment, and they do not lie near feed lanes in streams because they have little interest in food. In a nutshell, we find them not in slack water or in fast water, but in the "in between water" of moderate flow. The importance of "edges," the water in between the fast and the slow. It is helpful to review this because, under normal conditions, one should find salmon where he finds edges.

Normal conditions are when water is "right"; when streams are of average height, fairly clear, and not over 70 degrees in temperature. Salmon lie in different sorts of places when water is abnormal; that is, when it is too high and perhaps too cold, or when it is too low and too warm. Let's discuss normal conditions first, because they are most common during the fishing season, and most productive of good fishing.

When rivers are clear, moderately flowing, and under 70 degrees, the best places to find salmon are along the edges. A big rock in a stream forms two edges, the deeper one usually being the better one. Any obstruction along a stream's bank, such as a rock, a log or a ledge, forms an edge, even very close to shore. An edge is formed where water flowing over a bar merges with the main current. Salmon often rest in the deeper water just over the bar. Edges also are formed where a side channel meets the main channel, where a brook enters a stream, and where two parts of a stream merge just below an island. Salmon may lie in flat gravelly stretches if they contain potholes to break the flow. These stretches may contain fairly fast water, but the current is slower below the surface because of such depressions. These often are found in the narrowing tail of a pool. Wherever there is a major obstruction in the stream, such as a low dam or a waterfall, salmon should collect in the first good pool below it. They approach the obstruction and then return to the pool to rest before attempting to surmount it.

When rivers are high, and probably discolored, the problem becomes more complex. Larger and brighter flies are necessary for increased visibility. Fish should be in the "ponds," which is the flat water below pools, water that is too shallow ordinarily. The fish may be very near shore, where the flow is less swift; or they may be on bars which are too shallow when the flow is moderate.

Another problem is presented in the opposite situation, when water is low, clear, and over 70 degrees. Then, the salmon seek colder water with more oxygen, often in cold-water pools below entering brooks and streams. In fact, salmon may seek colder water by going up into the tributaries. These cald-water pools may be very shallow, but the fish have to put up with them. They often lie in plain sight-a tempting prey for poachers. Fishing for them is difficult because they usually ignore orthodox methods. A whipping fly, a rapidly retrieved fly, even a gaudy streamer or bucktail, may get results.

It is a maxim in salmon fishing that a fish can be caught if he rises to a fly or otherwise shows any interest in it, even so little as an increased quivering of his fins. The fish belongs to the person who attracted him, until the angler wishes to give up, or to offer the salmon to someone else. After resting the fish for a few minutes, the same fly, a smaller pattern, or a different fly may tempt a take. If that fails, a skated fly, dry or wet, or some other unusual method, may hook the fish.

WET FLY FISHING FOR SALMON

Under conditions of moderate flow the water in a salmon pool usually is only a few feet deep. Since the pool's bottom is rocky, salmon may be lying anywhere in it, and may not be seen. When we can't cast to individual fish, the usual custom is to cover all the good parts of the pool as shown in the accompanying drawing. In this stream, the current flows evenly; both sides are worth fishing. The stream can be covered from bank to bank by an angler in a boat in the middle. After fishing the head of the pool, the angler has moved to position A. The angler, standing if it is safe, makes quartering downstream casts, starting with short ones very near the boat. Salmon sometimes are hooked very close by.

It is of primary importance, regardless of the length of the cast, to make the fly swing on a tight line without any belly in it. A bellied line causes the fly to whip. Its speed usually should be moderate, which means that it is cast more nearly cross-stream in very slow currents and more nearly downstream in faster ones. When the swing is nearly completed, a few added feet can be obtained by transferring the rod to the other hand and moving it outward. At completion of the swing the fly can be bucktailed a bit in case a fish has followed it and may be reluctant to take. Some anglers bucktail the fly slightly during the swing; others use a dead drift. If one method doesn't work, try the other.

If the water looks good on both sides, alternate the casts from one to the other, extending each cast a foot or so for maximum water coverage. Of course, pay special attention to good lies, like the rock jutting into fairly deep water from the shore, the rock near midstream, and the edge formed by the current of the entering brook. When the casts have been extended as far as the angler wishes, he usually sits down, thus indicating to the guide that he wants to be moved to a new position where short casts can be resumed just below the previous longest one. By this means all the good water in the pool can be covered.

Salmon often can be seen lying in a pool. Select one and cast to the fish so the fly swings just above 'him or at the same depth. If he shows interest but doesn't take, he probably can be hooked by a smaller or different fly, or another method of presentation.

DRY-FLY FISHING FOR SALMON

Dry-fly fishing is not entirely a low-water method. It is very practical, and often superior to wet-fly fishing, under conditions of moderate flow. Seeing a salmon of any size rise to suck in a floating fly probably is the supreme thrill of angling. Using the dry fly for salmon isn't particularly difficult.

Salmon ordinarily take large and bushy dry flies, which fortunately are very good floaters. Being quite wind-resistant, they can be cast most effectively with a weight-forward floating line. Two efficient dry flies have been recommended: the Bomber in size 2 or 4 and the Wulffs in 4, 6 and 8. Trout patterns are acceptable in the rare cases when smaller dries are needed.

The dry fly usually is cast upstream and across, the rod being pulled back with a slight jerk to obtain a wavy line for a longer float. The rod usually is held parallel to the water, pointing at the fly.

Downstream or quartering downstream casts are used when needed. The rod is jerked back sharply just before the fly lands to provide the same snakelike effect for a long drift. Salmon often hook themselves, but the angler should tighten on feeling the fish.

These large dry flies are effective in themselves, and they also are used as locators. If a salmon rises to one without taking it, at least we know where he lies. The trick then is to tempt him with a smaller dry fly, or perhaps with a wet one.

Another method is to cast above and beyond the fish and to deliberately pull the dry fly under, trying to move it directly in front of his nose. Skittering a large dry fly over a reluctant fish often works. Cast the fly on a tight line ahead of him, then skitter it past his nose.

HOOKING, PLAYING AND LANDING SALMON

Trout and bass fishermen talk about "striking" their fish when they feel a connection. A salmon angler never strikes because this could take the fly away from the fish or break the tackle. When the salmon rises to the fly he usually will turn with it and take it down before spitting it out. When this connection is felt the angler merely tightens up. Ideally, this will sink the hook into the angle between the jaws, which is the most secure place. Some salmon take with such a savage strike that they are well hooked in any event.

When the fish has been hooked there is scant time to put loose line, if any, back on the reel, but the salmon must be played off the reel. He'll make a fast run, probably followed by a spectacular jump. On the jump, the rod tip should be lowered to give slack line, thus diminishing the chance of the fish breaking the leader with his tail. As the fish reenters the water, tighten the line again. Hold the rod nearly vertically so its spring can be used to the fullest. Always try to keep a tight line with as much tension as safety permits.

After the first run the angler usually is put ashore so he can handle his fish from the beach. When the salmon seems ready to be led in, a suitable spot on the beach is selected. An angler fishing alone can use a tailing device, which is a wire-spring noose attached to a handle that is slipped over the tail and pulled tight. An alternative is to beach the fish by leading him into shallow water and pulling him on land. Only moderate tension is needed because his flopping aids the process. When the fish is partly ashore, the angler (still holding it by the spring of the rod) can get behind it, grab it by the tail, and push it to safety.

A guide or someone else with a net can make all this easier. A salmon net has a long handle, a wide hoop and a deep bag. The angler indicates where he wants the fish netted, and the net handler rests the hoop on the bottom in about a foot of water. The angler leads the tired fish over the hoop, and the guide raises the net. All this is done quietly, without any jabbing. Jabbing at a salmon with a net is an excellent way to lose it.

ARE BLACK SALMON SPORTING?

When Atlantic salmon spawn late in the fall they often are imprisoned for the winter in pools under the ice. These are called "black salmon" or "kelts"-thin, ugly creatures bearing scant resemblance to the noble fish they once were.

Freed in spring by the breaking up of the ice, the starving fish feed and slowly grow stronger, taking on a glimmering of their former silvery sheen. During the high and cold spring runoff, they can be caught on rod and reel in areas where fishing for them is allowed.

Catching kelts is favored by some, violently opposed by others. The principal reason for permitting it seems to be the tourist revenue obtained.

Proponents of the practice think it is exciting to fish from boats in cold and swollen streams and to hook several big fish per day on large streamer flies and bucktails, even though patterns don't matter much and it only is necessary to drift the fly in the current to catch fish. These anglers maintain the salmon are edible -and that they release most of them, anyway.

Opponents scorn the practice, feeling that it is not sporting, that it requires no skill, and that it does great harm to the fishery. They say the slightly improved appearance of the fish mostly is an illusion because their bodies are highly liquid and deflate after being killed. They consider the flesh so unpalatable that even a dog wouldn't touch it.

These unfortunate fish have two strikes against them before they hit a hook. Exhausted by the long journey upstream, and by spawning, many starve and die in the pools under the ice. The survivors have a chance to return to the sea, but not if they expend their remaining energy fighting strong tackle. Released fish almost always die.

Statistics indicate that, of the black salmon which live to go back to sea, about 10 or 12 percent return to their rivers, fully mended, fat and vigorous, and much bigger than be^fore. These are the trophy bright salmon we dream of catching. It seems that they, and we, should be given the chance.

RIVERS AND RESERVATIONS

Maine

The Atlantic salmon fishery is being restored in Maine and seems to be improving year by year. A license costs very little, and anglers need not hire guides. Fishing normally is good on the Penobscot, Dennys, East Machias and Narraguagus rivers from mid-May through the first two weeks of June. The most prolific runs are on the Penobscot and the best places to fish for salmon are at the formerly famous Bangor Pool, in the city of Bangor, and in the pools up-river as far as the Veasey Dam. Due to breaks in the Bangor Dam the fishing there isn't as good as it used to be, but it still is popular. The pools on the four mile up-river stretch are crowded with anglers in season, each patiently awaiting his turn to fish the water by wading near shore. For this reason, fishing from small boats is popular, but guides with boats are scarce. There is a launching spot midway between the two dams on the Brewer side of the river. The best runs in the Machias usually are from late June through the first two weeks of July, with smaller runs in the fall. The Sheepscott has a small run from early July through the season. When rivers hold fish, the banks of some of them may be crowded. Information can be obtained from Vacation Travel Dept., State House Annex, Augusta, Maine, 04330.

New Brunswick

The great Miramichi River system has many salmon streams. Parts of many of these are open to public fishing, and visitors can arrange to stay at various sporting camps to fish some of the private stretches. Guides are necessary, but the services of one can be shared by two anglers. Summer fish run the rivers from late June into July, with larger runs in September. For reasons previously mentioned, the runs in the Miramichi River system predominantly were grilse, but more salmon now are entering the streams due to the abolition of netting. Contact the Fish and Wildlife Branch, Dept. of Natural Resources, Fredericton, New Brunswick, for further information.

Quebec

The vast Province of Quebec offers an abundance of salmon rivers, many of which are available for fishing at small cost. Since various rivers have runs at different times, and since conditions frequently change, readers should write for up-to-date information. Quebec offers facilities for every pocketbook, from inexpensive camping on public waters to daily rod fee arrangements on restricted stretches of famous streams controlled by clubs or by the government. Fees range from cheap to expensive, and the employment of guides is not always necessary. Information can be obtained from Service des Parts, P. O. Box 639, New Carlisle, P. Q.

Nova Scotia

While this province provides some of the most inexpensive salmon fishing, its productive rivers are few; notably the Medway, the Margaree and the St. Mary's. License fees are low, no guides are necessary, and waters are open to the public. There are spring and fall runs of salmon, with good fishing in June (unless the season is late) and usually better action in September. In June many of the rivers have runs of large seatrout, which increases the fun. The Margaree Salmon Association has a small angling museum near Margaree Harbor, on Cape Breton Island (connected to the mainland by a causeway), which is well worth a visit. Address inquiries to the Nova Scotia Travel Bureau, Halifax, Nova Scotia.

Newfoundland and Labrador

With the possible exception of Quebec, Newfoundland and Labrador offer the most remote (and sometimes the best) salmon fishing still available on the North American continent. Many rivers can be reached by road, others by float plane or railroad. Salmon enter some rivers in May, others as late as June or early July. The peak period usually is the first part of July. The nonresident season's fishing permit is inexpensive. While visitors must employ guides, one can serve two people. Lists of fishing camps and other inforMation are available from the Newfoundland-Labrador Tourist Development Office, Confederation Building, St. John's, Newfoundland.

THE FUTURE

As this is being written, the future of Atlantic salmon fishing, in North America, at least, is in doubt. Indiscriminate netting and long lining of salmon on the high seas prevent too many from returning to their rivers. Large numbers of those that do return are poached in the rivers themselves or are prevented from spawning by dams and pollution. The good side of this glum situation (if there is one) is that alarmed anglers are beginning to force governments and politicians to remedy it before it is too late.

While governments include many a fisherman dedicated to the preservation and restoration of the salmon, they also include others who care only about which side their bread is being buttered. Even those short-sighted individuals are now being made to realize that salmon caught commercially are worth relatively little, while those taken on flies by sportsmen bring vast sums into regional economies.

With the new restrictions on commercial fishing, more and more Atlantic salmon are being allowed to return to their rivers. As this book goes to press the two past seasons have provided excellent fishing. Let's hope that this trend will continue to the extent that the King of Fishes no longer will be subjected to mass slaughter, but will be allowed to return home freely to propagate his kind and to provide, to a reasonable extent, peak sport to devotees of the fly rod and the artificial fly.

13

PACIFIC SALMON

Although five species of salmon are common to northern Pacific waters, sport fishermen are mainly concerned with two. Mention either and watch their eyes light up! Then you'll hear tall tales of big ones landed by methods as diverse as trolling deep with "cannon balls" to laying an artificial fly across an estuarine pool.

The two featured actors in this annual angling bonanza are the spectacular chinook, or king salmon, and the smaller but no less sporty coho, or silver salmon. When the chinook reaches 30 pounds or more it is called a "tyee." These behemoths of the deep are sought in tidewaters and the lower reaches of great rivers such as the famous Campbell in British Columbia, where dedicated anglers try for them on light tackle with considerable ritual and frequent success.

The three others, more valuable commercially than for sport, are the sockeye (or blueback salmon), the pink (or humpback salmon) and the chum (or dog salmon). Although the sockeye has been ignored mainly because it is known as "the fish anglers can't catch," it is gaining the interest of a growing number of rodsmen who are learning how. This species has a nonmigratory freshwater kin fish known as the kokanee, a landlocked miniature which is fun on tiny tackle and will be discussed later.

Since the tackle and tactics for the chinook and the coho are very varied and more or less interlocking, we'll discuss them together. Most anglers can't tell a small chinook from a large coho, and they fish for both species in about the same ways. While very similar in appearance, the coho, or silver salmon, has a grayish-white mouth and gums, while those of the chinook, or king salmon, are black around the base of the teeth.

THE CHINOOK (OR KING) SALMON

After birth and youth in a river (similar to Atlantic salmon), young chinooks travel over a thousand miles northward, spending up to five years at sea before returning home again. Adult chinooks average about 20 pounds, but those that have remained longer at sea may be double and triple that size. The rod and reel record is 93 pounds; a 126V2-pounder was taken from a fish trap in Alaska in 1939.

From places as far away as the Aleutian Islands chinooks return southward along the Pacific coast to reach their rivers. Fish from these schools are caught by deep trolling at depths of from 50 to 150 feet. In estuaries and tidal pools they occasionally may be taken by near-surface methods and even on flies, particularly between dawn and daylight or when baitfish rise to the surface in the evening. Trolling depth, which is the cruising depth of the fish, depends largely on water temperatures. The favorite is about 54 degrees in a range between 50 and 65. Here again, a temperature-depth probe is usfful, and schools of fish can be located by electronic devices.

Chinooks enter their rivers between early spring and late fall. Peaks vary with the rivers and water conditions, but good runs can be expected in April and May and in August and September. July is a hot month in some areas. Spawning takes place, usually far upstream, between summer and early winter. Before entering fresh water the fish are of a bright silvery color, with dark blue to greenish backs prominently marked by irregular black spots. They turn darker and gradually deteriorate as they move upstream, dying after spawning. While there is good fishing when schools are spotted along the coast, most of it is in the estuaries and tidewater because the fish mill around, in and out with the tides, while acclimating themselves to the transition between salt and fresh water.

Chinooks (and cohos) have been transplanted very successfully into some of the Great Lakes. Deprived of the greater bounty of the sea, these are smaller fish, but they are lively fighters which often tip the scales at 20 pounds.

Since chinooks evidently stop feeding on entering fresh water, the best fishing for them is in the salt. However, they do strike at a variety of spoons, spinners, plugs, baits and flies, perhaps because of instinct, or from anger or curiosity. Lures should be fished slowly.

THE COHO (OR SILVER) SALMON

The appearance and habits of the coho are very similar to those of the chinook, but the coho is a favorite of light-tackle hardware heavers and fly fishermen because it often feeds near the surface and takes lures readily, sometimes accompanying this by spectacular jumps.

While coho grow to adulthood in the ocean they do not move far from their rivers, into which the schools begin to enter in early summer (earlier in northern California, and a bit later in Oregon and Washington). Entry time depends on water temperature; their favorite is between 45 and 60 degrees, with an optimum of 54. This being about the same as for the chinook, the schools of both species often are mixed. Average weights are between 10 and 15 pounds, with an occasional lunker going over 20.

Spawning takes place in the rivers between October and February, the fish dying thereafter, as all other Pacific salmons do. Like chinooks, the cohos drift in and out of the estuaries with the tides, becoming acclimated to fresh water and waiting for heavy rains to raise the rivers for the upstream journey. The fish are bright and silvery in salt water, but start to put on their reddish spawning colors as soon as they enter the fresh, this due to bacteriological breakdown of their skin cells. They are easy to catch in salt water; much less so as they move upstream.

Cohos are taken with live and cut baits such as anchovies, smelt, herring, squid, prawns, cut sardines and mackerel. While it is necessary to get the baits down to the level of the fish, schools often are seen feeding on top, usually marked by the wheeling and diving of shearwaters, cormorants and other birds. In such cases trolling around schools or across their path (but not through them) can produce quick results. Casting into such schools with spinners, plugs and spoons can bring smashing strikes at nearly every cast. Fly fishing is popular, and growing more so, during periods when fish are showing on the surface.

PARTY BOAT FISHING

In the pitch black of pre-dawn the boats stream out from their harbors, seemingly in endless procession; lightless (to aid night vision), they chug and hum along precarious channels, over treacherous bars, bucking peeling swells to open water in search of schools of salmon in the sea. On commercial boats, party boats, private ones, big and small, crude and elaborate, murmuring crews are busily preparing gear while sportsmen are sipping coffee or rigging tackle as these invisible armadas plow out of scores of ports from San Francisco Bay northward to brave clashing tides and currents, some in hope of profit, others in search of sport. Some of the paying guests are expert repeaters, but neophytes in Pacific salmon fishing can get their sea legs and learn the essentials by going along. Lacking the advice of old hands at the game, discuss your desires with proprietors of dockside tackle shops to obtain a list of skippers. Interview them and arrange with one for a booking. If the skipper is popular, the date may have to be made well in advance. Make sure he is agreeable to the kind of fishing you want to do. Discuss tackle; whether you will use that provided by the boat (which may be too strong to be very sporty) or whether you will bring your own. Discuss baits and lures, equipment and clothing, food and beverages. You want to be warm and comfortable, equipped with correct gear, but there's no sense in lugging things the boat carries for your use. It also may save later displeasure to agree in advance about who owns the fish you catch. Some skippers insist on a share.

Tackle and methods used on party boats vary so much under different fishing conditions that anything said here may be inappropriate, but an example or two may be helpful. In spring, salmon usually cruise deep, working shallower as the season progresses. August is the month for peak activity. The most active times are between first light and sunrise, the hours around dusk, and during tide changes. Let's assume here that the salmon are deep, because other conditions will be discussed later. Some anglers think deep fishing isn't very sporty, but it may be the only way to take fish.

Let's use a medium- to heavy-action two-handed boat rod about 8 feet long. On this put a metal-spooled star drag reel, such as a 3/0, loaded with between 275 and 350 yards of from 30- to 50-pound Dacron or monofilament, or perhaps metal line. This strong tackle is necessary because weights are used to slowly troll the lures at proper depths, which may be between 30 and 100 feet, or more, just above the bottom.

Newcomers to this form of fishing will find the weights and their rigging interesting. Called "bombs," or "salmon balls" or "cannon balls," these are from 1- to 3-pound balls of molded lead or scrap metal into which is set a brass screw-eye for attaching the tackle. On a strike, a spring-triggered device releases the bomb, allowing the fish to be handled weightlessly. These bombs are not very expensive, and many usually are used during a good day's fishing.

The connecting device is a metal tube about the size of a cigarette, with a swivel on both ends. In this is a spring-loaded slide-pin, or trigger, to which the bomb's eye-ring is clipped. This opens automatically to drop the weight when a fish strikes the lure.

The lure is connected to the swivel of the connecting device and its bomb by about 9 feet of strong monofilament. Baits usually are whole fish, such as an anchovy or a herring, carefully rigged on a special bait harness which runs through the body of the fish to keep it from spinning or rolling excessively while being trolled.

Artificial lures for chinooks for deep trolling usually are brightly finished metal spoons in brass and copper, brass and chrome or gold and bronze. Coho take the same lures, but also go for colored ones such as chartreuse with red dots, yellow-green, bright red, or cerise with black dots. They like brilliant colors, actively fished. Anglers on party boats usually ask the skipper to choose the lures.

Thus rigged, the rods are set in outrigger rod holders spaced along the boat's rail. A swift jerk of the rod tip signals a strike and a dropped bomb. While the angler handles his fish, one of the crew usually arrives with a yard-wide deeply bagged net. The action gets even more exciting when several men line the rail, all with fish on!

After dropping several of these bombs in only a single day's fishing, boat captains began to wonder if there might be a better way - one by which only one weight could be used again and again without having to jettison it on hooking a fish. There was, and the answer was the downrigger. Thus the expendable salmon balls have passed into history, or nearly so, but it seems worthwhile to make note of them because they evidently are forerunners of the downrigger. As this note is added in 1983 to the second edition of this book downriggers are familiar equipment for deep trolling in nearly all suitable large lakes from coast to coast.

Another popular method of fishing for Pacific salmon is with a plastic or metal diving plane which keeps the bait at the desired depth. A strike levels the planer, which offers little resistance while the angler handles the fish. Some planers are painted in bright colors which seem to attract salmon to the bait.

Another type of weight is crescent shaped with linkages at both ends for joining to line and leader. These are called drails, or keel leads, and are valuable in preventing line twist.

Party boats of different sizes offer varied facilities which newcomers may wish to consider. Large ones may be more stable, if one is inclined to seasickness. While they may be more comfortable, they also may be more crowded. Good skippers don't crowd their boats because many of their customers won't go out again after experiencing tangled lines and insufficient cient attention. Fish taken by each angler are strung on a ring or chain or something similar with his name or number on it. Crew members are careful to clean them and to see that everyone aboard gets what he catches. Sea birds, wheeling, screaming and diving for scraps tossed over the side, add to the fun of the return trip.

Smaller party boats offer more flexible facilities. One or more groups of fishermen, with the same or similar goals in mind, can agree to leave the dock when they wish, fish where and as they want to, and return whenever they desire, subject only to the captain's judgment of tidal and weather conditions.

PRIVATE BOAT FISHING

Small-skiff fishing is a growing sport along the northwestern coast. It offers excitement as well as solitude, but anglers unused to the capriciousness of the sea should be prepared for the worst. Dense fogs sweeping in from offshore can engulf small boats suddenly, and dangerous winds can come up. A compass, a chart and life preservers arc basic necessities. Small boats are on their own, without radar, direction finder or radio. If prudent, however, the small-skiff salmon fisherman can enjoy the sport at its best.

Small-skiff anglers usually set up their tackle in the same manner that fishermen do aboard party boats, except that somewhat lighter gear is the rule. Knowing that salmon surface and enter shallow water more frequently as the season goes on, there may be no need for the deep equipment which has been discussed. In fact, schools of salmon may be in such shallow water that they could be reached from shore.

When fish are near the surface, medium-size spinning gear with lines testing between 10 and 20 pounds should be sufficient. One may need to get the lure down by using a keel type of trolling lead of from 1 to 3 ounces, to which is joined about 3 feet of 40-pound monofilament, used as a shock leader to prevent gill cuts. A good trolling speed is between 3 and 4 miles per hour, using rigged bait, diving plugs or spoons. Since cohos usually take flies readily, try double-hooked patterns about 4 inches long in colors as recommended on page 496. A spinner or two ahead of the fly usually adds greatly to its effectiveness. Varying trolling speed, or increasing it to 5 miles per hour, often helps.

A hooked salmon may take the boat far from the school, and it's difficult to find it again. It helps to drop a marker when a fish is hooked. This can be a plastic bottle or any other sort of small buoy with a flag attached to it, a line wound around it, and a small anchor. Without something of this sort to identify the general area, one could troll in circles for hours without finding the school.

When cohos are showing on top, some fishermen like to troll with fly rods, using a high-density line, or a front section with a lead core, to get the lure down to the fish.

TROLLING WTIH FLASHERS AND DODGERS

Flashers and dodgers are slightly curved, nearly flat, rectangular polished metal attractors which are tied at both their ends into trolling tackle between lead and lure, thus providing added flash and action for attracting fish from long distances. The lure usually is bait, but it can be a wobbler, a plug, or something else. To provide best action, Luhr Jensen & Sons, Inc., whose flashers and dodgers are famous along the Pacific coast, recommends definite distances between the flasher or dodger and the lure and between the flasher or dodger and the lead.

These rigs usually are trolled with boat rods and conventional revolvingspool reels filled with line testing about 30 pounds. The line strength can vary according to the amount of lead used and the size of the dodger or flasher and the lure.

HOW TO LOCATE SALMON IN SALT WATER

When salmon are not feeding deep (usually around August and later) they often drive bait to the surface, whitening and disturbing it in an acre or more of fine showers of spray. Sharp eyes, perhaps aided by binoculars, can spot this activity, and really sharp ones can see it from a long distance.

Birds look for such activity and spot it immediately, a few alerting others, all flocking to the scene. So if anglers don't notice the surface-driven bait, the birds will point it out. One by one and in flocks they fold their wings and plummet into the sea, usually emerging with small fish in their beaks. They adroitly turn these, head first, into their gullets, letting each fish slide down before immediately attacking another. Birds on a feeding spree often are so gluttonous that the food they have eaten prevents them from flying. They have to sit on the water to digest their meals until they are light enough to take off. Thus, a flock of birds resting on the water may mean that fish are in the vicinity and boats should troll there.

The color of the water also may indicate fish. Green often indicates a shoal area. Black water is best for salmon, although anglers aren't quite sure why. Some think the color is due to dense swarms of plankton, which baitfish feed on, thereby attracting the big ones.

When a school of surface-feeding fish is spotted, a boat never should run into it. Slow down as the school is approached and watch the frantic bait as the silvery salmon slash through it. One can drift with the school and fish into it, or troll around it, or spot its direction of travel and fish just ahead of it. A drifting boat usually won't disturb the fish, but the noise of a motor is sure to put them down.

SALMON IN THE GREAT LAKES

We have noted that both coho and chinook salmon have been transplanted successfully into some of the Great Lakes. This is largely due to the abundance of alewives there. Deep trolling is the usual method of fishing for them, although the fish can be taken on top at times, and even off piers and breakwaters. The downrigger, or undertroll, method seems to be common. I haven't seen this done along the Pacific, and haven't noted any introduction of Pacific "bomb" methods into the Great Lakes, but this may happen at any time. Of the two methods, the downrigger seems superior, but it is less adaptable to large party boats.

The secret of finding salmon in the Great Lakes is in water temperatures, because the fish should be at the 50- to 55-degree level in summer. This may be around 100 feet deep. It depends somewhat on the thermocline level, but wind action on big lakes causes thermoclines to vary in depth, and may disperse them entirely. Downriggers (readily available commercially) are used with 8 pounds or so of weight. The break-away snaps allow use of any tackle of reasonable strength; even fly tackle with between 500 and 750 yards of Dacron line on the spool and a monofilament leader about 40 feet in length.

Bait and fluorescent spoons (usually red) and plugs of various kinds are all productive. Burke's pale-green soft plastic ones seem to be current favorites, but "hot" lures change so frequently that popular ones one season are forgotten during another.

Salmon fishing in the Great Lakes may become a grab-bag of other species, including big rainbow, brown, brook and lake trout-substitutions anglers shouldn't mind at all! Flurries of surface and near-surface fishing sometimes occur in June and perhaps into the summer, but winds can change favorable surface temperatures quickly.

MOOCHING

"Mooching" means fishing near bottom in moderate depths, usually of not over 50 feet, while roving about in a skiff or other small boat in bays and estuaries where schools of salmon should be.

The usual tackle is an 8- or 9-foot medium-action boat rod equipped with a star-drag reel loaded with 250 yards or more of monofilament testing between 15 and 30 pounds. Spinning tackle of comparable strength also is popular. Lures can be spoons or plugs, but whole herring or anchovy hookups or plug cuts are popular.

Fishermen use various mooching techniques, the following being fairly typical. Stop the boat in a selected area and strip out line, letting the bait (usually unleaded) sink straight down. A "strip" is the amount of line that can be comfortably pulled from the reel by grasping the line near the reel and pulling it out to arm's length. This usually is about 2 feet; so 15 counted strips, for example, would let out about 30 feet of line.

Let's assume that bottom depth is somewhat over 50 feet and that two men are fishing. One might make 15 strips and the other 25 strips, to fish the two baits at about 30 and 50 feet. Counting the strips tells the amount of line out so that the proper amount can be put out again after a salmon has been taken.

With the lines down, the boat is started and the lines are trolled 25 feet or so. The boat is stopped to allow the lures to sink again, whereupon it again is started and run another 25 feet; this action being repeated constantly. The result is to provide an active "injured minnow" action to the bait while keeping it moving up and down and forward in a manner similar to exaggerated jigging.

Stopping and starting the boat is unnecessary when tide or wind will drift it at a reasonable rate of speed, which should be between 2 and 4 miles an hour. Under such conditions, rods can be pumped slowly to keep the bait active. A sea anchor may be helpful in slowing a boat which is drifting too fast. A pail attached to the stern with 10 feet or so of rope will serve as a sea anchor.

Local anglers usually know good mooching areas, which of course change with seasonal and tidal conditions. It helps to watch the activity of other boats, the color of the water, and the actions of birds. We can guess that salmon may be deeper on bright days and nearer to the surface on dull ones, or around dawn and dusk, because baitfish seek these levels at such times. We know that changes of the tides are good fishing hours because bait is more active then.

When plug-cut baits are used they should be prepared with a sharp knife, and all innards should be removed. Monofilament should be checked constantly for abrasions, and knots should be examined and retied when they show any sign of weakness. Hooks must be kept needle-sharp. Big salmon may mouth a bait very lightly, perhaps giving the impression that it's a small fish. Let the fish take the bait until it pulls the rod tip down. Then strike - very hard!

Artificials are made to resemble plug-cuts, and other fast-sinking plugs often are useful. Local tackle shops may have up-to-date information. Bright spoons in mother-of-pearl usually are effective.

A hooked salmon never should be horsed. When it feels the steel it will run, and the reel's brake should have been adjusted properly to allow it to do so. Green fish should not be brought to boat because a longer line provides more spring in the tackle. Fish shouldn't be netted until they have given up and are lying on their sides. A salmon should be netted head first. If it feels the net near its tail it will try to move away -fast.

Two fishermen mooching in a boat often hook fish at the same time. To avoid tangled lines, they should decide who will bring his fish in first. The other fish can be held, well out on a tight line, until this has been done.

FLY FISHING

Hooks used in salt water should be of noncorrosive metal or should be plated to avoid rusting, which ruins sharpness and strength and which discolors fur and feathers. Those for coho usually are between sizes 3/0 and 1 in patterns such as the Siwash, but those for chinooks can be as large as 6/0 or 7/0. Flies vary in length up to about 5 inches; smaller patterns of course being used for casting. When prevalent baitfish, such as candlefish, anchovies or smelt, are in evidence they are imitated in size and colors as closely as possible.

Flies for trolling usually are tandem rigged, the tail hook frequently pointing upward. Hook shanks and the connection often are encased in mylar tubing, or the body of the front hook is elongated and wound with silver tinsel or folded foil. Thin mylar.strips can be added to wings for added flash. Flies for casting frequently are of the "Blonde" type, with the tail applied as a wing to prevent long wings from wrapping themselves around the hook shank when the fly is cast. Flies which are to be attached to spinners should have a straight ringed eye on the front hook. Some flies for Pacific salmon are well-known patterns; others are merely baitfish-imitating color combinations which anglers think should be lucky. Most are dressed with silver bodies, and usually with polar-bear hair, if it is obtainable. Here are a few popular color combinations:

Green over white
Blue over white
Orange over white
Brown (bucktail) over white
Green over yellow over white
Blue over green over white
Blue over red over white
Green over red over white
Gunmetal gray over medium green over fuchsia over white
Blue over green over yellow over white
Gray over green over peach over white

These flies usually are of hair for more pulsating action and because feathers are more perishable, although marabou often is used. Heads are built up, painted in an appropriate color with a large eye and pupil. Anglers think that a spinner (such as a number 3) ahead of the fly is an effective ruse. If the spinner causes line twist, a bead swivel 3 feet ahead of it should help.

In near-surface trolling best results are obtained when the fly is dropped back from 30 to 40 feet behind the wake of the motor. Trolling speed is important. Increase it until the fly bounces; then slow down until it is barely submerged. Run the motor at this speed. Cohos like a fast fly but chinooks want it trolled considerably slower.

When chinook and coho salmon are showing or are not very deep, flyrodders can enjoy so much fun that other types of tackle are forgotten. Systems 9, 10 or 11 are most practical. As in Atlantic salmon fishing, reels (usually single action) should have smooth, adjustable drags, with about 150 yards of backing for the line. Leaders, about as long as the rod, are tapered to between 10 and 15 pounds.

Since much of this fishing calls for deeply and quickly sunken flies, even weight-forward fast-sinking lines may not get them down fast enough. Such lines are available in slow, fast and extra-fast sinking types, of which the latter should do nicely for most purposes. If it doesn't, the answer is to use a "lead-head," which is a shooting head of between 25 and 30 feet of lead-core line attached to monofilament backing.

Lead-core line, marked by a different color every 10 yards, comes in 100 yard lengths of 10 colors, so one color is 30 feet long, and one color should be enough for a shooting head.

If we should decide on 25 feet of 25-pound-test line it would weigh 375 grains, which should be about right for average tackle in average water. A shorter length of heavier line could be substituted, and one weighing about 500 grains might be needed for fast, deep water. If a selected length and weight fails to suit the rod or fishing conditions it is easy to make adjustments.

Salmon often hold in deep pools and runs. Cast up and across to give the line time to sink. A slow, jerky retrieve usually works best, but this can be determined by experimentation. In addition to the bucktails previously mentioned, flies imitating shrimps can be effective, but tackle dealers and other fishermen will offer more up-to-date opinions than any book can provide. The Blonde method of dressing is somewhat shrimp-like and prevents the wing from wrapping itself around the hook shank. The Pink Blonde, or a red one, should do well. Larry Green, a noted Pacific salmon expert, says that each of his salmon has been taken on a different pattern. Loring Dodge, who transplanted himself from eastern Canada to British Columbia, likes flies made with artificial hair, which comes in all colors and shades, including hot ones. Since this is very fine hair it is good for adding various blends of color, but I like some stiffer material, such as polar-bear hair, mixed in with it.

When a salmon is hooked on a fly from a boat which is near land, it is best for the angler to be put ashore so he can handle the fish from the bank.

SOCKEYES AND KOKANEE

These two closely related salmons are also-rans in the estimation of most anglers, but a few notes on them may be appropriate.

Anglers usually happen upon schools of sockeye salmon by accident, and may ignore them because of their reputation for failing to take bait or lures. Those who know how to catch them say that fresh-run sockeyes are better fighters than cohos, and even more delicious to eat. This is the fish which often is referred to as the "blueback," or in Alaska, the "red salmon." Although I have had no experience with them, those who have have reported that they can be taken on spinning tackle with ¼-ounce wobblers in fluorescent red, hot pink, or red and white, and jigs in similar colors. It seems to help to put three thin tails on the hook, such as those which can be cut from the rubber or plastic skirts sometimes used on small plugs. Unlike other salmons, sockeyes won't hit on the retrieve, and they want lures fished very slowly. The idea seems to be to cast out the lure and to let it flutter down, jerking it up and repeating, as one would fish a jig. While these fish are reported to be plankton feeders, they will, for some strange reason, hit a lure that looks and acts like a swimming shrimp.

The kokanee is a very small landlocked salmon which rarely grows longer than between 8 and 12 inches; the record catch being about 4 pounds. Native to the Northwest, it has been introduced widely, even as far away as New England, because it is fun to catch on tiny tackle and because it provides valuable food for larger fish. For example, the giant Kamloops (rainbow) trout of Idaho grow to such size because of the abundance of kokanee in lakes such as Pend O'Reille.

Kokanee can be caught on small lures by trolling, fly casting or stillfishing. Try a small hook baited with a small worm trolled behind a redbeaded flasher. Flies in size 12 or smaller do well, especially shrimp imitations and little bucktails or streamers in combinations of white and pink or red. The little fish also take worms, single salmon eggs and maggots.

SALMON OF SUICIDE ROW

One of my favorite steelhead streams is the famous and beautiful Klamath, of northern California, also renowned in season for coho and chinook salmon. I never have fished Suicide Row, but Larry Green has, and his account of it seems worth repeating:

The mouth of the river can be reached by boat. A short distance upstream from the little town of Requa is a place where the treacherous curling breakers of the Pacific smash into the outward-flowing Klamath, and it is here, in early September, that migrating hordes of salmon pour into the mouth of the river to acclimate themselves to the fresh water before proceeding upstream. Here, a strong cable has been stretched across the river, so small boat fishermen can tie up to it to fish their lures in the fast current. In season boats are moored closely together along the cable, this being referred to as Suicide Row.

Incoming or milling salmon eagerly strike at baited anchovies or flashing blades. When a fish is hooked the angler must quickly start his motor; untie his bow line, and drift free before becoming entangled in hundreds of lines. Then he must get to shore to battle his fish. If a motor doesn't start immediately, the boat will be swept to the mouth of the river into the dangerous breakers and undertow. In earlier days many anglers lost their lives, thus giving the place its name. More recently the Coast Guard has stationed a special rescue unit here, standing constantly by during the salmon season. Always alert for trouble, the Guardsmen swoop out in a fast boat and tow hapless anglers to shore. In spite of the danger and confusion, hundreds of brightly silvered chinook are taken here almost daily. Some of them weigh over 50 pounds!

14

DIFFERENT NETS

The most important raw materials for nets in European fisheries have been vegetable fibres-cotton, flax, hemp, sisal and manila. When immersed in water, they are exposed to cellulose-digesting micro-organisms, especially bacteria. The resistance of vegetable fibres increases in the following order: flax, hemp, ramie, jute, cotton, sisal, manila, coir. Factors affecting the durability of the net of vegetable fibres, and of more importance than this resistance are:

Duration of Immersion in Water

Fishing gear left in water for a long time are more liable to rotting than those used only temporarily. Rotting is stopped only when the nets are dried completely, also inside the knots.

Water Temperature

The warmer the water, the quicker the rate of rotting'.

Kind of Water

Due to intensive rotting nets will be destroyed considerably sooner in eutrophic than in oligotrophic or dystrophic waters.

In middle European eutrophic water, with higher temperature, unpreserved cotton nets may become useless after 7 to 10 days. Only very intensive preservation methods can prolong their durability. Preservation is most effective for cotton nets whereas efficiency is poor with bast fibre or hard fibre nets.

The durability of fishing gear is primarily affected by rotting. The original cost is increased disproportionately owing to frequent renewals and preservation expenses. The smaller the business, the greater are the relative expenses for nets.

In a report on nylon as net material, Needham writes: "Nylon brings to one of man's oldest occupations the miracle of science and, in doing so, provides easier living for the fisherman . . . "This statement can apply to all synthetic fibres, for this miracle is apparent in their proof against rot. The first synthetic fibre, known as PeCe (polyvinyl chloride), was developed by the former. Some years later, twines made of these fibres began to be used for smaller gears in German inland fisheries, and its rot-proof quality became apparent. This meant much greater durability of gears (fyke nets made of PeCe, have been in continuous use for over 15 years), less work and lower preservation expenses, obviating the time consuming drying of nets and thus ensuring longer catching periods. Apart from this rot-proofness which is the most important characteristic of synthetic fibres, other important properties are: breaking strength, extensibility, elasticity, abrasion resistance, diameter, weight, stiffness, resistance to weathering, knot stability and behaviour in water, including change of length, sinking speed and visibility.

TYPES OF FISHING GEAR, MATERIAL AND KINDS OF SYNTHETIC FIBERS USED

A great variety of gear is used in fishing, differing very much in size, structure, method and purpose of use. They have been classified by v. Brandt. According to strain on the net-material German fishing gear can be divided into three groups[11]:

Low Strain

Fine gillnets belong to this first group. The traditional material is fine cotton twine of metric counts 270/6 to 70/9 (English counts 160/6 to 40/9) or similar thickness, with a diameter of less than 0.40 mm. and a breaking load (wet) not higher than about 3 kg.

Medium Strain

This group is formed by the various fishing gears of the German inland, coast and deep-sea fisheries which formerly were made of cotton twines, of metric counts 20 and 50 (English counts 12 and 30). Fine twines of flax or hemp were occasionally used. This includes:

most fishing lines

baskets, also trap nets and box nets

scoops

gape nets (without those mentioned above)

dragged gears also bottom trawls of smaller vessels floating trawls of cutters seines and purse seines dip nets and lift nets

drift nets

Heavy Strain

Bottom trawls of large vessels (trawlers and cutters) and gape nets set in fast running rivers belong to this group. They were made of thick manila, sisal or hemp twines, with diameters up to 3.9 mm. and breaking loads up to nearly 200 kg.

Table 14.1

Fibre	*Trade Names*		*Characteristics*	*Suitable for use in*
Polyvinyl Chloride	F: G: J:	Rhovyl PCU, Envilon, Teviron	medium breaking strength, Rhovyl medium abrasion resist., very good resistance to weathering	group 2
Polyvinyl Alcohol Kuralon,	J: G:	Vinylon, Manryo, resistance, Trawlon, Mewlon PVA	low price, medium breaking strength, medium abrasion good resistance to weathering Cremona,	group 2
Polyester	GB: USA: G: F: I: J:	Terylene Dacron Diolen, Trevira Tergal Terital Tetoron	very high breaking strength, low extensibility, medium abrasion riesistance, relatively good resistance to weathering	groups 1 and 2, group 3?
Polyethylene	GB: N: G:	Courlene X3, Drylene Nymplex Polyathylen-Hoechst	high breaking strength, very high abrasion resistance, wiry, swimming in water	group 2, group 3?
Polypropy-lene	I:	Meraklon USA	low price, medium to high breaking strength, low resistance to weathering, swimming in water	group 2, gr.la.3?
Polyamide mono-filament	Nylon-monofil., Perlon-monofil., Platil		transparency (little visible-ness in water), very good resistance to weathering, wiry fyke nets	group 1, group 2:

Fibre	*Trade Names*	*Characteristics*	*Suitable for use in*
Polyamide	Nylon-staple, Perlon-staple	high breaking strength, staple high abrasion resistance, good knot stability, very high extensibility, low resistance to weathering	group 2
Polyamide continuous filament	Nylon-filament Perlon-filament	very high breaking strength, very high abrasion resistance, high elasticity, low resistance to weathering	all 3 groups of gear

F=France, GB=Great Britain, G=Germany, I=Italy, J=Japan, N=Netherlands.

The characteristics relevant to fisheries purposes for the most important synthetic fibres, and shows for which of the three groups these fibres could be used. In the column "trade names", only the best-known are listed. Only nylon and Perlon are given for the polyamides, although fibres of this group and of equal value to fishing gear are produced in different countries under various names, such as: Amilan, Anzalon, Dayan, Dederon, Ducilon, Enkalon, Fefesa, Forlion, Grilon, Kapron, Kenlon, Knoxlock, Nylock, Rilsan, Silon, Steelon and Tynex.

SYNTHETIC FIBRES USED IN FISHING

Properties of primary importance in the evaluation of net twine are rot-proofness, breaking strength, extensibility and abrasion resistance. Resistance to weathering may be of primary importance for fishing gears, which are used just below the surface or partially out of water.

Based on experience in German fisheries and on investigations by the Institut fur Netz-und Materialforschung, an attempt is made below to evaluate these special properties.

Breaking Strength, in Knotted, Wet Condition

Twines of cotton, hemp and manila are strongest in wet condition. With synthetic fibres there are groups which have the same or nearly the same strength wet or dry: Polyvinyl chloride (PeCe, PCU, Rhovyl), Polyester (Terylene, Dacron, Trevira, Diolen), polyethylene poly propylene and some copolymers (Saran, Vinyon, Dynel, Acrylan). Other groups show a loss of strength when wet, such as: polyamide

(Nylon, Perlon), polyacrylonitrile (Orlon, PAN), polyvinyl alcohol (PVA, Manryo, Kuralon, Vinylon). Diminution of strength occurs as soon as the net twines come in contact with water. After a 24 hour immersion the following approximate losses of strength were observed: twines made from polyamide staple and cont. filament 13 to 17 per cent., polyamide monofilament nearly 20 per cent., twines made from polyvinyl alcohol staple about 25 per cent.

During prolonged immersion in water extending over several months, a further gradual decrease of strength takes place,[12] though by no means comparable to that of even well-preserved net twines of vegetable fibres.

It is of importance to know the decrease in strength caused by knotting. This is connected closely with the properties of the fibre substance. Continuous filaments of polyamide and polyester are of high tenacity (highly stretched) in order to obtain as high a strength as possible, and have a lower extensibility. Unfortunately, they show a diminution of knot strength too. Losses of strength of wet twines by knotting (in the average):

Twines made from		**losses in per cent**
manila, medium fine tw.		30
manila, thick twines		40
hemp		25
cotton		30
Perlon staple		35
Nylon contin. fil., fine tw.		33
Nylon contin. fil., medium fine tw.		43
Nylon contin. fil., thick tw.		50
Perlon contin., twisted	(trawl twines)	50
Perlon contin., braided	(trawl twines)	42
Peilon monofilament	(trawl twines)	30
Polyester contin.,	(trawl twines)	50
Polyethylene		36

Net material, therefore, has to be evaluated according to its *strength in knotted and wet condition.*

such an evaluation for net twines from group 2. Based on tests of a great number of net twines of different thickness, the breaking length in km. and tensile strength have been calculated. For calculating the

Table 14.2 : Breaking strength of net twines, single knot and wet

Twines made from	*Breaking Length km.*	*Tensile Strength kg./mm.²*
Polyvinyl chloride		
Rhovyl-Fibre (staple)	6.4	9.0
Rhovyl, continuous filament	7.6	10.6
PCU, staple	8.2	11.5
Polyvinyl alcohol, staple	10.2	13.3
Polyacrylonitrile, continuous filament	14.5	17.0
Polyester, continuous filament	26.3	36.3
Polyamide, staple	15.7	17.9
Polyamide, continuous filament	30.1	34.3
Cotton	12.2	18.5
Hemp	28.9	42.8

tensile strength (kg./mm.^{2}) the following specific gravities (which are mostly taken from the list of Grünsteidl and Preussler[16]) have been used:

Polyvinyl chloride	....	....	1.40 g./cm.^{3}
Polyacrylonitrile	....	....	1.17 "
Polyvinyl alcohol	....	....	1.30 "
Polyester	....	....	1.38 "
Polyamide	....	....	1.14 "
Cotton	....	....	1.52 "
Hemp	....	....	1.48 "

Hence of the synthetic net twines considered here, those of continuous polyamide and polyester are by far the strongest. Their wet knot-strength is twice that of cotton twines.

Extensibility and Elasticity

The extensibility of net twines consists of several components: that of the fibre, the single yarn, and finally extensibility of the end product or twine. The inherent extensibility of synthetic fibres varies, being large for the polyamide group and comparatively small for the polyester group. Short staple fibres produce a higher extensibility than continuous filaments. Hard twisted yarn is more extensible than a soft twisted one. The manufacturing process of the final products is also of great importance. The more the twine is twisted or the tighter the braid is plaited, the greater the extensibility. To obtain a high extensibility, it is better to use fibres with a large original extensibility than to use hard twisting of yarn and twine to attain this high extensibility.

Total extensibility comprises a part of elastic extension and the permanent elongation. Apart from the nature of the fibre and the manufacturing process, their value is dependent on the applied load. Where the total extension results mainly from the manufacturing process, i.e., from twist of yarn and twine, elasticity will be small, and the permanent elongation, which remains after unloading, will be comparatively great.

Net twine has a greater efficiency, if total extension and elasticity are high. It will be able to absorb kinetic energy and will take up

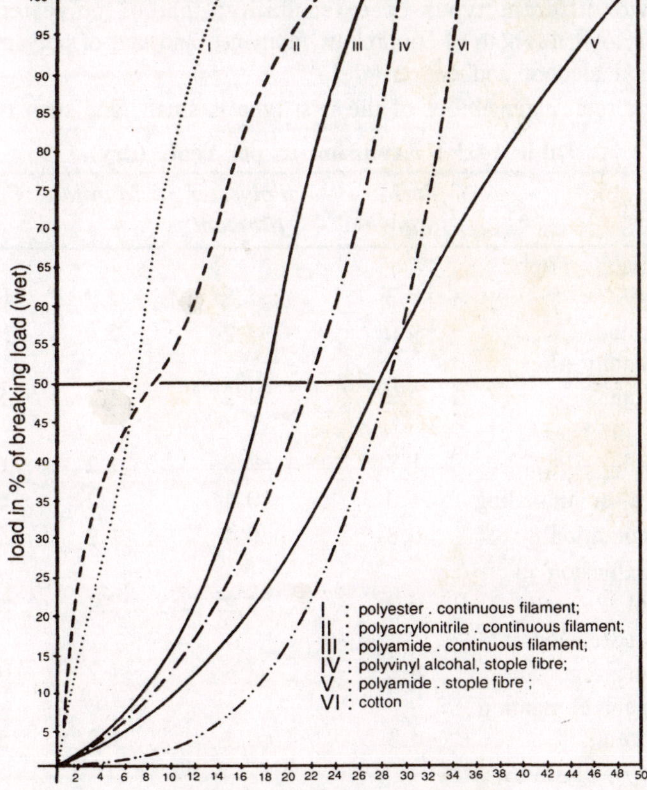

Figure 14.1 : Load-extension curves.

shock loads better than net twines of small extensibility. This problem was first examined thoroughly with nylon ropes. It is of particular importance for gear under heavy strain where one has to reckon with sudden strong loads.

It is difficult to evaluate the breaking extension of nets, as it must be expected that, for high-class fibres, knot-strength reaches only about 50 to 60 per cent. of the breaking load. Knowledge of breaking extension, is, therefore, not very important. The degree of extension with ascending load which, at the most, may amount to 50 per cent. of the breaking load is more important. The relation between load and extension, up to this limit, has to be observed, and an example is represented by the load-extension curves for net twines of material group 2. The curves have been drawn from average values which were calculated from the results of several tests of twines.

Two different types of extensibility; that of polyester and polyacrylonitrile (both of continuous filament), and that of polyamide., polyvinyl alcohol and cotton.

The total extensibility of the first type is small, and such twines

Table 14.3 : Extension in per cent. (dry).

Twine:	*Pcrlon filament*	*Polyester filament*	*Manila*	*Cotton*
Immediately after loading	8.3	3.8	2.2	14.2
1 hourloaded	9.0	4.2	2.4	15.0
Immediately after unloading	2.2	1.2	1.4	8.0
1 hour after unloading	1.7	0.8	1.2	6.0
1 day after unloading	1.0	0.4	1.2	5.0
Final condition	0.8	0.4	1.2	5.0
Total extension in per cent	9.0	4.2	2.4	15.0
Elastic extension in per cent	8.2	3.8	1.2	10.0
Permanent elongation in per cent.	0.8	0.4	1.2	5.0
Degree of elasticity in per cent	91	95	50	67

offer strong resistance, even at small loads. There is a steep ascent of the curves, as for those of manila twines.

The second type is affected at low tension or pressure, where the increase of extension already exceeds the increase of load. All curves apply to wet twines. Twines of staple fibres are much more extensible than those of continuous filaments, especially if the latter, as in the case in question, are of high tenacity. When cotton twines are wetted they show a shrinkage of nearly 10 per cent. With a small load, the extension of wet cotton twines is very great; with low tension pressure, shrinkage will be compensated by extension. Twines of fibres which have small shrinkage in water, such as twines of polyester or polyamide filament, will also show small differences between dry-extension and wet-extension.

Extension in connection with elasticity is of special interest. Elasticity regarding the stronger net twines, which are used for large bottom trawls. The twines have been loaded in dry condition for one hour by 30 per cent. of their breaking load.

The thick cotton twine shows the highest extension at load. After unloading, a rather high permanent elongation remains. For manila, the total extension is very small. Therefore, the absolute measurement of the permanent elongation is also small, though it comprises 50 per cent. of the total extension. Polyester twine is very elastic, its extensibility, however, is very small. Perlon filament twine has a proportionately high extensibility, together with high elasticity.

What significance does this difference in behaviour have for fishing gear?

A fixed extensibility cannot be set for net twines, as the requirements differ according to the type of gear.

In drift nets, for example, the fish will be caught in the meshes. The mesh sizes, which must be adapted to the size of the fish, should not alter, so the permanent elongation must be small. The twine should yield to the pressure of the fish caught in the mesh. Extension and elasticity, however, must not be too great, otherwise the fish would be squeezed in too tight, the quality of the fish would suffer, and it would be also very difficult to release it from the net. The very extensible twines of polyamide staple fibres are therefore less suitable for drift nets. Twines of polyester (Terylene, Dacron, Trevira, Diolen) would, however, be especially qualified for these gears, since their permanent elongation is small, but their relative elasticity great.

The requirements of gear exposed to strong tension or pressure are very different. This applies to trawls, especially the codends, when they are rushed up from a great depth to the surface or while heaving a large catch on deck. These nets were made of strong manila or sisal twines, in order to stand sudden shock loads. In consequence of their very small extensibility, hard fibres are not able to absorb kinetic energy. Best suited for large bottom trawls are twisted or plaited twines of polyamide filament (e.g. Perlon, nylon). Their extensibility is great enough to stand strong shock loads, and the high degree of elasticity guarantees a good constancy of mesh size. In the German trawl fishery most of the lighter bottom trawls and floating trawls of cutters are now also made of Perlon (in this case from staple fibres) because its high elastic extension gives the necessary strength for handling, also big catches. Perlon staple fibres have also proved suitable for otterboard stow nets.

As regards extensibility, drift and trawl nets are examples of extreme types of gear. Between them there is a great range of other types of gear with different requirements of extensibility and elasticity. For synthetic fibres, polyester filament and polyamide staple fibre are the extremes as regards extensibility.

Abrasion Resistance

In practice, it is scarcely possible to obtain exact knowledge of the relative abrasion resistance of nets made of different kinds of fibres. One is dependent on laboratory test methods, and only relative and never absolute values are obtained and these, moreover, are dependent on the test method employed.

For the following tests, the machine for testing abrasion resistance after Sander was used, in which the wet net twines are chafed across a bar of carborundum[19]. The counts or runnages of the various types of twines under comparison were taken as equal as possible. Thicker twines for large trawl nets were compared with one another, and also twines for gears of material group 2. If in the first case, abrasion resistance of thick cotton twine is equal to 100, the following approximate relation can be established:

Cotton .... 100 Manila .... 130 Hemp .... 280 Polyester cont. filament = 230 Polyamide cont. filament= 460

These figures refer to new, unused and wet twines, with a runnage of above 350 m./kg. The abrasion resistance of manila twines is very low. As some tests have shown, it is about the same as for sisal twines. As the breaking strength of nets of natural fibres is reduced by rotting, abrasion resistance will also be reduced. For twines of manila and hemp, the following relations have been established:

	Diminution of *abrasion resistance*	
breaking strength in per cent.	***manila, per cent.***	***hemp, per cent.***
10	2	5
20	3	9
30	5	14
40	6	19
50	15	23
60	26	27
70	37	34
80	48	50

Trawl nets are very much exposed to abrasion, which largely explains the short durability of manila nets. Polyamide fibres (Perlon, nylon) show the best abrasion resistance, and are by far superior to those of natural fibres. Trawl nets of Perlon filament used in the German trawl fishery, show about ten times the durability of manila nets, this is due not only to their rot-proofness, but also to the greater abrasion resistance of polyamides. Twines of polyester filament have only half the abrasion resistance of twines of polyamide filament, but their resistance is better than that of cotton and manila twines.

Abrasion resistance depends not only on the type of fibres and on the diameter, but also on the manufacturing process of the twine[18]. Hard twisted twines of polyamide filament have less abrasion resistance than soft twisted twines. Corresponding features apply to braids of polyamide filament.

The finer twines for gears of material group 2 have the following abrasion resistance (wet), when cotton is again equal to 100:

cotton	....	....	=	100
polyvinyl chloride Rhovyl-staple PCU-staple	....	....	=	50-55
polyvinyl alcohol				
staple	....	....	=	50-60
filament	....	....	=	110
polyacrylonitrile	....			
cont. filament	....	....	=	70

polyamide ....

staple .... = 170-250

cont. filament .... = 400-500

These relative figures correspond to twines of the same runnage or count. If the calculation is based on the same diameter, the values will be somewhat displaced. The third possibility would be a compa-rison of abrasion resistance of twines of the same breaking strength. In that case twines of high tenacity fibres, such as polyamide filament and polyester, do not show such a great superiority, since they are much finer than twines of weaker types of fibres.

Resistance to Weathering

There are few experiments on the resistance to weathering of nets made of vegetable fibres. These fibres are injured by sunlight. But such damage is small compared with that caused by micro-organisms in the water. Resistance to light and weathering is about the same in all vegetable fibres.

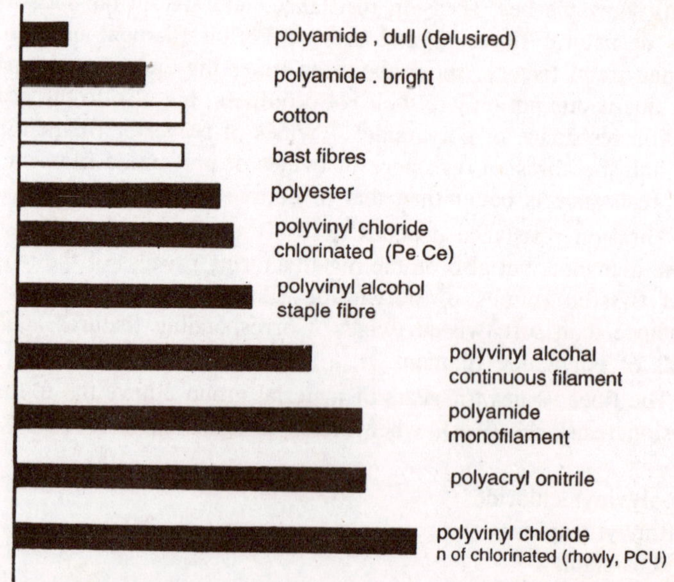

Figure 14.2 : Relative resistance to weathering.

Synthetic fibres show very great differences in their degree of resistance to light and weathering. Demonstrates their relative resistance to weathering. Of the fibres represented here, delustred polyamide offers least resistance, so that dull polyamide twines are not suitable for fish nets, since their resistance to light is too small.

It could be mentioned here that the USA are said to have succeeded in producing a dull nylon type with the same resistance to weathering as bright nylon. The resistance of twines made of bright polyamide (staple and filament) is still below that of cotton and hemp twines. Polyamide monofilament and fibres of polyacrylonitrile and non-chlorinated polyvinyl chloride have a very great resistance to weathering.

However, the other excellent physical properties of twines made of polyamide filament and staple more than compensate for their smaller resistance to light, especially for fishing gear of special economic significance, such as trawl nets.

As an example: in the Portuguese purse seine fishery, nets of unprotected (uncoloured) Perlon-staple twines, have been used for 1,306 working days and are still serviceable, whereas preserved cotton nets usually last for no more than about 500 working days.

For trap nets and other baskets, which stand partly out of the water or close below the surface, colouring of the material as a protection against light will often help. In this case nets of polyvinyl chloride, polyacrylonitrile and polyamide monofilament would be rather suitable.

SYNTHETIC FIBRES FOR VARIOUS FISHING GEAR

May now serve to answer the question as to the suitability of certain synthetic fibres for certain gear used in Germany.

Material Group

This comprises the different types of set gillnets, floating gillnets and trammel nets used in inland fisheries. Also fine herring nets of the Baltic fishery belong to this group. Gillnets are passive gears into which migrating fish swim by accident and become stuck in the meshes. Here, invisibility is of prime importance. The twines must therefore be of small diameter of just sufficient breaking strength, this depending on the species of fish to be captured.

Accordingly, it would be advantageous to use very fine twines from strong types of fibres. Two types of synthetic fibres have a particularly high strength: *polyester filament* and *polyamide filament*. Very fine twines made of the latter have stood the test very well in inland fisheries, and also for herring gillnets in the Baltic. They catch more fish than cotton gillnets, since they are finer and softer and less visible. The best material for fine gillnets is, undoubtedly *monofile polyamide-wire.* Its translucency makes it nearly invisible in water. The catch of gillnets of this material is many times that of cotton nets, as many fishing countries can testify.

Material Group

German fishing gear belonging to this group were all formerly made of cotton twines. For this, twines of breaking strength of 3 kg. up to about 50 kg. were used.

The largest gear of this group, the herring drift nets of the North Sea drifter, showing why twines of polyester filament are regarded as particularly suitable. Extension and elasticity of twines of polyamide filament are also suitable, but not twines of polyamide staple.

For gear of such large dimensions, the initial costs must especially be taken into consideration. The complete string of herring drift nets (called "fleet") often consists of more than 1,500 kg. cotton twine. The much greater strength of the synthetic fibres just mentioned cannot be fully utilised. It is not possible to use twines as fine as the breaking strength would permit, because of the damage that could be caused to the fish. Therefore, the weight advantage as compared to cotton nets and, consequently, the lower costs, will not be as great as for other gears. Thus, drift nets of polyester or polyamide are much more expensive than cotton nets. Perhaps it would be better, in this case, to use polyvinyl alcohol which is the cheapest synthetic fibre. Drift nets made of that fibre are scarcely more expensive than cotton nets and may be completely suitable, if a good stiffening process is used. The production of knotless drift nets of polyvinyl alcohol fibres would be a great advantage. Satisfactory results have been obtained in a first test with such nets made of Japanese Manryo.

The German Baltic Fishery recorded very good results with *salmon drift nets* made of polyamide filament, and this material will probably completely displace the traditional hemp nets during the next few years.

Passive gears, such as *baskets,* with the exception perhaps of the large trap nets (box nets, stake nets, fixed nets) of the coast fishery, do not generally require a very strong material. If fibres such as polyamide and polyester filament are used, then here too, may be a chance of catching more fish with a finer twine. Since all synthetic fibres are rot-proof, there is a very great choice of suitable materials. For trap nets, such as the Danish "Bundgarn", which stand partly out of the water,' fibres of high resistance to light would be recommendable, e.g.,)polyvinyl chloride fibres PCU and Rhovyl and als the new Japanese Teviron or Envilon.

Translucency, a special property of polyamide monofilament, has proved to be of primary importance in catching more fish, also with baskets. Eel fyke nets, made of monofilament, caught about twice as much as those made of cotton; that also would be true of other baskets. But this would not seem profitable, if the long leaders and wings of trap nets, were also made of monofilament. They would not lead the fish into the trap net (net proper), but act as gillnets and the fish would be caught in the meshes[22].

Dip nets, lift nets and falling nets (lantern nets, cast nets) are of little importance in the German fishery, but what has been said about material for baskets will mostly be applicable.

The most important gears of German rivers are *gape nets.* According to the flowing speed of the water, they require material of great to very great strength, of relative high extension and elasticity and of great abrasion resistance. Polyamides have these requirements. For swing nets (stow nets) at anchor (also for otter-board stow nets), with the smaller mesh sizes or such nets which are in rivers with a flowing speed of less than 5 km./hr., twines of Perlon-staple have lasted very well for more than six years. For stiffening, they were treated with "Black Varnish". Gape nets with large mesh sizes from rivers with strong current (e.g. the Rhine), are made either wholly, or in their most forward part, of plaited or twisted twines of polyamide filament, belonging to group 3.

Two more gear types of material group 2 are seines and light cutter trawls.

For *seines,* and also for *purse seines,* there are many possibilities of choice of synthetic twines. During the first years of introduction of PeCe into the German inland fishery there were, for instance, boat seines made of these fibres, although breaking strength and abrasion resistance were less than those of cotton twines. Since the physical properties of the non-chlorinated polyvinyl chloride fibres are superior, these can also be used for seines. The same goes for polyvinyl alcohol twines which are also inexpensive. These fibres, however, do not have a very high absolute strength in wet and knotted condition. Therefore, one cannot count on the gear being lighter than a cotton net of the same size. For polyester and polyamide, however, these relations are more favourable. If the breaking strength, wet and knotted, is taken as a basis, then gear of polyamide filament is only half as heavy as that of cotton.

Material Group

The requirements of the material of *trawls*. They are the same as for large gape nets; very high breaking strength and abrasion resistance and comparatively great extension and elasticity, and possibly the smallest weight. That goes both for lighter bottom trawls and floating trawls of cutters which still belong to material group 2, and for the large bottom trawls of deep sea trawlers belonging to *material group* 3. There are differences only in the degree of strain and not in principle. By far the best material for these purposes seems to be polyamides. For cutter trawls, besides twines of polyamide filament of titer 210 denier, weaker twines of polyamide staple may also be used. For heavy and large trawls, plaited and laid twines of polyamide filament have stood the test so excellently that the German trawl fishery works now mostly with Perlon nets for the catch of herring and of roundd fish.

Especially in the Japanese fishery, trawl nets of polyvinyl alcohol are frequently used. But their advantage seems to be in the low price only, as rot-proofness is peculiar to all synthetic fibres. They may be perfectly suitable for lighter trawl nets of smaller fishing boats, but for gears of large trawlers, twines of polyvinyl alcohol fibres do not seem to be very suitable. The test figures for one Japanese Manryo twine and for twines of manila and Perlon-filament, with nearly the same runnage, given in Table elsewhere in this chapter, may serve as an illustration.

Table 14.4

	Manyro 20'S/120 twine	*Manila Nm 0. 7/3*	*Perlon Nm. 3/12*
Runnage, m/kg.	239	~230	242
Breaking strength, wet, kg.	79.2	~95	156
Strength, wet, knotted, kg.	35.7	~60	99

The lighter a trawl net is, the easier its handling and the lower its trawling resistance. These aspects are particularly important for large gear. It can be seen that trawl nets of Manryo become much heavier and must have thicker twines, if they are to be as strong as manila trawl nets. The great advantages of trawl nets of polyamide filament in this respect are apparent.

SYNTHETIC FIBRES IN THE FISHING INDUSTRY

While Du Pont 66 nylon was first tested experimentally in the fishing industry in the United States for gillnets in 1939, military demand for nylon during the war delayed further work in the fishing industry. The use of nylon in gillnetting has grown rapidly. Recent estimates show that well over 90 per cent. of all Great Lakes' gillnets are now made of nylon.

More recently, nylon has taken over most of the salmon gillnetting in the Pacific. This rapid acceptance has been due primarily to the great increase in numbers of fish caught by nylon gillnets compared to the nets they replaced, as well as their lower maintenance costs, longer life, and easier care. Gillnets made of nylon twines have reportedly caught from 3 to 12 times more fish per net than those used previously, although the reasons for this phenomenon are not thoroughly understood.

In the heavier types of netting which used cotton twines in the past, nylon has also gained the approval of American fishermen. The first menhaden seine made entirely of nylon was placed in service in June 1954, and proved to be far more durable than nets previously used. In addition nylon nets have proved more economical in the long run. Significant growth has also been noted for nylon in lobster and shrimp gear and in tuna seining. Nylon is also being used in longlining for tuna and halibut.

In addition to the wide, general use of nylon twines for nets, the use of ropes of nylon and "Dacron" polyester fibre for hanging and purse lines has become quite common. Present industry sales of nylon in the U.S.A. for fish netting and twines amount to over 1,000,000 lb. annually. It is expected that in about five years' time at least 75 per cent. of all nets used in this country will be made of nylon.

FIBRE PROPERTIES

Du Pont 66 nylon has gained acceptance in netting because it offers an outstanding coordination of properties and price. Of these properties, high strength, wet or dry; abrasion resistance, complete resistance to rotting caused by marine organisms and high impact strength are the most important. Some specialised applications that call for low stretch under load have resulted in the use of "Dacron" polyester fibre in both net and rope forms.

Continuous filament nylon is the predominant synthetic type of raw material used by netting and rope manufacturers in the U.S.A., because it offers the highest breaking strength and maximum abrasion resistance. Spun nylon and combination twines are also being used successfully. Early problems of knot slippage or knot loosening have been largely overcome in nets made from continuous filament twines by the use of special resin-treated twines, special knots, or heat treatment of the finished nets. Spun twines of 100 per cent. nylon or twines. made of filament nylon combined with spun fibres are also used as a means of overcoming knot slippage without the need for any special twine treatments. Combination twines of filament nylon and spun fibres, in addition to good knot holding properties, offer the advantage of increased bulk at lower cost for certain types of netting where handling of fine twines might be a problem.

The following table gives some of the most important specification properties of nylon and "Dacron" used in netting:

Property	*Du Pont 66 nylon*		*"Dacron"*
	Type 300	*Type 700*	*Type 51*
Denier	210	840	220
Filaments	34	140	50
Denier per filament	6.2	6.0	4.3
Tenacity Dry (grams/denier)	8.0	8.7	6.4
Tenacity Wet (grams/denier)	6.8	8.1	6.4
Tenacity Loop (grams/denier)	5.4	7.1	3.9
Total extension dry (%)	17	17	10
Total extension wet (%)	22	24	10

These properties combine to produce more economical or more efficient nets.

GILLNETS

The high strength of nylon allows the use of thinner continuous filament twines which gill fish more effectively. Thus, the higher initial cost of nylon is more than offset by the greater number of fish caught per net. In addition, some fishermen feel that the elasticity of the nylon allows fish to force the twines apart and become caught. This may account for the fact that larger sized fish have been caught with a given mesh size than previously. The abrasion resistance of nylon is less important in this type of netting, but its rot resistance allows continued use of nets without the need for drying and treating.

PURSE SEINES

Abrasion resistance and resistance to rotting due to micro-organisms in sea water are the most important requirements for this type of net. It is also important to use twines of sufficient size to make handling easy. One hundred per cent. filament nylon twines are being used in menhaden purse seines to give maximum abrasion resistance as are combination twines of filament nylon and spun acetate, which give the required bulk and still achieve satisfactory strength and abrasion resistance.

To illustrate the unusually high abrasion resistance of nylon, a cotton "bunt" or centre section of a menhaden seine used in one area lasts an average of 5,000,000 fish caught. A 100 per cent. nylon menhaden net placed in service during June 1954 has caught a total of 50,000,000 fish; the net was still in service at the last report with the original bunt section intact.

The largest nylon purse seine ever manufactured in the U.S.A. was produced. Because of the greater strength of nylon, slightly finer twines were used which resulted in a weight saving of of approximately 2,400 lb., as compared to an all-cotton tuna seine. The nylon tuna seine weighed a total of 10,000 lb. and was 410 fm. long and 34 fm. deep. When wet, this nylon net weighs only half as much as a wet cotton net of the same size.

The Pacific fishermen, after using this net for nearly two years, are highly enthusiastic about its light weight and easier handling characteristics. In addition, the net shows little wear and the fishermen estimate it will be in service for a number of seasons to come.

In order to achieve optimum handling characteristics fishermen in the U.S.A. ordinarily dip nylon purse seines in a stiffening agent, such as an asphalt base tar, to impart required stiffness. Also, it has been found that slightly heavier weights should be used on the leadlines to provide the desired sinking qualities to the net.

TRAP NETS

The use of nylon in trap nets is growing slowly, although acceptance in the sardine industry has been widespread. Since these nets are stationary, rot resistance and abrasion resistance (particularly in the bottom section of the net) are important properties. Because of nylon's natural rot resistance, fishermen have found they can leave their nylon nets in the water almost indefinitely without fear of damage. One hundred per cent. filament nylon or combination twines of nylon and spun acetate are being used in trap netting. In the sardine fisheries experience has been good with a net of filament nylon, produced on a high speed knitting machine. This type of webbing has no knots and the mesh formation is made by proper settings on the knitting machine.

TRAWLS

In shrimp trawls where the sea bottoms that are fished are

moderately level, nylon nets perform well and are being accepted. In the larger nets for bottom fish where equipment is frequently snagged and either lost or badly torn, the extra cost of nylon may often not be justified.

FISH NET ROPES

With the increased use of synthetic fibres in fish netting, ropes and lines with the same strength, elasticity and rot resistant properties as the net material itself are preferable in most cases. For this reason, the use of ropes of nylon and "Dacron" polyester fibre has grown considerably as more fishermen have switched to synthetic netting. Nylon ropes require no preservative treatment and give outstanding wear life compared to ropes made of natural fibres. A tuna fisherman on the Pacific coast, for example, has reported that his nylon net lines have been in service for over five years, and are still in good condition.

A certain amount of difficulty was experienced at first in hanging nets on ropes of nylon or "Dacron". This was due primarily to the difference in elongation and shrinkage properties of ropes of these fibres as compared with natural fibre ropes. As a result of long experience with natural fibre ropes, the number of meshes of netting to hang per foot of rope had become well known and were correct for the stretch and shrinkage of such ropes. Fishermen have now learned that properly made ropes of nylon or "Dacron" stretch somewhat more than manila but do not shrink when wet. They have modified their hanging techniques to compensate for these differences.

Treated nylon is being used in longlining for ground fish such as halibut. Similar lines are being used for lobster pot warps, as well as for headers in the pots. In these cases, the shock absorbency resulting from the high elasticity of nylon is a definite advantage, particularly in rough weather.

NEW DEVELOPMENTS

Improvement of the nylon yarn as produced by Du Pont, as well as improved preparation of the twine by the net manufacturer continues.

An example is the recent introduction by Du Pont of Type 330 nylon yarn. This yarn has better resistance to sunlight than either cotton or previously available types of nylon. This yarn is now available to fish net manufacturers and should insure longer life for nets when sunlight is a significant factor.

A new development in the manufacture of nets of continuous filament yarn has been the use of "Taslan" textured nylon for netting twines. The texturing process creates tiny loops in the individual filaments, imparting new surface characteristics. Thus, good knot firmness can be obtained by the use of twines made from "Taslan" yarns. In addition, the texturing process is useful for combining two or more continuous filament yarns intimately without the need for twisting.

Such developments-and many others-are typical of the avenues which will continue to be explored as means of improving gear for the fishing industry.

"AMILAN" FISHING NETS

Synthetic fibre nets are now popularly used and their efficiency is highly appreciated by fishermen all over the world.

Amilan is the trade name of nylon produced by the Toyo Rayon Co. under licence and with the technical cooperation of E.I. du Pont de Nemours and Co., U.S.A. Amilan fishing nets are very well known today among fishermen in Japan, and are also exported to many fishing countries.

In the following sections the use of Amilan for nets in various types of fishing is described.

SALMON AND TROUT GILLNETS IN THE NORTHERN PACIFIC

The Northern Pacific fishing has become vitally important to Japan, as sea products are indispensable to feed Japan's ever-growing population. In pre-war days, the main practice in the Northern Pacific was large scale drift net fishing near to the coast of Kamchatka Peninsula. under agreement with the U.S.S.R. At present, however. the chief fishing is based on mother-ship operation.

When the Pacific War ended, production of Amilan was started and fishermen were using both conventional ramie and Amilan nets in the same quantity.

A series of tests clearly indicated the superior catching ability of nylon nets which in addition are claimed to stand about three years of consecutive use as compared to half a season for ramie. At that time, the price of Amilan nets was twice as high as that of ramie nets, but taking into account the durability and larger catches, they proved. however, more economical.

The catch figures illustrate this point. The first fishing fleets after the war caught 37,000 salmon and trout in the Northern Pacific. The catch was twice as much, at least partly due to the increased catchability of Amilan nets. Since then, more and more boats have adopted the nets and they are now used by even- fishing vessel.

SANMAI NET (TRAMMEL NET)

This net is known by different names in Japan, such as Jigoku-Net (Hell Net), Sanzyu-Net (Three-fold Net), etc., and is widely used for inshore fishing. Before nylon made its appearance, the nets were made of natural fibres such as cotton, ramie and silk. Today, the nets are made of Amilan and about 80 or 100 times more of such nets are being operated.

PURSE SEINE NET

The use of nylon for the construction of these nets has great advantages owing to its increased tensile strength per weight unit, as compared to cotton. When Amilan is used the boat and crew can work a much larger net, or less crew is necessary to operate a purse seine of the size a cotton net would require. Amilan is therefore now used widely in the tuna and bonito fisheries of Japan.

DRAG NET

The strength, lightness, small moisture absorption and durability of Amilan trawl nets allow for increased trawling speed. Fukuoka Fisheries Experiment Station showed that the nets produced higher catches. Some small fishermen, however, still hesitate to use them, because of the cost, but it is considered that their advantages outweigh this factor.

DEVELOPMENT OF SYNTHETIC NETTING AND ITS EFFECT ON THE INDUSTRY

The Japanese fishing net manufacturing industry has made considerable progress in the last few years, due largely to the development of synthetic fibres, such as nylon, vinylon, vinylidene, vinyl chloride, and combination synthetics. The export of nets, twines and ropes of all types from Japan, including natural fibres, totalled 7,163,000 lbs. in 1956, a 22 per cent. increase over the previous year's figure, compared with an increase of 9½ per cent. in exports of natural fibre fishing nets, twines, and ropes.

The following are the highlights in a survey made by the Japanese Government's Fisheries Agency of export trends in fishing gears, classified according to natural and man-made fibres.

NATURAL FIBRES

The exports of cotton fishing gear continue to lead all. other kinds of fishing nets, totalling 5,220,000 lbs. in 1956. This accounts for 73 per cent. of the total exports; 460,000 lbs. of manila hemp were exported in 1956; 336,500 lbs. of sisal and a total of 22,000 lbs. of other natural fibres (flax, coir, ramie and silk).

SYNTHETIC FIBRES

The exports of nylon nets, twines and ropes amounted to 683,000 lbs. and increased 158 per cent. in 1956 over 1955. More than 91 per cent. were nylon fishing nets, while 8 per cent. were twines; ropes constituted I per cent, of this total. The export of other synthetic fibres in 1956 was as follows; vinylon 221,000 lbs. (increased 2.8 times over the previous year); vinylidene 63,500 lbs.; vinyl chloride (which is a comparatively new synthetic fibre) 3,000 lbs.; combination synthetics 165,000 lbs. (increased 3.8 times in 1956 and are expected to exceed 420,000 lbs. in 1957).

DIFFERENT FIBRES FOR DIFFERENT FISHING PURPOSES

Synthetic fibre fishing gear has contributed materially toward the stabilisation of the fishing industry, but because of the industry's complexity, no synthetic fibre developed to date is ideal for all types of fisheries. At present, gillnet fisheries find nylon a satisfactory replacement for linen netting. The purse seine fisheries have accepted Marlon (nylon/vinylon) as a replacement for cotton netting, while vinylon may prove an acceptable replacement for cotton and manila for fishing lines and ropes. For trawl fishing however, no particular synthetic stands out as acceptable to replace natural fibres. Approximately 75 per cent. of the nylon netting exported from Japan is used in the gillnet fisheries of various countries. A great variety of nylon types can be produced by changing its chemical construction, production method, yarn size, degree of twist, amount and method of relieving the stress in the twine, and the different processes for setting the knots.

It should be borne in mind that when selecting the appropriate nylon twine to replace linen for a gillnet, wet knot strength of nylon is less than that of first grade linen. So, where it is necessary to maintain the same strength a slightly heavier nylon twine must be chosen.

Nylon exposed continually to sunlight of normal intensity for two months can lose as much as 40 per cent. of its tensile strength. It must therefore be kept away from direct sunlight.

Nylon is not weakened by bacteria, but a nylon net should be kept clean, as acid caused by fish slime can be harmful.

DEEP WATER GILLNETS AND SEINES

To determine the twine diameter most suitable for deep water gillnets, consideration must be given to:

a. Strength of the twine
b. Fishability
c. Form of the net in the water when fishing
d. Durability and
e. Initial cost of the net.

Factors a and b are opposite in function, namely thicker twine means higher strength, but lower catching ability, therefore, a compromise has to be found.

The specific gravity of the material influences the form of the net when fishing and it was found that pure nylon nets were too light. Attempts were made to overcome this deficiency by hanging a strip of vinylidene net 360d/30 ply on a ratio of nylon 4 : vinylidene I under the nylon. This method was not entirely satisfactory.

The later method, which proved satisfactory, was to combine nylon and vinylidene fibre during manufacture of the yarn and twine. This product called Livlon has considerable merit for deep sea gillnetting operations.

It is especially important to determine accurately the most suitable twine size for deep water purse seine because the webbing is subject to much water resistance while sinking. The specific gravity of a fibre can be determined by the laws of physics, but accurate determination of the resistance is complicated by various factors.

For a comparative study, the unit of resistance $K = D/L$ can be used where D is the diameter of the twine and L the meshsize (stretched).

The unit of resistance of a standard sardine purse seine made of cotton twine 20/15 ply with a mesh size of 20 m/m would be 0.05.

If the influence of the surface condition of the twine on the resistance is called "R", the numerical values of "R" for different types of twine are as follows:

Cotton	...	...	1.00
Nylon	...	...	0.75
Vinyl Chloride	...	...	0.70
Vinylon	....	....	1.00

These factors "K" and "R" influence the shape of the seine in operation, sinking speed, etc.

It may be assumed that a set is made with a cotton seine 20/15 ply, 20 m/m stretched mesh, 300 fathoms long by 10 fathoms deep, on a sardine school of 100 tons, 20 fathoms in width, that the travelling speed of the school is 0.5 fathoms/second, and that it takes 2 minutes for the sardines of the lower part of the school to arrive at the skirt of the net. In this case, the seine must be pursed within two minutes to catch the entire school, thus the net must sink at a rate of 1 fathom per second, which is quite difficult to accomplish with cotton netting. However, Livion seine netting of smaller diameter twine, manufactured with a special twisting method equal in specific gravity to cotton, is able to meet the requirements.

Marlon netting was first put to commercial use in 1951, and many of the original nets are still giving good service. Its most notable characteristics are derived from the combination of the best features of nylon and vinylon fibres.

Some of the advantages of such a combination are: increased specific gravity of the twine, maintenance of strength by the predominance of nylon fibre in the twine and decreased knot slippage.

Marlon netting is available in a variety of colours, twine and mesh sizes, and completely made up purse seines of standard or controlled specifications are available.

Like most synthetics, it requires little maintenance, but reasonable precaution should be taken to prevent undue exposure to sunlight.

CONCLUSIONS

Synthetic fibres are rapidly replacing natural fibres in the construction of all kinds of fishing gear, but an ideal synthetic fibre suitable to all types of gears is not yet available to the fishing industry.

At present the synthetic fibres most commonly used in the main types of fishing gear can be listed as follows

Type of Gear	***Synthetic most used***
Gillnets, trammel nets, tangle nets	nylon
purse seines, beach seines, lamparas, trap nets	Marlon and other combination synthetics
beam trawl, otter trawl, balloon trawl	various but none entirely satisfactory
twines and ropes	vinylon. nylon and combinations

NYLON FISHING TACKLE IN SWEDISH INLAND FISHERIES

This paper gives a description of tests which were begun in 1947 by the Institute of Freshwater Research (Sotvattenslaboratoriet), on the design of nets made of nylon.

One of the greatest disadvantages, particularly of continuous filament or monofilament nylon, ways the difficulty of making knots that would not slip. This was gradually eliminated by using various methods, such as the treatment e.g. chemical treatment of the material before or after knotting, as well as the use of double knots. In cases where the nets are exposed to great stress, the last-mentioned method is still the most generally used.

Certain difficulties are still encountered in the manufacture of monofilament nylon nets. Apart from the use of double knots, the chemical treatment has proved to be most effective.

One more disadvantage of nylon nets for fishing purposes is their relatively high sensitivity to ultra-violet rays. Tests carried out with both normal sunshine and quartz-lamps have proved multifilament nylon to be most sensitive to such radiation, whereas the monofilament nylon is more resistant. While it is true that special impregnation of multifilament nylon with catechu offers a certain protection it cannot prevent the gradual reduction in strength. From the practical point of view this sensitivity to light is not an important disadvantage to the fisherman. If he knows of the destructive influence of light on his fishing tackle, he can take care that his nets and tackle are not exposed to sunshine more than necessary. Practical experience gained by Swedish fresh-water fishermen has also proved that nets that must be discarded after 3 to 4 years have not become unserviceable due to the destructive action of sunshine, but to damage through tearing during intensive fishing.

Quite different are, on the other hand, the conditions prevailing in connection with the use of certain types of fixed fishing tackle such as bottom nets and fish baskets (bow nets), when part of the tackle often remains standing above the water level. In such cases the sensitivity of nylon to sunshine necessitates frequent replacement of the destroyed parts. It would be practical to use for these parts Terylene or Saran fibres, which are extremely resistant to U.V. radiation.

It must be mentioned here that monofilament nylon very easily becomes slimy through the adherence of microscopic particles whirling about in the water, particularly when the wind is strong. Since the comparatively high catching ability of this type of net is mainly dependent on its relative invisibility in water, the slimy coating of the net reduces its efficiency which, in certain cases, becomes less than that of spun nylon. This slimishness is particularly noticeable when the catch is checked without lifting the net from the water, and the net is thus left in a submerged position, sometimes for several days. The thorough rinsing of the nets is absolutely necessary to avoid a reduction of the catching ability. This disadvantage is responsible for the relatively limited use of monofilament nets in many flat-country lakes, where the water often has a high slime content.

COMPARATIVE FISHING TEST

In order to estimate the catching ability of continuous multifilament nylon compared with cotton, fishing tests were carried out, using nets of equal size. Separate tests have proved the catching ability of nylon nets to be twice that of cotton nets. Other tests showed that the excess catch made with nylon nets varied between 1.75 to 3 times.

Apart from the tests carried out by the Institute, supervised tests were made by professional fishermen and gave, on the whole, the same results. All tested nylon nets were manufactured from continuous-multifilament twine as staple nylon twine is unsuitable for gillnets. The reason for the high catching ability of nylon is its low water absorption, relatively high elasticity and high breaking strength. These qualities permit the use of thinner twine than is possible with the weaker cotton. This rule is valid for all kinds of net fishing: the catching ability of nets increases with a reduced twine diameter. The choice of the proper diameter is naturally merely a question of experience. The twine must not, however, be so fine as to cause very quick wear and consequent need for frequent repairs.

Since 1951 the Institute has carried out catching tests with nets made from monofilament nylon. This is an almost ideal material for certain fishing tackle as, due to its transparency, it is practically invisible in water. It was found that the monofilament gillnets caught not less than seven times as much fish as cotton nets, and approximately four times as much as nets of continuousmultifilament nylon.

The tests, which were carried out during both summer and autumn, also proved that the catching ability of nets made of monofilament nylon are to a certain extent dependent on light conditions. Thus, the excess catches were higher during the light season when the advantage of transparency is greater. It has also proved possible to achieve good results with these nets in daytime in lakes with clear water, when cotton or continuous-multifilament nylon nets were quite unsuccessful.

A test was carried out during the same season in two different lakes. The degree of visibility in these lakes was 13 yards and 13 yards respectively. In the former lake, the ratio of the catches made with nets of cotton, continuous-multifilament and monofilament nylon was 1:2:14. The corresponding figures for the lake with reduced degree of visibility were 1:2.5:11. The fact that the catches made with monofilament nets in the latter lake were unexpectedly high indicates that apart from the positive influence of the invisibility in water on the fishing results, this material must possess additional advantages on account of its structure.

Continuous-multifilament nylon proved to be more efficient than cotton quite irrespective of the size of gillnets used-whether high or low (superiority varying between 1.75 and 3).

When higher nets made of monofilament nylon came into use, it was discovered that the superiority of this material compared with continuous-multifilament nylon disappeared and that the catch, in some cases, was even less. Professional fishermen using such high nets had the same experience. The reason most probably lies in the increased stiffness of the monofilament nylon used for these nets. This may have an adverse influence on the flexibility of the net surface, which is of importance when fish jostle against the net. Nevertheless, singular positive results were achieved and one of these cases may be quoted.

In a fishing test a net of monofilament nylon caught 60 per cent. more white fish than a corresponding net of continuous-multifilament nylon. The net used for this species of fish is 20 ft. deep, provided with extremely light weights, and has its floats on the water level, connected to the net by lines by which the net position between bottom and surface can be varied.

THE SCOPE OF APPLICATION OF NYLON NETS

The change from cotton to nylon nets among professional fishermen, despite their conservative attitude to modernisation, is practically 100 per cent. They soon realised the great advantages of nylon, thanks to its high catching ability and absolute resistance to rot as compared with cotton. It is also undeniable that the introduction of the new material for fishing tackle has brought about a considerable improvement in the economic situation of professional fishermen. Today, the tackle of professional fishermen is mostly made of continuous multifilament or staple nylon. The shallow nets used by the non-professional and casual fishermen are made of monofilament nylon.

EXPERIENCE IN THE USE OF NYLON NETS

In order to secure the greatest possible catches when using nets made of monofilament nylon, it is necessary to follow certain rules. Perhaps the choice of the right diameter in relation both to the mesh size and to the species of fish for which the net is intended, is the most important. There are quite a number of different species of fish in the same lake and consequently the material must be strong enough for the strongest species of fish. The following standard diameters are used for certain mesh sizes in Swedish lakes: 0.12 to 0.15 mm. for 42 to 55 mm. stretched mesh; 0.15 to 0.20 mm. for 60 to 75 mm.; 0.20 to 0.25 mm. for 85 to 100 mm.; and 0.22 to 0.30 mm. for 110 to 150 mm. In many lakes in northern Sweden, where the catches often consist mainly of white fish, thinner material is used for the respective mesh sizes, as such fish does not tear the nets to the same extent as pike, pike-perche, trout or char.

Unfortunately, nets made of monofilament nylon earned a bad reputation, because the firms delivered nets where the material was too thick in relation to the mesh size, which resulted in inferior catching ability. Another reason was that with some of the nets the knots were slipping.

The nets made of continuous-multifilament nylon at present on the market are almost without exception of good quality with good resistance against slipping.

Fishermen using fixed fishing tackle have to a great extent changed over to the use of nets made of staple nylon twine, particularly in lakes where this kind of fishing is carried on practically the year round and where the rot-problem in connection with cotton becomes extremely serious. As already stated, the parts of the tackle remaining above water level must be renewed now and then, due to deterioration by sunshine.

The staple nylon twine used for this type of tackle is always treated with a stiffening agent which also makes it less elastic. As a rule coal-tar or bituminous varnish, dissolved in an equal quantity of benzol, is used.

The maintenance of fishing tackle made of nylon is extremely simple. The nets must not be exposed to sunshine for long periods of time. The repairs are, however, somewhat complicated. So far as continuousmultifilament twine or monofilament is concerned, special knots must be made since usual netting knots, such as those used with cotton twine, are not firm enough.

SYNTHETIC MATERIALS IN THE NORWEGIAN FISHERIES

Synthetic materials are mainly used in Norway for catching cod and coalfish with gillnets, and have mostly replaced the conventional cotton and hemp fibres. Nylon 66, nylon 6 and Terylene are chiefly used, the twine being imported from Great Britain, Canada, U.S.A. and the continent, although a small amount is produced in Norway. Synthetic materials are also used for snoods, lines and ropes.

The use of synthetic materials by Norwegian fishermen has gradually increased since 1955, following experiments with gillnets of nylon and Perlon which were started by the Directorate of Fisheries in 1951 and continued in 1952, 1953, 1954 and 1955. On the condition, that reports were made on catchability, etc., certain fishermen along the coast were given a quantity of nets of nylon and perlon to use alongside the conventional gillnets made from cotton and hemp. The number of fish caught by each type of net was counted in every haul and a report sent to the Directorate of Fisheries when fishing was finished. The following example is typical of the reports received (overleaf).

As will be noticed from the report, the catchability of the nylon nets was approximately twice that of the conventional. nets. This figure is also typical as an average, although according to several reports, catches were sometimes seven times higher. The types of twine used in these experiments were comparable as to wet knot strength and, according to the reports, durability and strength were satisfactory.

When the results of these promising experiments were published, the interest in the synthetic material grew rapidly among the fishermen and production of nylon gillnets was taken up by the Norwegian manufacturers to meet the increasing demand. The nets produced in Norway are mainly the single knot type.

The following figures illustrate the growth of imports of synthetic material for the fisheries, including twine, filaments and nets (a large part of the nets are of the double knot type):

1954/55	approx.	15 to 20 tons
1955/56	,,	290 ,,
1956/57	,,	400 ,,

1954/55 was the first season when nylon came into use.

The strength of nylon is often overestimated when first used for

	65 ordinary codnets (cotton and hemp)		*10 nylon nets*	
1953 Date	***Total catch per day***	***Average catch per net per day***	***Total catch per day***	***Average catch per net per day***
19/2	40	0.62	19	1.9
21/2	108	1.66	132	13.2
23/2	168	2.58	170	17.0
24/2	29	0.44	17	1.7
26/2	21	0.32	14	1.4
27/2	27	0.42	9	0.9
28/2	71	1.09	11	1.1
2/3	45	0.69	9	0.9
3/3	371	5.70	112	11.2
4/3	87	1.33	26	2.6
5/3	234	3.60	47	4.7
7/3	319	4.90	111	11.1
10/3	323	4.96	133	13.3
11/3	235	3.61	93	9.3
12/3	140	2.15	21	2.1
14/3	314	4.83	71	7.1
16/3	221	3.40	83	8.3
17/3	307	3.18	93	9.3
18/3	59	0.90	12	1.2
19/3	76	1.16	31	3.1
20/3	231	3.55	52	5.2
24/3	498	7.66	129	12.9
25/3	87	1.33	14	1.4
26/3	14	0.22	9	0.9
27/3	73	1.12	18	1.8
28/3	366	5.63	57	5.7
30/3	82	1.26	27	2.7
31/3	140	2.17	16	1.6
1/4	270	4.15	28	2.8

Total 1,885 nets	4,956 fish	290 nets 1,564 fish
Average=2.63 fish per net per day		Average=5.38 fish per net per day.

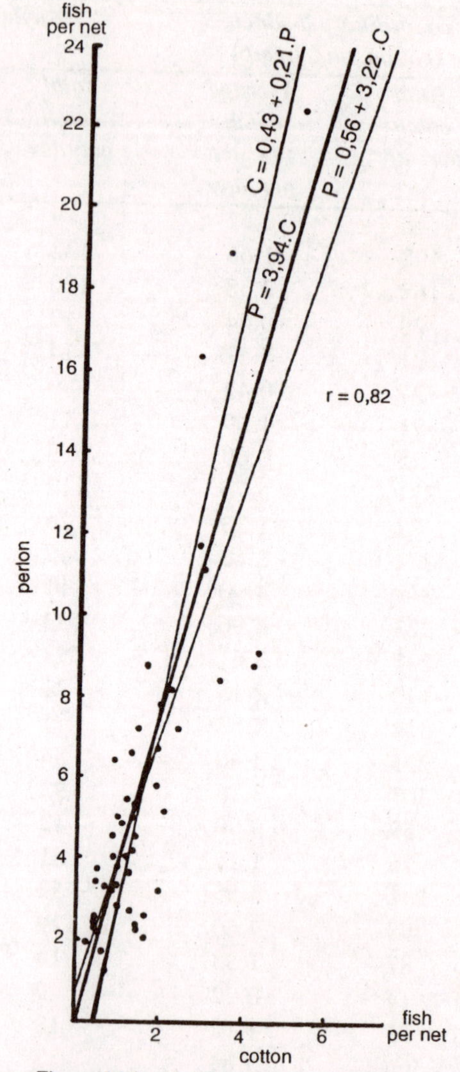

Figure 14.3 : C = cotton. P = perlon.

fishing gear. The dry tensile strength of nylon shows a great superiority when compared to that of natural fibres, but it sustains a great reduction of strength when wetted, whereas the tensile strength of cotton increases in wet condition. The strength of nylon is further reduced by knotting. This has often led to the wrong selection in size of nylon twines in replacing cotton and hemp twines.

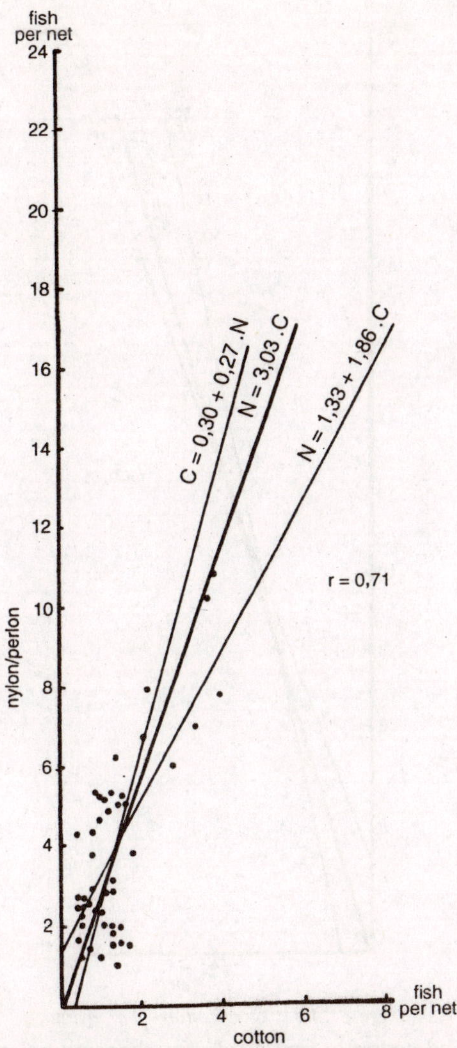

Figure 14.4 : C = cotton, N = nylon.

For example, nylon twine, chosen on the basis of dry tensile strength, produced a fine netting with good catchability but resulted in much damage to nets in hauling the catch in the cod fishing season 1954/55. An investigation showed that the twine was too fine and the wet mesh strength of the nets was lower than that of similar nets of hemp and cotton.

Tests, wet and dry, with and without knots, were carried out with

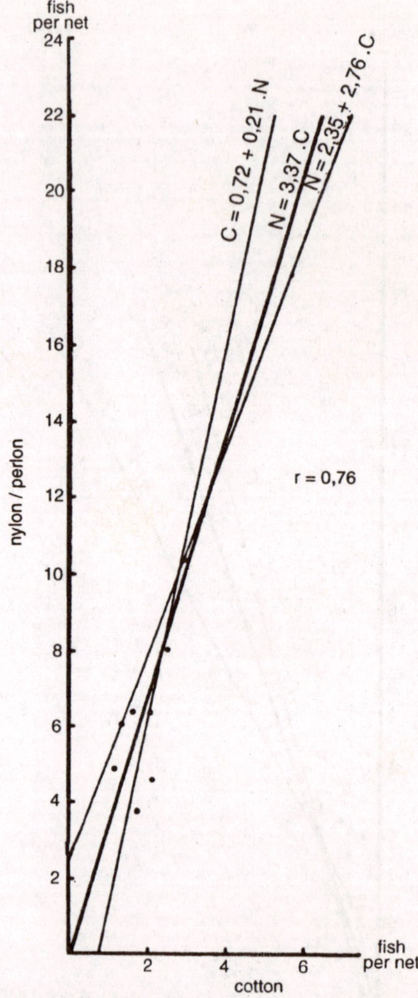

Figure 14.5 : C = cotton. N = nylon.

nylon, hemp and cotton to find a rule for choosing nylon as a substitute for cotton and hemp for a particular net.

As a result of these tests it was found that a nylon twine with 50 per cent. more runnage than the cotton it would replace, or which was 50 per cent. longer per unit weight, had a satisfactory wet knot strength.

Knot slippage in single knot nylon netting has also created problems, but experiments and experience have led to methods being evolved to overcome this difficulty. The application of heat and/or bonding agents has made it possible to produce a single knot webbing which is satisfactory as regards knot slippage and stiffness of the netting.

The degree of knot slippage depends on the construction of the knot and double knots show less tendency to slip than single knots. On the other hand, mending double knot webbing requires more work.

Polyester fibre nets of Terylene are used for catching cod and coalfish. Their catchability and durability seem to be similar to those of nets made from the polyamide fibres.

In 1954/55 experiments with codtraps made from the Japanese polyvinyl alcohol fibre Kuralon were carried out by the Directorate of Fisheries. As this type of gear is almost continuously in the sea until it is renewed, traps made from hemp or cotton have a relatively short life, being destroyed by micro-organisms. A first report from fishermen said that Kuralon traps had been in the sea continuously for 20 weeks while, during this same period, traps from hemp and cotton had to be renewed every seventh week. Up to April 25th, 1957, seven reports said that Kuralon traps had been in the sea continuously from 3 to 6 months, averaging 4.5 months, and there were no signs of reduction in the strength of the twine.

Experiments are still being carried out with most types of fishing gear, such as purse seines, herring nets, etc. made of synthetic material. Despite the high cost of these materials, there seems to be a great future for their use in the various types of fishing gear.

ON THE FISHING POWER OF NYLON GILLNETS

On the provision of the Fisheries Director, Bergen, a few fishing experiments for cod with gillnets made of artificial fibres were carried out in northern Norway in 1952 and 1953. The results obtained were promising and for the season 1954 a large series of experiments was planned in order to ascertain the fishing power and the practical usefulness of such nets compared with conventional ones.

The experiments were planned and carried out under the leadership of Fisheries Consultant M. Halaas, Fisheries Directorate, Bergen. An agreement was made with a number of fishermen from different parts of the coast that they should include some nets of artificial fibres in their sets of conventional nets and report on the results. As a whole this arrangement proved to be successful. Most of the fishermen continued this work also in the season 1955.

SPECIFICATIONS OF THE NETS

The experiments include the species cod, coalfish and mackerel. In the Norwegian gillnet fisheries for cod and coalfish the mesh size and the dimensions of the nets vary in the various coastal districts. In order to obtain comparable results the nylon nets were in each case made to match these different specifications as far as possible.

The following types of artificial fibre nets were used:

Mackerel: nylon, 21.0/2 × 3, 600 m./kg., breaking strength 8 kg. 18 knots per ell (24 inch).

Cod: nylon, 210/5 × 3, 2,500 m./kg. breaking strength 20 kg. 6¼ and 7 knots per ell (24 inch).

Perlon No. 26, continuous multi-filament, 2,200 m./kg., breaking strength 19 kg., 6¼, 6½ and 7½; knots per ell (24 inch).

Coalfish: nylon, 210/6×3, 2,000 m./kg. breaking, strength 24 *kg.*, 7½, 8 and 8¼ knots per ell (24 inch).

The knots in the Perlon nets were single, in the nylon nets double.

The conventional nets used for comparison were of the following types:

Mackerel: cotton, 18 knots per ell (24 inch).

Cod: cotton, 12/12 and 12/15, 6¼ and 6½ knots per ell (24 inch).

Coalfish: hemp, 7/3, and 8/3, 8 and 8¼ knots per ell (24 inch).

FISHING RESULTS

In the reports from the fishermen the following data were given for each lift: number of conventional nets used, number of synthetic fibre nets used, and total number of fish caught in each type of net. The distribution of the fish in the single nets is thus not available, but each lift can be described by a paired mean.

In most cases the numbers of nylon nets used were small compared with the numbers of other nets in the sets, causing a large spread in the distribution of the paired means. This is, however, partly compensated by the large number of lifts in most of the series.

The distribution of the paired means from one of the experimental series. In this case the number of both types of nets were almost equal. The regression lines have been fitted by the method of least squares. (The fact that there was a small variation in the number of nets used in some of the lifts has not been accounted for in this calculation.)

In the series numbered 2 to 6 in this table both nylon and Perlon nets were used in the same sets together with the conventional cotton nets. Records of the catch by each of these types of artificial fibres are available only from the series 5 and 6. They show slightly lower values for Perlon than for nylon, but the number of observations is too small to allow conclusions.

In our analysis there is no logical reason to prefer one of these regression lines to the other. They run almost symmetrically relative to the origin, and it seems fair to assume that the true relation is a simple proportionality The best description of this relationship is therefore a straight line through the origin and the paired mean of all the observations shown as the line P=3.94 C.

Figures given the chapter elsewhere show the results of two more series of cod catches with regression lines calculated in the same. The spread is larger, but the regression lines are still fairly symmetrical around the origin.

Similar cluster diagrams were made for all the series or smaller groups of experimental series, and they all show a more or less pronounced trend in the distribution of the paired means. In Table I the results have been listed as the probable proportional relationship between the fishing powers of the two types of nets used, represented by the straight line through the origin and the paired mean of all the observations.

In the series numbered 2 to 6 in this table both nylon and Perlon nets were used in the same sets together with the conventional cotton net. Records of the catch by each of these types of artificial fibres are available only from the series 5 and 6. They show slightly lower values for Perlon than for nylon, but the number of observations is too small to allow conclusions.

CONCLUSIONS

In the case of the cod it is seen that the observed fishing power of the nets made of artificial fibres varies from 2.5 to 4.4 times that of the cotton nets. The corresponding range for the coalfish is from 1.4 to 2.3 and for the mackerel 1.2 and 1. 3.

The variation "within the species" may partly be chance variation, partly due to different experimental conditions such as the rigging of the nets, the arrangement of the two types of nets in the sets, etc. The variation "between the species" is, however, probably a true one, and must in some way be related to the nature of the higher efficiency of the artificial fibre nets.

The observer of experiment No. 6 remarked in his report that the nylon nets seemed to catch a wider range of fish sizes than usual. A sampling was planned for the season 1955/56 to ascertain whether such a difference in the selectivity really existed, but sufficient data was unfortunately not obtained. Such a difference could be caused by the higher elasticity of the nylon twine. Another practical experience which may touch on this point is that the quality of the cod caught on the nylon nets seems to be inferior to that of the fish caught on cotton nets, presumably because the fish die more quickly in the nylon nets.

If the higher fishing power of the nylon nets is connected with the ability of the nets to catch a wider range of fish sizes, then this would offer us an explanation of the different efficiency of the nylon nets towards the different species found in these experiments. As a result of the pronounced schooling behaviour of the coalfish, the range of fish sizes present in a concentration of coalfish is usually considerably smaller than in a concentration of cod. This phenomenon is even more striking in the mackerel. In agreement with this hypothesis is v. Brandt's report of only up to 35 per cent. increase of catches with Perlon nets in herring drifting.

RELATIVE EFFICIENCIES OF NETS MADE OF DIFFERENT MATERIALS

Mr. B. B. Parrish (U.K.) Rapporteur. When judging the relative efficiencies of gear made of different materials, selectivity must be considered. By selectivity we mean how much will a certain gear catch, what species and sizes of fish?

All fishery workers are concerned with this question in one form or another. The fisherman usually wants from each of his operations the greatest possible catch, but not necessarily the greatest amount of any or all kinds of fish. He tries to catch fish of a species and size for which there is a market in his particular aiea. The technologist sees in the catch an indication of the efficiency of the gear used, as compared to another.

The fishery economist looks at the catch as an income resulting from a certain expenditure.

The fishery biologist looks at the catch from the point of view of its significance with regard to the fish stock from which it has been taken, as that stock determines what can, and in part what will, be taken in the future. For him the catch is a proportion of the stock and the removal of this proportion signifies certain effects on the stock-changes in its size and other properties.

Let us first of all define what selectivity really is. We can say that every individual in a fish population of a certain area has an equal chance of being caught, then the catch will reflect the composition of the whole population. When the proportions in the catch show a different composition, then the gear may be said to be selective. This can be due to the biological characteristics of the fish, to the type of the fishing operation, to the behaviour of the fish in relation to the operation or to the rig and design of the gear so that selection is not attributable to the characteristics of the gear alone.

Another type of selection will operate when the fishing boats of a certain area fish in regions where certain high quality fish are to be found, to satisfy the market demands for a particular species.

Perhaps the most important factor, determining the extent to which this type of selection process will operate, is the knowledge of the distribution of the species wanted. The collection of such knowledge should benefit from a close cooperation between fishery biologists and the fisherman. So we can say that selection begins as soon as the fishing fleet puts to sea and concentrates in places where the most desirable catch is to be found. This selection process becomes more complex where several types of gear operate at the same time to supply different market demands. A second stage of selection, which we can call gear selection, starts when the gear is put into operation.

Gear selection can be divided into two components: that due to the avoidance of the gear by certain species and that due to the ability or tendencies to escape from the gear when the fish come into contact with it. If the magnitude of these components can be identified and measured, then the gear designer, the gear manufacturer and fishermen can take account of them in the construction and operation of the gear, or the fishery administrator in his fish conservation practices.

In gillnet fishing, for instance, the catch depends on fish swimming into and being meshed or entangled by a wall of net. The influence which biological factors may have in this action is illustrated very clearly by Nomura who discusses the factors affecting the selectivity and fishing ability of gillnets in Japanese waters. This paper demonstrates that the behaviour of the fish in relation to the visibility of the nets, plays a very important part. Saetersdal and Mugaas also mention data which point to the inter-action of the fish behaviour to, and the gear selectivity of, different types of material used for gillnets, as being important in determining the size of catch and its composition.

In line fishing, too, there are data which demonstrate the importance of these factors; fish of some sizes or ages, for example, may avoid baits of a particular shape or size, while others will take them. A further stage at which selection takes place is the escapement of fish from the gear once they have contacted it, such as through the meshes of the net.

This mesh escapement, of course, is a process which is well known to technologists, fishermen and fishery administrators as well as fishery biologists. It is, in fact, the selection process which is utilised in effecting fishery conservation practices in some important regional fisheries.

For example, we have learned a great deal in recent years about the behaviour of fish in trawl codends and seine codends by the adoption of television techniques, under-water observations by frogmen, by measuring devices attached to nets, by controlled experiments with gears and observations on the catches in different parts of the net. This has shown that mesh escapement is effected by features of the gear as well as by the behaviour of the fish. One of the most important of the gear factors which affects escapement is the material from which the net is made. The results of experiments have shown conclusively that the escape of fish differs between nets having the same mesh size but made of different materials. Some quality of the material which is being identified by many workers with the property of flexibility and stretchability of the twines is responsible for this. Mesh escapement is also affected by the size of the catch as fish in the codend will tend to block meshes and prevent further escapement.

I do not propose to go into the methods by which the selection processes can be studied and examined yet I must draw attention to the method known as comparative fishing. Comparative fishing has been used almost exclusively hitherto to study selectivity and other features of fish operations, such as determining the fishing power of nets of different structures, different materials and so on, and in four of the present papers there are examples of the use of the comparative fishing technique. That the method must be worked under strictly controlled conditions is apparent from the experiments referred to in the papers on the efficiency of gillnets made of natural and synthetic fibres. Whereas these papers conclude that the synthetic fibres are catching much more efficiently than the natural fibres. the results of similar experiments conducted elsewhere show very little superiority of synthetic fibres. The higher catches of synthetic fibre nets may be due to the decreased visibility when they are fished in clear water. On the other hand, one would not expect to find a marked difference in very turbid water.

Mr. S. Holt (FAO). In measuring how effective one gear is by comparison with another slightly different gear, we use the "comparative fishing method". Although it essentially means fishing the two gears side by side and comparing the results, the difficulty has been that results are very variable and therefore rather complex statistical methods have to be used if we are to make the correct interpretation of the results of such trials. The trouble is that, although we can measure the differences between gears, we do not know if that difference will be true when we make the same experiment on another day, or in a different place, a different season, or at a different time of day.

Comparative fishing experiments can give us measurements, they cannot tell us the reasons why. The problem of interpretation of the results is a matter for both biological and technical study. There are very complex problems involved, as Mr. Parrish has said, but somehow they have got to be solved if we are to try to devise gears rationally. If we are going to accelerate the rate at which we can improve fishing gears, we must do more than just make trials working in the dark the whole time. To do this at all we will need to understand a lot more about the reactions of fish, particularly to different characteristics of fishing gear such as its visibility, the speed at which it moves, and so on.

At a meeting of biologists in Lisbon this year, it was suggested that this Congress was the appropriate place to tell technologists and fishermen what kind of information biologists expect from them.

Fishery biologists are in general responsible for measuring the properties of the fish which determine its liability to capture. They believe it is the responsibility of gear technologists to analyse and measure the corresponding properties of the gear. They believe that this work would be greatly assisted if a world catalogue and description of fishing gears could be compiled and if a classification of gears based on toxonomic principles could be devised. They believe that a simple classification of gear would serve the following purposes. Firstly, it would stabilise the nomenclature of gears with special reference to synonymous terms used in various countries. Secondly, it would assist the population dynamic worker by indicating the essential properties of gears with which they are working, and thirdly it would facilitate the correct comparison of results obtained with various gears in different circumstances.

The biologist cannot advise on the correct gear to use, unless he knows what is the composition of the fish stock on the ground. If that changes, then the choice of gear must also change, i.e. another gear may be more economical. The trouble is, we cannot wholly measure the composition of the fish on the ground without also knowing a great deal about the selectivity of the gears which are actually being used by the fishermen. The only solution is by very close co-operation between the fishermen, the gear technologist and the biologist.

Mr. S. Springer (U.S.A.). In the Gulf of Mexico we use shrimp trawls which run from 12 ft. across the mouth to 125 ft. and all of these are used to a very great extentsome on the same boats. The smaller nets are used for try nets to test the ground and the larger nets are used for production. The smaller nets are usually a lot faster than the larger nets. I presume that all of these nets work well enough but one thing that we have found in analysing the catches of the nets of different sizes is that the smaller nets catch smaller fish than the larger nets. Yet it appears that, owing to the greater speed of the smaller net, this should not be so. One explanation of why this may happen could be seen in the television film showing haddock inside the codend. In general, these haddock seemed to move and swim forward to a certain extent before they actually go into the tail of the codend. It would seem to me that we might infer from this that the net has to have a certain length in order to retain the large fishthat is, a short net with a short codend would catch more small fish because the larger fish, being faster swimmers, can escape. I would be interested to know if there is any experience like this in other fisheries and whether there is any relation between the length of the throat and the codend and the size composition of the catch.

Dr. Went (Ireland). From a point raised this morning it would appear that the fisherman in general is not convinced of the practical value of selectivity and comparative fishing for the industry. I think we must stress the very great importance of this subject to the practical fisherman who, of course. is interested in getting fish. However, if he is to continue obtaining good catches, fish conservation regulations must be made to preserve the stock. I think the experiments on selectivity of fishing gear and comparative fishing will give us the basis on which to draw up such international regulations.

Mr. M. Ben-Yami (Israel). I would like to emphasise the importance of comparative fishing in assessing the relative value of fishing gear. We met this problem in Israel in connection with the high opening trawl we had developed. We were aware of the difficulties in determining the actual value of differences in catching efficiency between two types of gear accurately. But, for the commercial fishermen only very obvious differences count. 1, therefore, think that accurate statistical evaluation, as applied in selection experiments, is not needed for such practical purposes. We actually used two techniques of comparative fishing: one and the same trawler fishing with the new gear and the conventional gear alternatively, and two boats, one with the new gear and one with conventional gear, towing side by side. These experiments were conducted for several months in commercial operation in various areas. After eliminating all not strictly comparable hauls, some 60 comparative tows remained, the evaluation of which indicated that the new gear gave about 20 per cent. better catches. For such commercial comparative fishing it is advisable to put a very experienced and very conservative, but good, fisherman in charge of the conventional gear and make the designer of the new gear compete against him. If a new design passes such a test successfully, I think one can be sure and the fishermen can be convinced, that something worth while has been achieved.

15

NETS

Continuous filament yarns of nylon and Terylene are extremely "lively" when twisted so that when twine is cut the yarns promptly untwist. Heat setting of the twine during manufacture reduces this liveliness by altering the linear molecular structure of the fibre. If this process is carried out when the twine is in relaxed state contraction takes place, and influences the length/weight ratio and the extensibility. The extent of the contraction depends upon the temperature and time the twine is heated. The nylon and Terylene twines are treated at or below 100 deg. C. for up to 10 minutes.

If the twine is permitted to contract during this process a greater length will be necessary to make any particular specification of net and the raw material cost is therefore increased. The elasticity of the twine is affected and if it were allowed to contract during the heat setting process, extension under load would be increased; much depends on the temperature and time. For this reason the Kenlon brand twines are heat-set by methods, devised and developed within the Organisation and contraction during heat-setting is avoided so that the length/weight ratio is improved. These twines show about 16 per cent. up to 25 per cent. extension at break whereas twines which have been heat-set in a relaxed state show about 21 per cent. up to 30 per cent. extension at break. The difference in length/weight ratio may be anything from 5 per cent. up to 10 per cent. in the case of nylon 66 of 210 denier.

IMPREGNATION

The low coefficient of friction against itself caused some difficulty when synthetic fibre nets were first introduced and necessitated the use of double knots to prevent knot slippage. If suitable impregnating materials and methods are employed, it is possible to increase the coefficient of friction so that secure single knots may be achieved with both machine-made or hand-made nets. It goes without saying that coating of the twine with materials which are incompatible with nylon or Terylene does not give satisfactory results. Reasonable resistance to flaking off during rubbing or surface abrasion and rapid leaching out in water are two of the points to be considered in this respect.

KNOTS

Where double knots are made, satisfactory results may be obtained with thermoset type twines, but where single knots are made it is advisable to use bonded type twines.

In either case heat setting of the net is usual and this must be done at a temperature higher than that used for the heat setting of the twine. It is therefore an advantage if the twine is heat set at a temperature less than 100 deg. C., so that boiling water or low pressure steam may be used for heat setting the net. Where possible, the net should be kept on tension during heat setting, otherwise some contraction will take place, thus affecting mesh size.

To further avoid knot slippage the net can be impregnated with suitable materials. When nets are so treated with a recent development of our own it is practically impossible to move the knot in any way, under any conditions, wet or dry.

SHRINKAGE IN WATER

Where an exact match of mesh size is required a check should be made when both natural fibre and synthetic fibre nets are saturated in water. This is necessary because of the different water absorption characteristics.

ROPES

From the remarks above regarding rot-proofing, it will be obvious that, in order to gain the maximum advantage, synthetic fibre mounting ropes should be used where possible. It is sometimes found that ropes made from staple fibre yarns (instead of continuous filament yarns) are the most suitable as this avoids any possibility of the tying cords gliding on the head rope. On the other hand, ropes made from continuous filament yarn are much stronger and this consideration may have a bearing on the size of rope required and therefore on the cost.

SOME CONSIDERATIONS ON NETMAKING

A Few decades ago the construction of fishing nets, from the braiding of the webbing to the assembly of the whole gear was the task of the fishermen themselves. The boat owner was only interested in the cost and, in many cases, the fishermen had to help pay for the gear or even owned a part of it. In the driftnet and line fisheries, for instance, each fisherman owned a certain number of the nets that went to make up the set. The cost of materials was high in relation to earnings from the catch and very often more consideration was given to the price and durability of the material than to its catchability and efficiency as a part of the gear.

The operating range of the vessels was small and the , nets used were largely the same for each home port as the gear was adjusted to the fishing conditions prevailing on the grounds within reach of that port. Large differences were, however, to be noted between different fishing areas, in boats, in gear and even in the type of operation.

The use of larger vessels, engine power and better material has brought many changes; fishing gear is now constructed to give a high degree of efficiency in a certain fishing method over a much greater range of grounds. Higher speeds of operation, with resulting higher strains on the nets, and the use of new materials and improved methods of netmaking, have brought changes to net manufacture. While the cost of material in relation to returns from the catch has decreased, the cost of labour has increased and the new nonrotting fibres have made preservative treatment practically unnecessary, so that the conditions governing the cost of fishing gear are now totally different. Today the nets are increasingly planned and constructed ashore and the fishermen are relying more and more on the netmaker to make and improve his gear.

Although some observations of gear in action have been made by frogmen and results are being obtained by underwater television, the netmaker himself does not normally see how the net he has made, functions. He still has to evaluate the performance of the net by interpreting various telltale signs which are only apparent when the net is back on board, such as localised tightening of knots, deformation of meshes, gilling of fish, wear on knots, etc. Many of these signs only become apparent after some time so that the fisherman must still, as of old, keep his gear under close observation. His criticisms and suggestions, based on such observations, are of the greatest help to the netmaker.

GENERAL

When making a new design, the aim of the netmaker is to produce a net which will give the largest catching potential for the smallest cost in materials and the least energy output. Depending on the type of gear, many different factors have to be taken into account, such as resilience of webbing, strength and elasticity, resistance to flow of water, weight and bulk, speed of operation, cost of materials, condition of fishing ground, etc. However, in most cases, the most important factor is the shape the completed net will take when in action.

Gillnets and trammel nets need no fashioning as their shape in operation is determined by the looseness given to the webbing when it is hung between the float- and leadlines. This not only determines the opening of the meshes but also the resilience and the bulge the webbing can take during certain types of operation. The amount of resilience given to these nets differs according to the type of operation, the net height and fish species to be caught. It is regulated by the amount of, and relation, between, the buoyancy and weight of floats and sinkers.

A certain amount of shaping is necessary during the construction of roundhaul nets to give the webbing its bulge during operation and to take up the various strains on the webbing during shooting and hauling. This is achieved partly by using the appropriate hang-in coefficient but mainly by adjustment of take-up and/or tuck-in between the strips of webbing that go to compose the complete net. With trawl nets and other types of bagnets, the netmaker has to tailor the webbing to make individual net sections fit the curve of the lines to which they are hung and to obtain the required bagshape. The problem here is made more difficult by the fact that in some cases the operational conditions are liable to change; variation in resistance, or drag due to change in the flow of water through the net, influence the performance of the whole gear. Points and direction of forces acting on the net may change, with a resulting change in the shape of the net in operation.

The strains on the net must be spread as evenly as possible and, in general, should be across the knots. The twine must everywhere be strong enough to withstand the strains acting upon it, yet, on the other hand, its weight, resistance to flow of water and cost have to be taken into account.

In the case of gillnets, the mesh size must be appropriate for gilling and/or entangling the particular fish species caught; with other nets, it must prevent such gilling. When deciding on the mesh size for bagnets, the netmaker has to take into consideration, apart from the resistance, the water release so as prevent "whirling" or backwash within the net.

Apart from the above considerations, which are connected with the catchability of the net, the netmaker has to give thought to the handling of the net during the shooting and hauling operations. For instance, very fine twines may make hauling difficult and too much loose webbing may slow down the speed of operation as in the case of roundhaul nets.

There is no such thing as an ideal net. All nets are a compromise to fit a certain set or collection of fishing conditions; all are subject to controversy, being praised by some fishermen and condemned by others. Much depends, of course, on how the net is used, both as to assembly of the net to other gear parts and the skill with which it is operated. A well constructed net may give unsatisfactory results because of defective auxiliary gear, such as floats and kites, sinkers and anchors, trawl boards, etc. All such' gear must be of the right type and fitted in the correct position; they help the net to take its correct shape in operation and are therefore as important as the net itself. The fisherman must realise that he can only get maximum efficiency from a given net if he uses it under the conditions and in the manner for which it was constructed. Finally, much can be done to adapt a net to the conditions prevailing, if the netmaker is made aware of those conditions.

STRAINS AND STRESSES

Although certain studies have been made on the forces acting on the webbing of a net, most netmakers still use empirical values when planning and constructing new nets. There is a complex stress-strain reaction in webbing, very much like that in an elastic fabric, i.e. it elongates in the direction and at the point of stress, while contracting in the lateral direction. In fact, the webbing takes up a part of the stress by yielding to it.

The deformation of webbing due to a stress acting on it depends on the position of the stress point in relation to the whole net and the magnitude and the direction of the stress. Normally the stress acts "across the knots" and in that direction a small strain will be taken up completely by the partial closing of the meshes directly below the point of stress. Larger stresses will cause an increased strain reaction along the "halvers" until the selvedge is reached.

Further increase of strain will result in a lateral closing of the webbing section.

The amount of reaction to a given stress in influenced by the hang-in coefficient and the mesh size, increasing as these values increase. In general, individual stresses cause positive strain reaction in the area of webbing enclosed by those halvers which run out from the point of stress; the meshes along these lines are opened and the knots tend to slip, while directly under the point of stress the meshes are closed and elongated. Above the "halver" lines, the webbing is loose and the area of loose webbing is in proportion to half the area under stress.

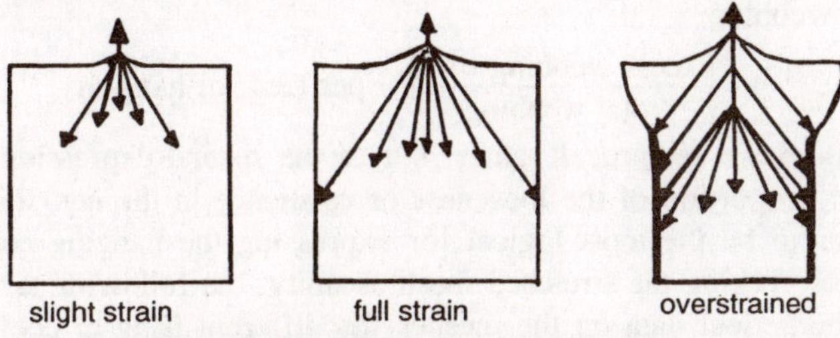

Figure 15.1 : Effect of stress on webbing.

When considering the effect of the direction of strains and forces acting on webbing, it can be said that the more the strain acts out of the vertical (across the knots), the easier the webbing will give to the strain, but the more deformation it will cause. The only remedy for diagonal stresses is the use of tension lines to take up the strain. An important fact is that, by proper adjustment of the hanging in stress points, the stress can be neutralised to a certain extent. This is done by allowing the meshes to elongate by close hanging at the point of stress, thereby spreading the strain.

For the rational design of fishing nets and, in particular, of trawl nets, more accurate knowledge of the stress-strain behaviour in webbing is needed to counteract the deformation it causes to the shape of the net and weaknesses which arises from it. This could well be a subject of practical research at one of the Research Institutes.

Because of this yielding characteristic of webbing it is difficult to foresee the actual shape of a new designed net. However, to the experienced netmaker, the condition of a used net gives much indication on its behaviour in operation; the tightening of knots below stress points, the swelling of twine in slack parts, the deformation of meshes, accumulation of scrap, wear on the knots, etc., all help to form a picture of the net's performance and its shape in action.

HANGING OF WEBBING

The correct hanging of the webbing to the framing or supporting lines is an important factor in all nets. At present there are two methods in use for expressing the hanging coefficient and this causes confusion when constructing nets from a plan. The first is to express the length of the line, to which the webbing is hung, in a percentage of the total stretched webbing:

$$\frac{\text{length of line} \times 100}{\text{length of stretched webbing}} = \text{per cent. of hanging}$$

The second method is to express the amount of excess or loose webbing (total webbing minus line length) as a percentage of the total webbing :

$$\frac{\text{excess webbing} \times 100}{\text{total webbing}} = \text{per cent. of hang-in}$$

Both are reciprocal values, but as the second expression gives a direct proportion of the looseness or resilience in the net, this would appear to be the most logical for expressing the hanging coefficient of nets. Taking the stretched mesh as unity, the following table gives the theoretical data on the meshes for different hang-in coefficients. It is clear that the mesh height, for instance, will depend to a certain extent on the strains acting on the webbing.

The resilience of a section of webbing is determined by the forces acting on it. With low strains, however, the hang-in coefficient has a big influence so that when hung squared, webbing has its greatest area but least flexibility. With increasing hang-in, it becomes looser in the lateral direction, whereas with less than 28 per cent. of hang-in the webbing has a certain vertical springiness.

In gillnets and entangle nets the hang-in determinA the looseness of the net, true gillnets being usually hung somewhat tighter than entangle nets.

The resilience and fishing height of these nets is, of course, largely influenced by the amount of sinker weight on buoyed-up nets, and the amount of buoyancy on bottom operating nets, which produces the strain in the webbing.

In roundhaul nets, the hang-in is adjusted to give depth and bulge during operation; in the wings, however, it is often curtailed as too much webbing affects the speed of operation.

With trawl nets, the hang-in to be given to wings, quarters and bosom, depends on the general shape and the cut of the sections of

Hang-in or Looseness percent	*Ratio of line to webbing of percent.*	*Angle of mesh*	*Height of mesh*	*Width of mesh*	*Filtering coefficient (areas of mesh)*
10	90	128 20'	0.436	0.90	0.785
12	88	123 20'	0.475	0.88	0.836
14	86	118 40'	0.510	0.86	0.877
16	84	114 20'	0.542	0.84	0.911
18	82	110 10'	0.572	0.82	0.938
20	80	106 20'	0.599	0.80	0.958
22	78	102 40'	0.626	0.78	0.975
24	76	99 —	0.649	0.76	0.986
26	74	95 30'	0.672	0.74	0.991
28	72	92 10'	0.693	0.72	0.998
30	70	89 —	0.713	0.70	0.998
32	68	85 40'	0.733	0.68	0.997
34	66	82 40'	0.751	0.66	0.991
36	64	79 30'	0.768	0.64	0.983
38	62	76 40'	0.784	0.62	0.972
40	60	73 40	0.801	0.60	0.961
42	58	70 50'	0.815	0.58	0.945
44	56	66 10'	0.829	0.56	0.928
46	54	65 20'	0.842	0.54	0.909
48	52	62 40'	0.854	0.52	0.888
50	50	60 —	0.866	0.50	0.866
52	48	57 20'	0.877	0.48	0842
54	46	54 50'	0.888	0.46	0.817
56	44	52 10'	0.898	0.44	0.790
58	42	49 40'	0.907	0.42	0.762
60	40	47 10'	0.916	0.40	0.733
62	38	44 40'	0.925	0.38	0.703
64	36	42 10'	0-933	0.36	0.672
66	34	39 50'	0.940	0.34	0.639
68	32	37 20'	0-947	0.32	0.606
70	30	34 50'	0.954	0.30	0.572

the net. It is difficult to determine accurately, and is usually based on previous experience and experimenting. It is, therefore, most important that the amount of hang-in on wings, quarters -and bosom is denoted on trawlnet plans.

The fastening of the webbing to the lines is done in different ways according to the gear. The main requirement is to' ensure that the hanging cannot shift over the line or become loose once the net is hung. The net-line hitch and rolling hitch are about the best knots to use when making hangings to carry two or more meshes. The clove hitch is to be discouraged unless locked by an additional half hitch, as it tends to become loose. In most cases, it is better to use a rather heavy twine for the hangings as it reduces wear on, and cutting of, the selvedge meshes. When the nets are hauled mechanically, it is best to keep the hangings rather short to avoid snagging while in operation.

The backhand marline hitch is still the most secure for rigid fixing of the webbing to the lines, but is also the most time-consuming to make. It has the further disadvantage of requiring renewal every time the selvedge meshes need repairing.

JOINING OF WEBBING

When sections of webbing of the same width are joined together, the angle of the meshes in both sections is the same. When the width of the sections differs, the angle of the meshes in the widest section will be smaller, so that this piece will have more hang-in and, by the same token, less lateral strain. Several combinations are possible and each is applied according to the purpose of the joint:

(a) the sections have the same width but different mesh size

(b) the sections have the same mesh size but differ in width

(c) the sections differ both in width and mesh size,

The first is a simple joint between sections of different mesh size and the take-up meshes are inserted at suitable intervals. It is, however, advisable to allow the large strip a little extra width to enable it to expand fully during operation.

In the cases of (b) and (c) the greater width of one of the sections is given to allow the webbing freedom to bulge out laterally, as in the case of roundhaul nets and some types of bagnets. Under (c) is included the special case where, although both width and mesh size differ, the sections are joined mesh to mesh, which is very common in trawl nets. It must be borne in mind that in the above case the effect is more a narrowing of the wide section instead of bulging. It results in the wider webbing section reacting as if it were baited at a faster rate; this, at the same time, helps to shift the stress towards the side or selvedge of the complete section.

In small-mesh webbing, it is better to make the takeups by creases rather than by baitings, especially in the case of joining two sections of equal width, as the baitings tend to form a contraction in the row they are made, which causes slipping of knots.

Where a tuck-in is needed in the lacing of vertical strips of webbing because of difference in depth or mesh size, the tucks should be made by half meshes or legs, rather than by full meshes. Full mesh tucks form strain points in the webbing, which soon become holes.

In calculating the width of the sections to be joined, the take-up or tuck-in rate should be rounded off to allow for an easy sequence. This not only facilitates the actual work of joining but allows the sequence to be expressed in a simple fraction form such as 5/6, for example, meaning 6 meshes are joined to 5, or one take-up every fifth mesh. Simple take-up sequences are, furthermore, easy to replace in mending and when new webbing sections have to be inserted during overhauling of nets.

TAILORING OF WEBBING

All bagnets need fashioning, either by braiding them in the round, with baitings inserted at appropriate intervals, or by assembly of shaped webbing sections. With the latter method, the sections can be braided to shape or they can be cut. to shape from marine-made strips of webbing. Theoretically, it is possible to cut or braid webbing to any shape; in practice, however, all shapes are obtained by decreasing the width of sections at certain baiting rates, as curves give rise to many difficulties both in construction and in mounting the webbing to its supporting lines. Curves, therefore, are always made by a series of straight lines, each at a different angle to the vertical or "across the knots" direction.

When drawing plans of nets, it is best to use a "1 to 2" ratio, i.e. all width measurements are drawn in as half values while heights are drawn to full value. It would be helpful when comparing net designs, if such a ratio could be adopted by all netmakers, as a comparable picture of the general net shape would then be apparent at a glance. Apart from representing a fair mesh opening, 53 degrees or about 56 per cent. hang-in, this ratio gives a fair picture of the net shape and has the advantage of being easy to work with on the drawing board. Furthermore, with this ratio the baiting rates can be taken directly from the drawings as they are equivalent to the cotangents of their angle of slope to the vertical. The angle of these slopes increases with the baiting ratio and shows the slopes of the most frequently used baiting ratios and illustrates the equivalent cutting rates.

In some net sections, a faster slope is necessary to fit the slope of curving supporting lines as, for instance, at the quarterpoints where the wings join the bosom in trawlnets. In such cases, the slope can be increased by inserting baitings under the selvedge of the wing which, by reason of its fly meshes, already has a slope of 1 : 2.

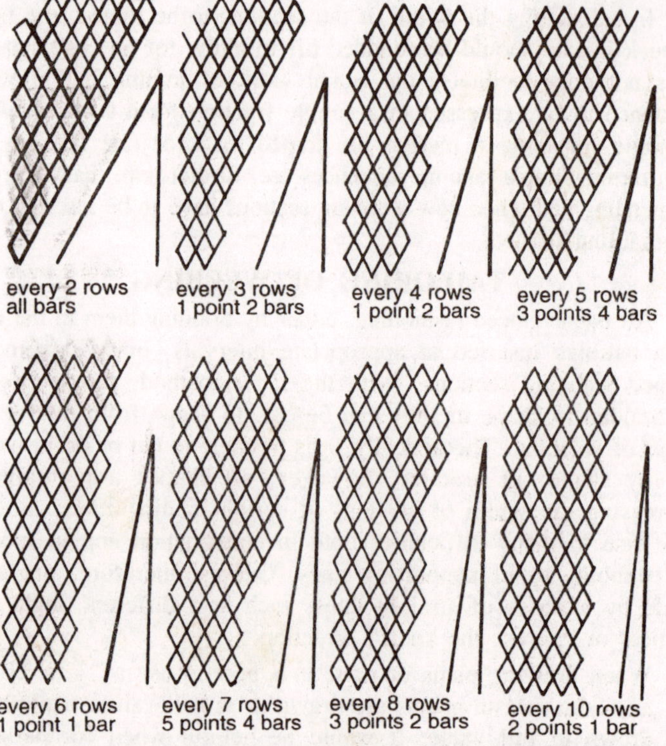

Figure 15.2 : Slope and cutting ratios of most frequently used baiting rates.

The additional baitings cause a localised closing of the meshes and a consequently greater height, which allows the selvedge to stand further off.

Where desired, the meshes can be kept lower down at the normal opening by inserting creases to equalise the number of meshes.

When considering the dimensions of new nets, it is always better to overestimate the width the net will have in action; this is especially true of all types of trawl nets. Too high assessment of the opening width of a trawlnet may result in the use of too much webbing, but the net will still operate properly and catch fish. Underestimating the width at the trawl mouth, however, will cause a change in the whole shape of the net. When the wings are pulled further open than calculated, the bosom is pulled forward while the sideseams fall back; this causes a contraction in the webbing of the throat with resultant bad water release. It may even lead to the cod-end turning over.

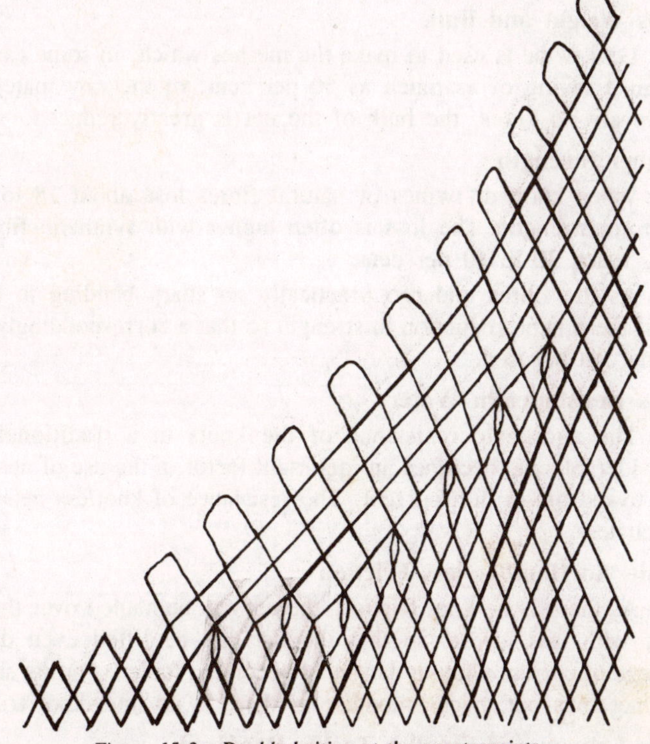

Figure 15.3 : Double baiting at the quarterpoints.

This desirability to over- rather than under-estimate net width does not mean that it is better to choose a bigger net, but that careful consideration should be given to the amount of webbing to be allowed to a headline. Indeed, where it is a question of the *size* of net to be used by a certain vessel the contrary is true: it is always better to underestimate.

THE KNOTLESS NET

In this type of webbing the twines are joined at the mesh corners by an interlacing of the two twine strands. Knotless nets did not come into general use while only natural fibres were available, but followed with the development of rotproof synthetic fibres and their increased use for fishing nets. Now this type of webbing is rapidly becoming popular for several kinds of nets.

SPECIAL ADVANTAGES

The rapid change-over to knotless nets in Japan is chiefly due to the following special features, of which the most important are:

Less Weight and Bulk

Less twine is used to make the meshes which, in some cases, can mean a saving of as much as 50 per cent. of the raw material. As there are no knots, the bulk of the net is greatly reduced.

Higher Strength

When knotted, twines of natural fibres lose about 18 to 20 per cent. in strength. The loss is often higher with synthetic fibres and may reach 30 to 40 per cent.

As the fibres undergo practically no sharp bending in knotless nets, there is no reduction in strength so that a correspondingly lighter twine can be used.

Less Resistance in Water

The aggregate resistance of the knots in a traditional net is considerable and becomes an important factor in the use of nets which are towed or set in a current. The resistance of knotless nets is very much less.

Easier to Handle—less Friction

As there is no knot friction, the net can be hauled over the ship's side with less effort, so that the net can be lifted even during a change of current without danger of becoming fouled. During shooting, the net runs out much smoother as there is no inter-knot friction.

Less Labour and Smaller Tackle Required

As the whole net is lighter and less bulky, time and labour is saved and tackle can be lighter, factors leading to a saving in manpower.

No Wearing Away of Knots

The damage caused by abrasion of knots in the belly of trawls, and other nets which are dragged over the sea bottom, is well known. In other types of nets the knots are worn away during operation by rubbing against the ship's side, net rollers or other gear parts.

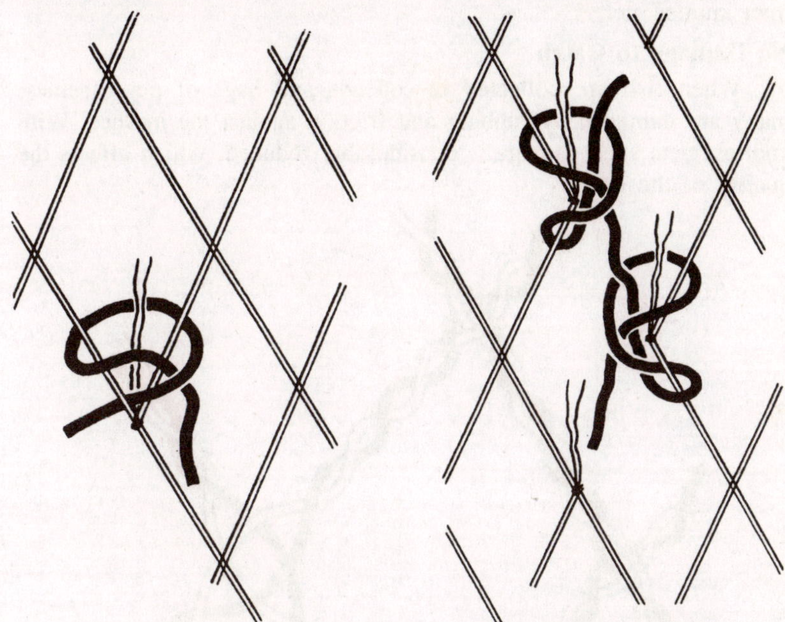

Figure 15.4 : Starting. *Figure 15.5 : Joining.*

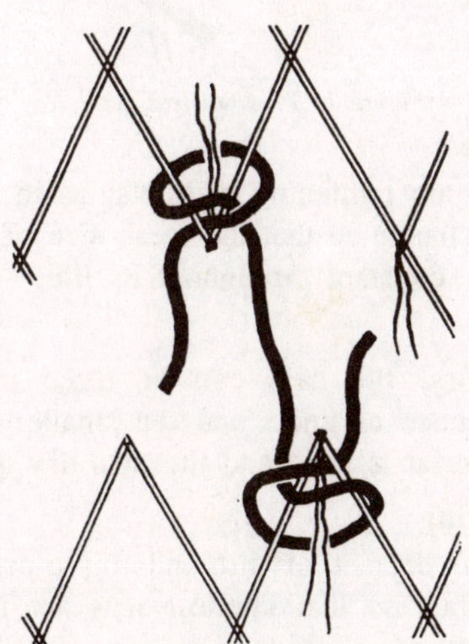

Figure 15.6 : Pick-up.

With synthetic fibres, which are rotproof, knotwear is very important as it shortens the useful life of the net.

Knotless nets, having no such "points" of wear, should last longer than knotted nets.

No Damage to Catch

When fish are collected in codends and bags of purse seines, many are damaged by rubbing and friction against the meshes. With knotless nets such damage is considerably reduced, which affects the quality of the catch.

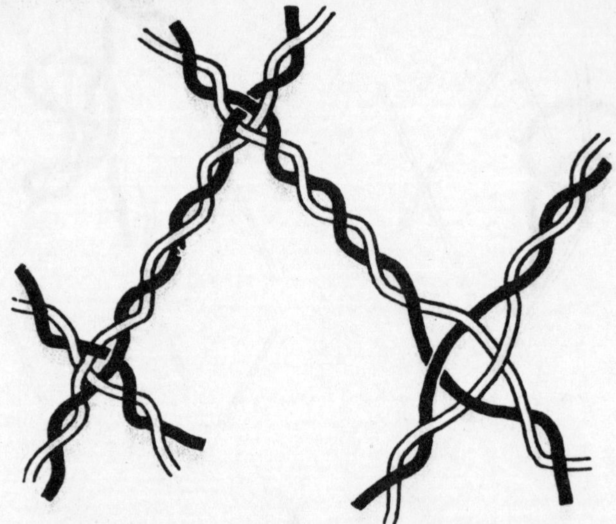

Figure 15.7 : Knotsless net.

Constant Mesh Size

Because there is no tightening of knots, as in knotted nets, the meshes undergo no change so that the mesh size of a knotless net is almost 100 per cent. constant throughout its life.

Easy to Dye

Having no knots, the nets can be dyed more easily and completely. The absence of knots and the smaller bulk also means that less dyeing material is used and the nets dry quicker.

Less Liable to Fouling

There can be no deposit of dirt and micro-organisms between the interstices of knots, so that knotless nets are much less fouled and need less washing.

MENDING OF KNOTLESS NETS

When knotless nets were first introduced, the fishermen were anxious about the mending of them. In fact, torn parts are mended as easily as in knotted nets. The only difference is that, when starting, the mending twine should be attached from one mesh outside and all ends of torn meshes should be bent back within the mending knot.

METHOD OF MANUFACTURE

The knotless net is manufactured in two stages. In the first stage the twine is prepared by doubling the appropriate number of single yarns together and then twisting two or more of such yarns into strands with a "Z" twist. The strand thus obtained is wound on the bobbins of the net making machine.

In the net making stage two such bobbins are attached to both sides of a special flat spindle. Turns of the spindle insert in the twine a second twisting operation in an "S" direction. When the necessary number of twists are made, the two bobbins move and exchange places with two neighbouring bobbins and the point of intersection of twine, which corresponds to a "knot" in the knotted net, is made accordingly.

Strands of both twines run across and through each other, and are twisted in opposite directions as shown. Thus a "knot". or intersection is made.

The twine in the knotless net has an "S" twist of two strands.

The webbing is examined, flaws and/or irregular twists are corrected and each web section is then completed by weaving on the top and bottom selvedges.

Any size of web section can be made to suit different types of gear.

The nets are heat-set so that the twist of the strands is fixed permanently. During this operation the tensile strength of the twine increases by about 15 per cent. By adjusting the process, the flexibility of the twines can be modified to suit particular purposes. Hard fibres can be made softer and soft fibres stiffened.

The heat setting process is necessary for nets of filament yarns such as nylon, Krehalon and Saran (vinylidene), Teviron (polyvinylchloride) and Kuralon 5 but not for cotton, Kuralon (vinylon) and the other spun yarns.

With the exception of materials belonging to the polyvinylchloride group, such as Krehalon and Saran, the nets made of nylon, Kuralon, etc. can be dyed with commercial dyes, pigment colours or coal tar. Heat setting is not generally required when coal tar is used.

RATIONAL MANUFACTURING OF FISHING GEAR

Mr. H. Warncke (Germany)-Rapporteur: Our problem is now to produce machine-made webbing on a more rational basis and so reduce costs. This aim could be achieved by:

1. Greater uniformity and standardisation of net types, and of the assembling components, which would gradually bring stock-carrying risks of manufacturers and dealers to a minimum;
2. A reasonable limitation of types of synthetic materials;
3. A general exchange of information on net-making problems.

Some progress has been achieved towards such a standardisation by the fishermen and gear technicians. This is illustrated by Barraclough and Needler in their description of the construction of a midwater trawl for a 62 ft. 175 h.p. trawler equipped with typical British Columbia gear and deck layout, and by du Plessis who describes the South African Pursed Lampara. These two examples show that by careful consideration of the many factors, fishing nets can be restricted to a few types. Standardisation must, however, be based on scientific study and practical experience.

The papers presented indicate that synthetic materials are becoming increasingly important to the fishing industry. It appears, however, that a fibre well suited for some types of net and gear lacks certain qualities for others. The net manufacturer must therefore produce and stock many types of gear in order to satisfy the requirements of the fishermen. A limitation in this ever increasing variety would enable the net manufacturer to reduce his stock-carrying overheads to a reasonable level, and so reduce production costs.

It was recommended that net manufacturers should exchange information so as to arrive at more rational production. However, owing to the big differences in production conditions between countries, created by subsidies, import rules, etc., most manufacturers would at present be reluctant to agree to such exchange as it would interfere with business competition.

It would seem, therefore, that the most likely methods of achieving rationalisation are:

(a) Limitation of mesh-sizes to a standard set of sizes, together with a standard method of measurement. This would reduce the need for frequent adjustments to the net-making machines and allow for simplifications in their construction;

(b) limitation of sizes and twists of twines. For instance, the denier system counts that are at present in production, should be sufficient to supply all the required sizes of twines. The number of different twists and plies could probably be reduced without disadvantage to the fishing industry.

Such rationalisation could be discussed by working groups formed of representatives of all parties concerned.

Problems of more general interest which might be considered are:

(a) Does dyeing really provide sufficient protection to the synthetic fibres against decay through exposure to ultra-violet rays, as seems to be widely accepted? Recommendations concerning the appropriate methods of dying synthetics are often difficult to interpret. Such recommendations should be consistent with the limited dyeing facilities of the fishermen.

(b) Which is the appropriate size of synthetic twines used in place of natural fibres? To replace 210/6 ply and coarser twines, synthetic fibres, 25 per cent. lighter in weight, are generally chosen. For finer twines, the criterion is the same running length as the natural fibre. This would mean that, in some cases, the main advantage of synthetics, i.e. their higher tensile strength, is not fully utilised and unnecessarily strong and heavy twines are used.

The Nippon Seimo Co. Ltd. deal with the construction of knotless nets designed and developed in Japan. They report that such nets, constructed of 2 strand twine, suffer no loss in strength, whereas natural fibres lose from 18 per cent. to 20 per cent. and synthetic fibres from 30 per cent. to 40 per cent. when knotted.

As the selvedges have to be woven separately, causing additional expense, it would be interesting to know if the initial financial advantage of lower weight is thus reduced? The relation between cost of production and price of material can perhaps explain why knotless webbing, which was not in demand when made of natural fibres, has increased in sales since it has been made of synthetic fibres.

Mr. D. McKee (Scotland): With regard to catering to the wishes of the fishermen, it is necessary for the manufacturers first to obtain, either from the fishermen or from fisheries management, information as to the thickness or weight by runnage of the twines or lines required.

A good illustration of this point is the fishing line situation. In inshore haddock line fishing, a man uses a line weighing approximately 1 lb./60 fin. Now he wants a synthetic line. We find the breaking strength of such a line in cotton or hemp is, say, 20 lb. The problem is-does he want a line of this strength or does he want the same thickness? Generally, in the case of light lines, the man wants the same body or feel in the new line, rather than the same strength. My firm which is mostly concerned with inshore herring fishing, gillnetting and that type of gear, has published a booklet for the industry, on the makeup of synthetic twines. This gives the dry and wet breaking strength of a cotton twine and recommended equivalent size in nylon or other synthetic twines in yds./lb., and m./kg. One omission is the breaking strength of knotted webbing, but information on this point can be obtained directly from our works. There is a certain risk in declaring breaking strengths and giving yardages as the figures depend on the testing method used. We, therefore, would claim at least 10 per cent. margin either way.

On the question of economy, I would like to quote an example. Recently, while abroad, I was confronted with general complaints about the economic failure of the fishing. Until a few years ago, the fishermen were using natural fibres and then changed to synthetics-which cost double the price. When natural fibres were in use and the net got damaged, it was thrown overboard. I tried to find out how much more service they obtained from a synthetic net. I got the amazing answer-little, if any. It appeared that when the synthetic fibre net became damaged the same procedure was followed, and the net was thrown aside and not used again. It is not the part of the manufacturer to work down to a.,onomy level of that type, although successful manufacturing depends on the success of fishing.

There are many dfferent types of mesh measurement. In Canada, for fine gillnets, you measure between knots; yet, in some parts of Canada, heavier nets are measured from the inside of one knot to the outside of another knot. The best suggestion I can make to the fisherman is to rely on the netting manufacturers, but to have a precise idea of requirements.

Mr. A. Robinson (U.K.): I fully agree with the comments made so far, particularly with those of Mr. McKee. Mr. Warncke also has made a very strong point regarding the difficulties which arise due to the large stocks which are necessary at present. My firm is essentially in the twine trade, but if the variety of sizes and special requirements can trouble us, how much more will they trouble the net manufacturer! Of course, these are difficulties inside the trade, perhaps not fully realised by the fishermen, but they have a very strong bearing indeed on the price that the fisherman is eventually asked to pay.

I also feel that the twine maker, and I am sure also the net maker, will, if only he is given full information, do all he can to help the fisherman. Net and twine factories get, daily, numerous inquiries and orders for materials, from which it is difficult to determine exactly what is required. Yet this is necessary because the customer does not have sufficient technical knowledge or information. However, if he states his problem exactly he will get reliable information in exchange.

With regard to the dyeing of synthetic fibres, nothing very definite has been proved so far as we are aware. Our tests have failed to prove that dyeing does give protection against deterioration due to sunlight.

Most firms would, I think, be willing to issue comparative lists showing the conversion from natural to synthetic fibres. We, ourselves, give such information quite freely and I believe the wet knot strength should be the criterion for such material. A comparison on any other basis, certainly so far as net twine is concerned, is apt to be misleading. At the same time, there are circumstances where other characteristics of synthetic fibres can be taken into consideration as, for instance, the greater elasticity which improves the energy absorption of the twine, quite apart from its merits as to breaking strength. It has been mentioned that synthetics may be as much as 25 per cent. lighter than natural fibres. On the basis of wet knot strength, and also taking into consideration the high tenacity yarns available today, a fairly reliable changeover to cotton can be made on the basis of a synthetic twine giving about 50 per cent. greater length per unit weight.

I would like to raise one further point, i.e. that of mutual understanding between fisheries workers all over the world. Apart from using the same standards and tests in our work, we should also be able to understand one another's writings. Translation of fisheries literature is extremely difficult because of the different meanings given to the terms used in different countries and areas.

Dr. E. Hess (FAO): As you probably know, FAO has for the last 8 years been publishing the World Fisheries Abstracts in English,

French and Spanish. We have, during these years accumulated quite a lot of terms and we are now in the process of preparing this material for publication as a dictionary in the three official FAO languages, English, French and Spanish, at least to begin with. The form of these dictionaries was agreed upon with UNESCO, who has done much work along similar lines.

The English dictionary will form the base and each term will have a definition. Each term is given a number and its meaning in another language can be found by simply referring to that same number in the other language part or book.

The whole field of fisheries technology will be covered, not just gear. The first section, on fish curing, is ready now. The section on fishing gear and boats we hoped to have ready for this Congress, but I am sorry to say we did not get that far. We have another section on refrigeration also in an advanced stage. Before long, we hope to have this published, but will first issue a mimeographed draft and send it to as many people as we can think of in various countries for their comments. After study of these comments and necessary adjustments we will then publish in book form.

Mr. O. Aagaard (Norway): Much has been said about standardisation as seen from the manufacturer's point of view. The request of the Fishermen's Organisation. However, when the same fishermen discovered that standardisation would mean that their individual taste and wishes as to twine and rope sizes, mesh sizes, etc. would be interfered with, the whole thing was dropped.

Before going over to any kind of standardisation, it may be useful to hear the fishermen's opinion.

Prof. S. Takayama (Japan): About half the knotless nets used at present in Japan, i.e. 3 million lb./year, are made of synthetic fibres. All Japanese knotless nets are made of two strand twines.

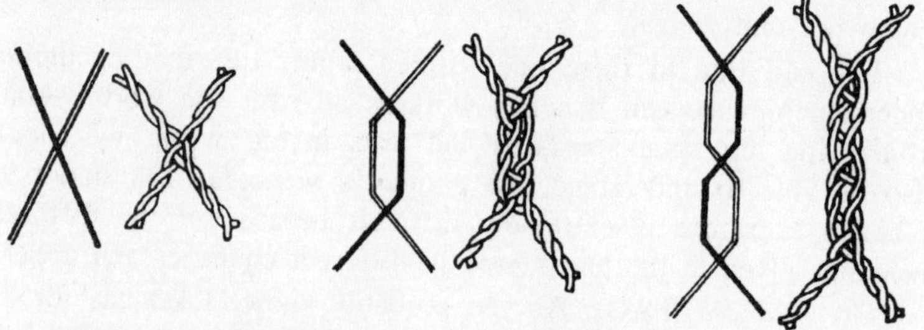

Figure 15.8 : Three types of joinings used in knotless nets.

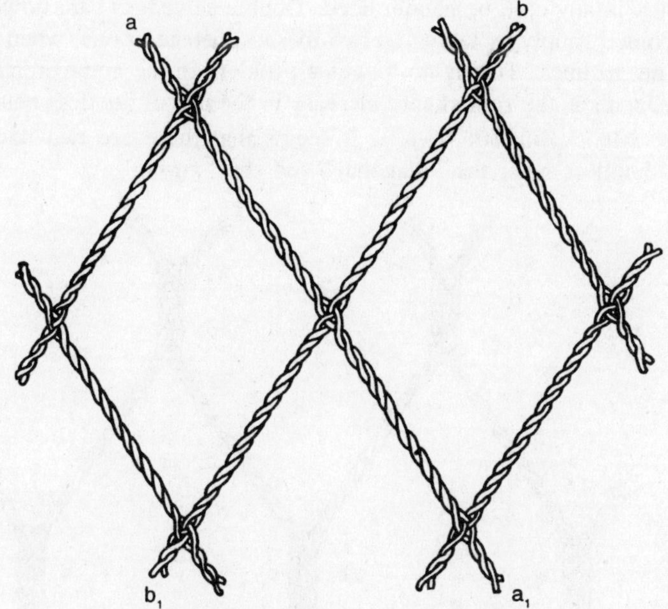

Figure 15.9 : Diagonally constructed knotless net.

There are two types of nets based on practically the same system of connecting. In the first type the twines run diagonally through the webbing, whilst in the second type they run in a zigzag line. These different directions depend on how many times the two twines are threaded through each other knotless nets made of braided twine are now being manufactured and tested for instance in U.S.A. and Beligum.

If they pass through only once, the direction of the twines is diagonal; if they are threaded twice, the zigzag line is obtained; if they pass through three times, a diagonal is obtained, and four times again a zigzag, i.e. uneven numbers of interweaving results in diagonal twine direction and even numbers in zigzag. The number of the interweavings can be built up until a quite unusual mesh form results which could be called the tortoise type.

Knotless nets can be adapted for practically all types of gear, but until now, they are mainly used for big setnets, leader nets and gillnets.

Mr. H. Kobayashi (Japan). As regards mending tears or connecting pieces of webbing, there is no remarkable difference between knotted and knotless nets, and consequently, there are practically no additional costs with regard to - selvedges. If double selvedges are needed, they must in any case be handbraided. Double selvedges can, however, be avoided simply by taking up two meshes instead of one, when hanging a net to lines. This is no longer a problem in the commercial fishery in Japan as the remarkable increase in the use of knotless nets shows.

Mr. J. Buchan (U.K.). It seems that there are two basic types of knotless nets, the "diagonal" and the "zigzag".

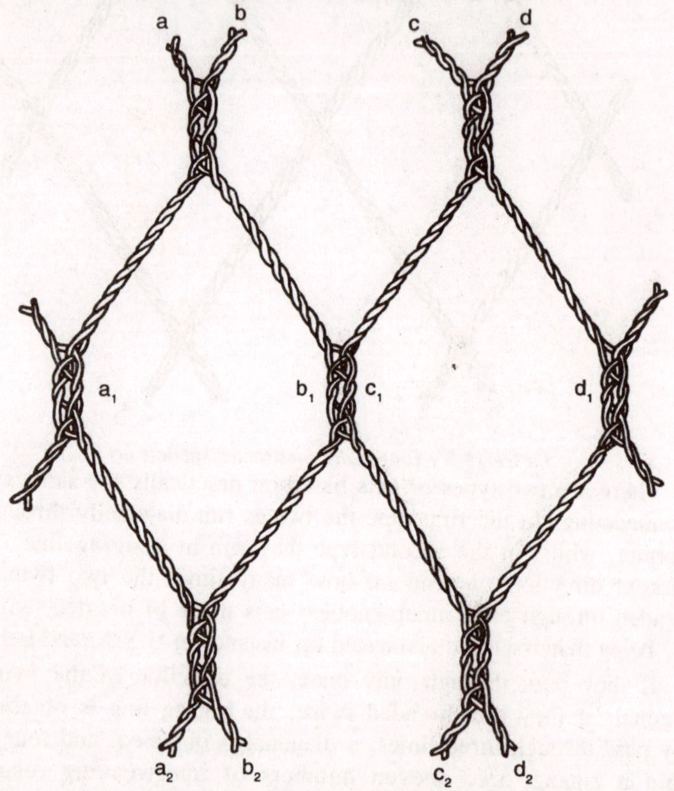

Figure 15.10 : Zigzag constructed knotless net.

The twines in a webbing made with conventional double English knot, run in a zigzag line. There are two ways of testing this knot, either by pulling the two bars belonging to one continuing twine against the two bars belonging to the other twine, or by testing the two bars belonging to different twines against the opposite pair. The same two testing directions can be applied for both types of knotless net.

We have found that the conventional double English knot will give an efficiency of 54·6 per cent., that is the breaking load of the knot is 54 per cent. of the total breaking load of the two twines in the first direction, and 51 per cent. in the second. In the first type of knotless net, that is the straight cross, you get 75 per cent. and 57 per cent., but the knot will break at the weakest point, i.e. 57 per cent.

With the second type of knotless nets (that made by double interweaving of twine), we obtained 91 per cent. and 51 per cent., showing that the actual lowest efficiency of the knotless connection is not, in fact, better than the conventional double English knot.

Knotless nets are made in a two-fold construction, if one strand breaks, the twine can untwist, causing break in the net running, which will not happen with an ordinary webbing braided with double English knot.

Prof. A. von Brandt (Germany). The fact that not only Japan but also Russia use knotless nets to a large extent indicates to me that there must be considerable advantages. Furthermore, the gillnetting experiments we have carried out with Japanese nets in Germany gave good results. We had no difficulties with "knot" slippage and repair, for which our fishermen found an even simpler way than the one recommended by the Japanese.

In addition to what we have been told until now, I believe that knotless nets should have great value for trawl nets. The limiting factor for size of the gear or towing speed is the relation between towing power and gear resistance. Dr. Scharfe has shown how the towing resistance can be decreased by using hydrofoil otter boards and thinner synthetic twine. The smaller area of knotless nets should further decrease the towing resistance. This would be a valuable gain in addition to the savings in material resulting in lower weight.

INDEX

I

J

K

L

M

N

Q

R

S

Z